MICROMECHANICS WITH *MATHEMATICA*

MICROMECHANICS WITH *MATHEMATICA*

Seiichi Nomura

Department of Mechanical and Aerospace Engineering
The University of Texas at Arlington
Arlington, TX
USA

This edition first published 2016
© 2016 John Wiley & Sons Ltd

Registered office
John Wiley & Sons Ltd, The Atrium, Southern Gate, Chichester, West Sussex, PO19 8SQ, United Kingdom

For details of our global editorial offices, for customer services and for information about how to apply for permission to reuse the copyright material in this book please see our website at www.wiley.com.

The right of the author to be identified as the author of this work has been asserted in accordance with the Copyright, Designs and Patents Act 1988.

All rights reserved. No part of this publication may be reproduced, stored in a retrieval system, or transmitted, in any form or by any means, electronic, mechanical, photocopying, recording or otherwise, except as permitted by the UK Copyright, Designs and Patents Act 1988, without the prior permission of the publisher.

Wiley also publishes its books in a variety of electronic formats. Some content that appears in print may not be available in electronic books.

Designations used by companies to distinguish their products are often claimed as trademarks. All brand names and product names used in this book are trade names, service marks, trademarks or registered trademarks of their respective owners. The publisher is not associated with any product or vendor mentioned in this book.

Limit of Liability/Disclaimer of Warranty: While the publisher and author have used their best efforts in preparing this book, they make no representations or warranties with respect to the accuracy or completeness of the contents of this book and specifically disclaim any implied warranties of merchantability or fitness for a particular purpose. It is sold on the understanding that the publisher is not engaged in rendering professional services and neither the publisher nor the author shall be liable for damages arising herefrom. If professional advice or other expert assistance is required, the services of a competent professional should be sought

Library of Congress Cataloging-in-Publication Data applied for

A catalogue record for this book is available from the British Library.

ISBN: 9781119945031

1 2016

Contents

Preface ix

About the Companion Website xi

1 Coordinate Transformation and Tensors 1
1.1 Index Notation 1
    1.1.1 Some Examples of Index Notation in 3-D 3
    1.1.2 Mathematica Implementation 3
    1.1.3 Kronecker Delta 6
    1.1.4 Permutation Symbols 9
    1.1.5 Product of Matrices 10
1.2 Coordinate Transformations (Cartesian Tensors) 11
1.3 Definition of Tensors 13
    1.3.1 Tensor of Rank 0 (Scalar) 13
    1.3.2 Tensor of Rank 1 (Vector) 14
    1.3.3 Tensor of Rank 2 15
    1.3.4 Tensor of Rank 3 17
    1.3.5 Tensor of Rank 4 17
    1.3.6 Differentiation 19
    1.3.7 Differentiation of Cartesian Tensors 20
1.4 Invariance of Tensor Equations 21
1.5 Quotient Rule 22
1.6 Exercises 23
    References 24

2 Field Equations 25
2.1 Concept of Stress 25
    2.1.1 Properties of Stress 29
    2.1.2 (Stress) Boundary Conditions 30
    2.1.3 Principal Stresses 31
    2.1.4 Stress Deviator 35
    2.1.5 Mohr's Circle 38

2.2	Strain		40
	2.2.1	Shear Deformation	47
2.3	Compatibility Condition		49
2.4	Constitutive Relation, Isotropy, Anisotropy		50
	2.4.1	Isotropy	52
	2.4.2	Elastic Modulus	54
	2.4.3	Orthotropy	56
	2.4.4	2-D Orthotropic Materials	57
	2.4.5	Transverse Isotropy	57
2.5	Constitutive Relation for Fluids		58
	2.5.1	Thermal Effect	58
2.6	Derivation of Field Equations		59
	2.6.1	Divergence Theorem (Gauss Theorem)	59
	2.6.2	Material Derivative	60
	2.6.3	Equation of Continuity	62
	2.6.4	Equation of Motion	62
	2.6.5	Equation of Energy	63
	2.6.6	Isotropic Solids	65
	2.6.7	Isotropic Fluids	65
	2.6.8	Thermal Effects	66
2.7	General Coordinate System		66
	2.7.1	Introduction to Tensor Analysis	66
	2.7.2	Definition of Tensors in Curvilinear Systems	68
	2.7.3	Metric Tensor[10], g_{ij}	69
	2.7.4	Covariant Derivatives	70
	2.7.5	Examples	73
	2.7.6	Vector Analysis	75
2.8	Exercises		77
	References		80
3	**Inclusions in Infinite Media**		**81**
3.1	Eshelby's Solution for an Ellipsoidal Inclusion Problem		82
	3.1.1	Eigenstrain Problem	85
	3.1.2	Eshelby Tensors for an Ellipsoidal Inclusion	87
	3.1.3	Inhomogeneity (Inclusion) Problem	95
3.2	Multilayered Inclusions		104
	3.2.1	Background	104
	3.2.2	Implementation of Index Manipulation in Mathematica	105
	3.2.3	General Formulation	108
	3.2.4	Exact Solution for Two-Phase Materials	116
	3.2.5	Exact Solution for Three-Phase Materials	123
	3.2.6	Exact Solution for Four-Phase Materials	132
	3.2.7	Exact Solution for 2-D Multiphase Materials	137
3.3	Thermal Stress		137
	3.3.1	Thermal Stress Due to Heat Source	138
	3.3.2	Thermal Stress Due to Heat Flow	146

3.4	Airy's Stress Function Approach	155
	3.4.1 Airy's Stress Function	156
	3.4.2 Mathematica Programming of Complex Variables	161
	3.4.3 Multiphase Inclusion Problems Using Airy's Stress Function	163
3.5	Effective Properties	172
	3.5.1 Upper and Lower Bounds of Effective Properties	173
	3.5.2 Self-Consistent Approximation	175
	3.5.3 Source Code for micromech.m	178
3.6	Exercises	188
	References	189

4 Inclusions in Finite Matrix — 191

4.1	General Approaches for Numerically Solving Boundary Value Problems	192
	4.1.1 Method of Weighted Residuals	192
	4.1.2 Rayleigh–Ritz Method	203
	4.1.3 Sturm–Liouville System	205
4.2	Steady-State Heat Conduction Equations	213
	4.2.1 Derivation of Permissible Functions	213
	4.2.2 Finding Temperature Field Using Permissible Functions	227
4.3	Elastic Fields with Bounded Boundaries	232
4.4	Numerical Examples	238
	4.4.1 Homogeneous Medium	238
	4.4.2 Single Inclusion	240
4.5	Exercises	251
	References	252

Appendix A Introduction to *Mathematica* — 253

A.1	Essential Commands/Statements	255
A.2	Equations	256
A.3	Differentiation/Integration	260
A.4	Matrices/Vectors/Tensors	260
A.5	Functions	262
A.6	Graphics	263
A.7	Other Useful Functions	265
A.8	Programming in *Mathematica*	267
	A.8.1 Control Statements	268
	A.8.2 Tensor Manipulations	270
	References	272

Index — 273

Preface

Micromechanics is a branch of applied mechanics that began with the celebrated paper of Eshelby published in 1957. It refers to analytical methods for solid mechanics that can describe deformations as functions of such microstructures as voids, cracks, inclusions, and dislocations. Micromechanics is an essential tool for obtaining mechanical fields analytically in modern materials including composite and nanomaterials that did not exist 50 years ago.

There exist a number of well-written books with a similar subject title to this book (micromechanics, continuum mechanics with computer algebra, etc.). However, many of them are written by mathematicians or theoretical physicists that follow the strict style of rigorous formality (theorem, corollary, etc.), which may easily discourage aspiring students without formal background in mathematics and physics yet who want to learn what micromechanics has to offer.

The threshold of micromechanics seems high because many formulas and derivations are based on tensor algebra and analysis that calls for a substantial amount of algebra. Although it is a routine type of work, evaluation of tensorial equations requires tedious manual calculations. This scheme all changed in the 1980s with the emergence of computer algebra systems that made it possible to crunch symbols instead of numbers. It is no longer necessary to spend endless time on algebra manually as symbolically capable software such as Maple and *Mathematica* can handle complex tensor equations

The aim of this book is to introduce the concept of micromechanics in plain terms without rigorousness yet still maintaining consistency with a target audience of those who want to actually use the result of micromechanics for multiphase/heterogeneous materials, taking advantage of a computer algebra system, *Mathematica*, rather than those who need formal and rigorous derivations of the equations in micromechanics. The author has been a fan of *Mathematica* since the 1990s and believes that it is the best tool for handling subjects in micromechanics that require both analytical and numerical computations. Unlike numerically oriented computer languages such as C and Fortran, *Mathematica* can process both symbols and numerics seamlessly, thus being capable of handling lengthy tensorial manipulations that can release mundane and tedious jobs by human beings. There have been intense debates in user communities about the difference and preference among *Mathematica* and other numerical software such as MATLAB, all of which are widely used in engineering and scientific communities. The major difference is that software such as MATLAB offers only a limited support for symbolic variables through licensing Maple and is not integrated in the system seamlessly, whereas in *Mathematica*, there is no distinction between symbolic and numerical variables; more importantly, it is not possible to derive and manipulate formulas employed in this book with MATLAB alone.

One of the unique features in this book is to introduce many examples in micromechanics that can be solved only through computer algebra systems. This includes stress analysis for multiinclusions and the use of the Airy stress function for inclusion problems.

Many of the subjects presented in this book may be classical that may have existed for the past 200 years. Nevertheless, those problems presented in this book would not have been possibly solved analytically had it not been for *Mathematica* or, for that matter, any computer algebra system, which, the author believes, is the raison d'être of this book.

This book consists of four chapters that cover a variety of topics in micromechanics. Each example problem is accompanied with corresponding *Mathematica* code. Chapter 1 introduces the basic concept of the coordinate transformations and the properties of Cartesian tensors that are needed to derive equations in continuum mechanics. In Chapter 2, based on the concepts introduced in Chapter 1, the field equations in continuum mechanics are derived. Coordinate transformations in general curvilinear coordinate systems are discussed. Chapter 3 presents a new paradigm for inclusion problems embedded in an infinite matrix. After a brief introduction of the Eshelby method, new analytical approaches to derive the stress fields for an inclusion and concentrically placed inclusions in an infinite matrix are discussed along with their implementations in *Mathematica*. Chapter 4 is devoted to the inclusion problems where the matrix is finite-sized. The classical Galerkin method is combined with *Mathematica* to derive the physical and mechanical fields semi-analytically. The Appendix is an introduction to *Mathematica* that provides sufficient background information in order to understand the *Mathematica* code presented in this book.

Seiichi Nomura
Arlington, Texas

About the Companion Website

This book is accompanied by a companion website:

www.wiley.com/go/nomura0615

This website includes:

- Listing of Selected Programs
- A Solutions Manual
- An Exercise Section

1

Coordinate Transformation and Tensors

To describe the state of the deformation for a deformable body, the coordinate transformation plays an important rule, and the most appropriate way to represent the coordinate transformation is to use tensors. In this chapter, the concept of coordinate transformations and the introduction to tensor algebra in the Cartesian coordinate system are presented along with their implementation in *Mathematica*. As this book is not meant to be a textbook on continuum mechanics, the readers are referred to some good reference books including Romano et al. (2006) and Fung (1965), among others. Manipulation involving indices requires a considerable amount of algebra work when the expressions become lengthy and complicated. It is not practical to properly handle and evaluate quantities that involve tensor manipulations by conventional scientific/engineering software such as FORTRAN, C, and MATLAB. Software packages capable of handling symbolic manipulations include *Mathematica* (Wolfram 1999), Maple (Garvan 2001), and others. In this book, *Mathematica* is exclusively used for implementation and evaluation of derived formulas. A brief introduction to the basic commands in *Mathematica* is found in the appendix, which should be appropriate to understand and execute the *Mathematica* code used in this book.

1.1 Index Notation

If one wants to properly express the deformation state of deformable bodies regardless of whether they are solids or fluids, the use of tensor equations is essential. There are several different ways to denote notations of tensors, one of which uses indices and others without using indices at all. In this book, the index notation is exclusively used throughout to avert unnecessary abstraction at the expense of mathematical sophistication.

The following are the main compelling reasons to mandate the use of tensor notations in order to describe the deformation state of bodies correctly.

1. The principle of physics stipulates that a physically meaningful object must be described independent of the frame of references.[1] If the equation for a physically meaningful object changes depending on the coordinate system used, that equation is no longer a correct equation.
2. Tensor equations can be shown to be invariant under the coordinate transformation. Tensor equations are thus defined as those equations that are unchanged from one coordinate system to another.

Hence, by combining the two aforementioned statements, it can be concluded that only tensor equations can describe the physical objects properly. In other words, if an equation is not in tensorial format, the equation does not represent the object physically.

The index notation, also known as the Einstein notation (Einstein et al. 1916)[2] or the summation convention, is the most widely used notation to represent tensor quantities, which will be used in this book. The index notation in the Cartesian coordinate system is summarized as follows:

1. For mathematical symbols that are referred to quantities in the x, y, and z directions, use subscripts, 1, 2, 3, as in x_1, x_2, x_3 or a_1, a_2, a_3, instead of x, y, z or a, b, c. The subscripted numbers 1, 2, and 3, refer to the x, y, and z directions, respectively. Obviously, the upper limit of the number is 2 for 2-D and 3 for 3-D.
2. If there are twice repeated indices in a term of products such as $a_i b_i$, the summation with respect to that index (i) is always assumed. For example,

$$a_i b_i \equiv \sum_{i=1}^{3} a_i b_i = a_1 b_1 + a_2 b_2 + a_3 b_3 \quad \text{(3-D)}.$$

There is no exception to this rule. An expression such as $a_i b_i c_i$ is not allowed as the number of repetitions is 3 instead of 2.

A repeated index is called the *dummy index* as it does not matter what letter is used, and an unrepeated index is called the *free index*.[3] For example,

$$x_i x_i = x_j x_j = x_\alpha x_\alpha,$$

all of which represent a summation ($= \sum_i x_i x_i$). An unrepeated index such as x_i (or x_j or x_α) stands for one of x_1, x_2, or x_3.

It should be noted that the notations and conventions introduced are valid for the Cartesian coordinate system only. In a curvilinear coordinate system such as the spherical coordinate system, the length of base vectors is not necessarily unity, and this mandates the aforementioned index notation to be modified to reflect the difference between the contravariant components and the covariant components, which will be discussed in Chapter 2.

---

[1] This "frame of references" refers to the Galilean transformation in classical mechanics, the Lorentz transformation in special relativity, and the general curvilinear transformation in general relativity.
[2] Albert Einstein introduced this notation in 1916.
[3] This is similar to definite integrals. The variable used in a definite integral does not matter as

$$\int_a^b f(x)dx = \int_a^b f(y)dy = \int_a^b f(z)dz.$$

The variables x, y, and z are called *dummy variables*.

1.1.1 Some Examples of Index Notation in 3-D

1. $x_i x_i$

   As the index i is repeated, the summation symbol, $\sum$, must be added in front, i.e.,

   $$x_i x_i = \sum_{i=1}^{3} x_i x_i$$
   $$= x_1 x_1 + x_2 x_2 + x_3 x_3$$
   $$= x^2 + y^2 + z^2.$$

   Note that $x_i x_i$ is different from $(x_i)^2$. While $x_i x_i$ represents a single expression with three terms, $(x_i)^2$ represents one of the three expressions $((x_1)^2, (x_2)^2, \text{ or } (x_3)^2)$.

2. $x_i x_i x_j$

   Note that the index i is a dummy index (repeated twice) while the index j is a free index (no repeat). Therefore,

   $$x_i x_i x_j = x_j \sum_{i=1}^{3} x_i x_i$$
   $$= x_j ((x_1)^2 + (x_2)^2 + (x_3)^2)$$
   $$= \begin{cases} x(x^2 + y^2 + z^2) \\ \quad \text{or} \\ y(x^2 + y^2 + z^2) \\ \quad \text{or} \\ z(x^2 + y^2 + z^2). \end{cases}$$

3. $x_i x_i x_i$

   This is not a valid tensor expression as the number of repeated indices must be 2.

1.1.2 Mathematica Implementation

As *Mathematica* itself does not support tensor manipulation natively, it is necessary to devise a way to handle index notation and tensor manipulation. In this book, a list or a list of lists (a nested list) is used to represent tensor quantities. Using a nested list to define a tensor of any rank is straightforward but at the same time limited to the Cartesian tensors. For tensors defined in a curvilinear coordinate system, a slightly different approach is needed.

When running *Mathematica* first time, a default directory should be selected so that all the notebook files can be saved and accessed in this directory. By default, *Mathematica* looks for all the files stored in `c:\users\<user>\` where `<user>` is the user's home directory.[4] The `SetDirectory` command can change this location. For example, if you want to change the default directory to `c:\tmp`, the `SetDirectory` command can specify the default directory as

---

[4] The Windows operating system uses "\" (backslash) as the directory delimiter while the Unix system uses "/" (forward slash) as the directory delimiter. However, the "/" in the SetDirectory command in *Mathematica* works for both.

In[1]:= `SetDirectory["c:/tmp"]`

Out[1]= `c:\tmp`

It is noted that the directory delimiter needs to be entered as "/" (forward slash) even though the Windows delimiter character is "\" (backslash).

To enter a three-dimensional vector, $\mathbf{v} = (x^2, y^2, z^2)$, the following *Mathematica* command can be entered to create a list with braces (curly brackets) as

In[2]:= `v = {x^2, y^2, z^2}`

Out[2]= $\{x^2, y^2, z^2\}$

An individual component of **v** can be referenced using double square brackets (`[[...]]`) as

In[3]:= `v[[1]]`

Out[3]= x^2

The partial derivative of **v** with respect to x can be entered as

In[4]:= `D[v, x]`

Out[4]= `{2 x, 0, 0}`

You can also differentiate an individual component as

In[5]:= `D[v[[2]], y]`

Out[5]= `2 y`

Implementation of the coordinate component, x_i, into *Mathematica* can be done by using the `Table` function. To define a position vector, **r**, whose components are (x_1, x_2, x_3), enter

In[1]:= `r = Table[x[i], {i, 1, 3}]`

Out[1]= `{x[1], x[2], x[3]}`

The given `Table` function generates a list of elements. For example, the following command generates a sequence of i^2 for $i = 1, \ldots, 10$.

In[2]:= `Table[i^2, {i, 1, 10}]`

Out[2]= `{1, 4, 9, 16, 25, 36, 49, 64, 81, 100}`

In the definition of the position vector, **r**, it is noted that the coordinate components, (x_1, x_2, x_3), are entered as `x[1]`, `x[2]`, `x[3]` instead of `x[[1]]`, `x[[2]]`, `x[[3]]`. It is important to distinguish a single square bracket (`[...]`) and a double square bracket (`[[...]]`). The single square bracket (`[...]`) is for a parameter used in a function. The quantity, `x[1]`, stands for a function, x, with the argument of 1. By using a single square bracket, the quantities such as `x[1]`, `x[2]` can stand for themselves, meaning that initial values do not have to be preassigned. On the other hand, if `x[[1]]` were used instead of `x[1]`, 0 would be returned unless a list x is previously defined.

To define a function in *Mathematica*, use the following syntax:

In[1]:= `f[x_] := x^2 + 1`

In[2]:= `f[6]`

Out[2]= `37`

In[3]:= `f[a^2]`

Out[3]= $1 + a^4$

In the aforementioned example code, a user-defined function, `f[x]`, that returns $x^2 + 1$ is defined. The syntax is such that variables of the function must be presented to the left of the equal sign with the underscore and the definition of the function is given to the right of a colon and an equal sign ($:=$). It is important to note that a function in *Mathematica* returns itself if no prior definition is given.

In[1]:= `f[x_, y_] := x^2 + y^2`

In[2]:= `f[a^3, b^2]`

Out[2]= $a^6 + b^4$

In[3]:= `f[1]`

Out[3]= `f[1]`

In[4]:= `g[a]`

Out[4]= `g[a]`

In the aforementioned example, `f[x, y]` is defined as a function that takes two variables returning $x^2 + y^2$. However, when `f` is called with only one variable, `1`, it returns itself, i.e., `f[1]`, as `f` with only one variable has not been defined. When `g[a]` is entered, it returns itself without evaluation as there is no prior definition of `g[x]` given. It is this property of a function in *Mathematica* that enables manipulating index notations.

As an example of using `x[i]` as the coordinate components, here is how to implement the summation convention. As *Mathematica* does not have support for the summation convention built in, if `x[i]x[i]` meant as $x_i x_i = x_1^2 + x_2^2 + x_3^2$ is entered as

In[1]:= `x[i] x[i]`

Out[1]= `x[i]`^{2}

Mathematica does not automatically expand $x_i x_i$ and reduce the result to r^2. Hence, it is necessary to explicitly use the `Sum[]` command as

In[6]:= `Sum[x[i] x[i], {i, 1, 3}]`

Out[6]= `x[1]`^{2} + `x[2]`^{2} + `x[3]`^{2}

for $x_i x_i$.

However, it is possible to implement the summation convention in *Mathematica* with the following procedure.

```
In[1]:= Unprotect[Times];
        x[i_Symbol] x[i_Symbol] := r^2;
        Protect[Times];
```

The aforementioned three-line code adds a new rule to automatically replace x[i]x[i] by r^2 through pattern matching. When *Mathematica* evaluates the product of two quantities, it calls its internal function, Times, which tries to simplify the result with various pattern-matching algorithms. It is not possible to modify the pattern-matching algorithms as they are part of the definition of the Times function and they are protected by default. Therefore, it is necessary to unprotect the multiplication operator, Times, in *Mathematica* with the Unprotect command so that a new rule for the summation convention can be added. The second line is to tell *Mathematica* a new rule that whenever a pattern of x[i]x[i] where i is any symbol but not a number is entered, it is automatically replaced by r^2. Finally, in the third line, the Times function is given back the Protect attribute so that any further modification is prevented. After these three lines are entered, the summation convention for x[i] x[i] is automatic as

```
In[4]:= x[j] x[j]
Out[4]= r²

In[5]:= x[3] x[3]
Out[5]= x[3]²

In[6]:= x[j] x[j] x[i]
Out[6]= r² x[i]

In[7]:= x[i] x[i] x[j] x[j]
Out[7]= r⁴
```

In the aforementioned examples, $x_j x_j$ is reduced to r^2 but $x_3 x_3$ remains unchanged. The expression, $x_j x_j x_i$ is simplified to $r^2 x_i$ and $x_i x_i x_j x_j$ is reduced to r^4. With this pattern-matching capability of *Mathematica*, manipulation of tensor quantities can be greatly simplified. More detailed examples will be shown later.

1.1.3 Kronecker Delta

One of the most important symbols in tensor algebra is the Kronecker delta, which is defined as

$$\delta_{ij} = \begin{cases} 1, & \text{if } (i,j) = (1,1), (2,2), \text{ or } (3,3) \\ 0, & \text{otherwise.} \end{cases}$$

Coordinate Transformation and Tensors

The Kronecker delta is what is equivalent to 1 in numbers or the identity matrix in linear algebra. It is also used to define the inverse of a tensor.

Examples
1. δ_{ii} (3-D)

$$\delta_{ii} = \sum_{i=1}^{3} \delta_{ii}$$
$$= \delta_{11} + \delta_{22} + \delta_{33}$$
$$= 3.$$

2. $\delta_{ij}\delta_{ij}$ (3-D)

$$\delta_{ij}\delta_{ij} = \sum_{i=1}^{3}\sum_{j=1}^{3} \delta_{ij}\delta_{ij}$$
$$= \delta_{11}\delta_{11} + \delta_{12}\delta_{12} + \delta_{13}\delta_{13}$$
$$+ \delta_{21}\delta_{21} + \delta_{22}\delta_{22} + \delta_{23}\delta_{23}$$
$$+ \delta_{31}\delta_{31} + \delta_{32}\delta_{32} + \delta_{33}\delta_{33}$$
$$= 3.$$

3. $\delta_{ii}\delta_{jj}$ (3-D)

$$\delta_{ii}\delta_{jj} = \sum_{i=1}^{3} \delta_{ii} \sum_{j=1}^{3} \delta_{jj}$$
$$= (\delta_{11} + \delta_{22} + \delta_{33})(\delta_{11} + \delta_{22} + \delta_{33})$$
$$= 9.$$

4. $\delta_{ij}x_i x_j$ (3-D)

$$\delta_{ij}x_i x_j = \sum_{i=1}^{3}\sum_{j=1}^{3} \delta_{ij}x_i x_j$$
$$= \delta_{11}x_1 x_1 + \delta_{12}x_1 x_2 + \delta_{13}x_1 x_3$$
$$+ \delta_{21}x_2 x_1 + \delta_{22}x_2 x_2 + \delta_{23}x_2 x_3$$
$$+ \delta_{31}x_3 x_1 + \delta_{32}x_3 x_2 + \delta_{33}x_3 x_3$$
$$= (x_1)^2 + (x_2)^2 + (x_3)^2.$$

The Kronecker delta can be implemented in *Mathematica* by several different ways. The following implementation is to use a function to define the Kronecker delta, which is valid for any number of dimensions.

```
In[1]:= delta[i_Integer, j_Integer] := If[i == j, 1, 0]

In[2]:= delta[1, 2]

Out[2]= 0

In[3]:= delta[i, j]

Out[3]= delta[i, j]

In[4]:= Sum[delta[i, i], {i, 3}]

Out[4]= 3
```

The underline "_" after the variable name is used by *Mathematica* to restrict the type of variable to be the type that follows _. In this case, delta[i, j] is evaluated only when both i and j are integers. The function If[condition, t, f] gives t if condition is true and f otherwise. Note that the two equal signs (==) in "i==j" mean equality, not an assignment as in most of the programming languages.

1.1.3.1 Summation Convention for δ_{ij}

Some of the properties of the Kronecker delta that involve the summation convention include:

$$\delta_{ii} = 3, \quad a_j \delta_{ij} = a_i, \quad a_{ij} \delta_{ik} = a_{kj}, \quad \delta_{ij} \delta_{ik} = \delta_{jk}.$$

The following *Mathematica* code implements the aforementioned rules so that summation convention that involves the Kronecker delta is always automatically executed.

```
In[10]:= SetAttribute[delta, Orderless];
         delta[i_Integer, j_Integer] := If[i == j, 1, 0];
         delta[i_Symbol, i_Symbol] := 3;

In[14]:= Unprotect[Times];
         Times[a_Symbol[j_Symbol], delta[i_, j_Symbol]] := a[i];
         Times[a_Symbol[i_Symbol, j_], delta[i_Symbol, k_]] := a[k, j];
         Times[delta[i_Symbol, j_], delta[i_Symbol, k_]] := delta[j, k];
         Protect[Times];
```

The SetAttribute[delta, Orderless] command sorts the variables in delta into standard order so that delta[j,i] is automatically changed to delta[i,j]. The part i_Integer specifies that the variable i must be an integer value, and the part i_Symbol specifies that the variable i must be a symbol, not a number. The subsequent code instructs *Mathematica* to add new rules that use the summation convention with delta[i,j]. After this *Mathematica* code is entered, the summation convention involving the Kronecker delta is automatically enforced.

```
In[9]:= delta[i, i]

Out[9]= 3

In[10]:= a[j] delta[i, j]

Out[10]= a[i]

In[11]:= a[i, j] delta[i, k]

Out[11]= a[k, j]

In[12]:= delta[i, j] delta[i, k]

Out[12]= delta[j, k]
```

The aforementioned automatic conversion makes manipulation of tensor products easier as will be seen in later chapters.

1.1.4 Permutation Symbols

The permutation symbol,[5] ϵ_{ijk}, is defined as

$$\epsilon_{ijk} = \begin{cases} 1, & (ijk) = (123), (231), (312), \\ -1, & (ijk) = (213), (321), (132), \\ 0, & \text{otherwise.} \end{cases}$$

The permutation symbol, ϵ_{ijk}, takes a value of 1 if (ijk) is an even permutation[6] of (123) and a value of 0 if (ijk) is an odd permutation of (123). Note that ϵ_{ijk} is 0 if any two indices are the same.

The permutation symbol is used for the vector product. The ith component of vector product, $\mathbf{a} \times \mathbf{b}$, is expressed as

$$(a \times b)_i = \epsilon_{ijk} a_j b_k.$$

In *Mathematica*, the permutation symbol is built-in and can be invoked by the Signature function as

```
In[1]:= Signature[{1, 2, 3}]

Out[1]= 1

In[2]:= Signature[{1, 1, 3}]

Out[2]= 0
```

---

[5] The permutation symbol is defined in 3-D only.
[6] If a sequence, (ijk), is obtained from (123) with an even number of permutations of two numbers, it is called an even permutation. Otherwise, it is called an odd permutation.

The vector cross product, **a** × **b**, can be implemented as

In[3]:= **Table[Sum[Signature[{i, j, k}] a[i] b[j], {i, 3}, {j, 3}], {k, 3}]**

Out[3]= {-a[3] b[2] + a[2] b[3], a[3] b[1] - a[1] b[3], -a[2] b[1] + a[1] b[2]}

The rotation operation, rot **v**, can be implemented as

In[4]:= **Table[Sum[Signature[{i, j, k}] Dt[v, x[k]], {j, 3}, {k, 3}], {i, 3}]**

Out[4]= {-Dt[v, x[2]] + Dt[v, x[3]], Dt[v, x[1]] - Dt[v, x[3]], -Dt[v, x[1]] + Dt[v, x[2]]}

where Dt[f, x] is the total derivative, df/dx.

1.1.5 Product of Matrices

As another example of summation convention, the product of $n \times n$ matrices, A and B, is expressed as

$$AB = C,$$

or

$$\begin{pmatrix} a_{11} & a_{12} & \cdots & a_{1n} \\ a_{21} & a_{22} & \cdots & a_{2n} \\ \cdots & \cdots & \cdots & \cdots \\ a_{n1} & a_{n2} & \cdots & a_{nn} \end{pmatrix} \begin{pmatrix} b_{11} & b_{12} & \cdots & b_{1n} \\ b_{21} & b_{22} & \cdots & b_{2n} \\ \cdots & \cdots & \cdots & \cdots \\ b_{n1} & b_{n2} & \cdots & b_{nn} \end{pmatrix} = \begin{pmatrix} c_{11} & c_{12} & \cdots & c_{1n} \\ c_{21} & c_{22} & \cdots & c_{2n} \\ \cdots & \cdots & \cdots & \cdots \\ c_{n1} & c_{n2} & \cdots & c_{nn} \end{pmatrix}.$$

The ij component of C is expressed as

$$c_{ij} = a_{i1}b_{1j} + a_{i2}b_{2j} + a_{i3}b_{3j} + \cdots + a_{in}b_{nj}$$

$$= \sum_{k=1}^{n} a_{ik}b_{kj}$$

$$= a_{ik}b_{kj}. \tag{1.1}$$

Equation (1.1) indicates that the (ij) component of C can be computed by the (ik) component of A and the (kj) component of B.

In *Mathematica*, a matrix with the components in symbolic form can be entered as a nested array as

In[2]:= **mat = Table[a[i, j], {i, 1, 3}, {j, 1, 3}]**

Out[2]= {{a[1, 1], a[1, 2], a[1, 3]},
{a[2, 1], a[2, 2], a[2, 3]}, {a[3, 1], a[3, 2], a[3, 3]}}

In[5]:= **MatrixForm[mat]**

Out[5]//MatrixForm=
$$\begin{pmatrix} a[1, 1] & a[1, 2] & a[1, 3] \\ a[2, 1] & a[2, 2] & a[2, 3] \\ a[3, 1] & a[3, 2] & a[3, 3] \end{pmatrix}$$

The multiplication between two matrices can be carried out by a dot (.).

In[6]:= **amat = Table[a[i, j], {i, 3}, {j, 3}]**

Out[6]= {{a[1, 1], a[1, 2], a[1, 3]},
        {a[2, 1], a[2, 2], a[2, 3]}, {a[3, 1], a[3, 2], a[3, 3]}}

In[7]:= **bmat = Table[b[i, j], {i, 3}, {j, 3}]**

Out[7]= {{b[1, 1], b[1, 2], b[1, 3]},
        {b[2, 1], b[2, 2], b[2, 3]}, {b[3, 1], b[3, 2], b[3, 3]}}

In[8]:= **cmat = amat.bmat**

Out[8]= {{a[1, 1] b[1, 1] + a[1, 2] b[2, 1] + a[1, 3] b[3, 1],
         a[1, 1] b[1, 2] + a[1, 2] b[2, 2] + a[1, 3] b[3, 2],
         a[1, 1] b[1, 3] + a[1, 2] b[2, 3] + a[1, 3] b[3, 3]},
        {a[2, 1] b[1, 1] + a[2, 2] b[2, 1] + a[2, 3] b[3, 1],
         a[2, 1] b[1, 2] + a[2, 2] b[2, 2] + a[2, 3] b[3, 2],
         a[2, 1] b[1, 3] + a[2, 2] b[2, 3] + a[2, 3] b[3, 3]},
        {a[3, 1] b[1, 1] + a[3, 2] b[2, 1] + a[3, 3] b[3, 1],
         a[3, 1] b[1, 2] + a[3, 2] b[2, 2] + a[3, 3] b[3, 2],
         a[3, 1] b[1, 3] + a[3, 2] b[2, 3] + a[3, 3] b[3, 3]}}

Note that ordinary multiplication in *Mathematica* is either an asterisk (∗) or a space.

1.2 Coordinate Transformations (Cartesian Tensors)

In this section, Cartesian tensors that are subject to transformational rules between two Cartesian coordinate systems are introduced. This automatically excludes such curvilinear coordinate systems as the polar coordinate system. Tensors that are used in general curvilinear coordinate systems will be discussed in Chapter 2.

Consider a rotation of a coordinate system from a Cartesian coordinate system, x_i, or (x, y), to another Cartesian coordinate system, $x_{\bar{i}}$, or $(\bar{x}, \bar{y})$ as shown in Figure 1.1.

Note that, although the coordinate systems are changed, the position vector, **r**, itself remains unchanged. Therefore, it follows

$$\mathbf{r} = x_1 \mathbf{e}_1 + x_2 \mathbf{e}_2$$
$$= x_{\bar{1}} \mathbf{e}_{\bar{1}} + x_{\bar{2}} \mathbf{e}_{\bar{2}},$$

or

$$x_{\bar{i}} \mathbf{e}_{\bar{i}} = x_i \mathbf{e}_i, \qquad (1.2)$$

where $(\mathbf{e}_1, \mathbf{e}_2)$ and $(\mathbf{e}_{\bar{1}}, \mathbf{e}_{\bar{2}})$ are the base vectors in the original coordinate system and the rotated coordinate system, respectively. It is noted that the base vectors are normalized and mutually orthogonal as

$$(\mathbf{e}_i, \mathbf{e}_j) = \delta_{ij},$$

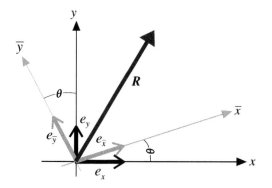

Figure 1.1 Two-dimensional coordinate transformation

where $(\mathbf{a}, \mathbf{b})$ is the inner product between two vectors, $\mathbf{a}$ and $\mathbf{b}$. By multiplying $\mathbf{e}_{\bar{j}}$ on both sides of Equation (1.2) in the inner product, one obtains

$$x_{\bar{i}}(\mathbf{e}_{\bar{i}}, \mathbf{e}_{\bar{j}}) = x_i(\mathbf{e}_i, \mathbf{e}_{\bar{j}}),^{7}$$

or

$$x_{\bar{i}} \delta_{\bar{i}\bar{j}} = x_i(\mathbf{e}_i, \mathbf{e}_{\bar{j}}),$$

or

$$x_{\bar{j}} = \beta_{\bar{j}i} x_i, \qquad (1.3)$$

where

$$\beta_{\bar{j}i} \equiv (\mathbf{e}_{\bar{j}}, \mathbf{e}_i).$$

For 2-D, Equation (1.3) is written in conventional form as

$$\begin{pmatrix} x_{\bar{1}} \\ x_{\bar{2}} \end{pmatrix} = \begin{pmatrix} \cos\theta, & \sin\theta \\ -\sin\theta, & \cos\theta \end{pmatrix} \begin{pmatrix} x_1 \\ x_2 \end{pmatrix}.^{8} \qquad (1.4)$$

The matrix, $\beta_{\bar{j}i}$, is called the transformation matrix and is a unitary matrix based on which the definition of the Cartesian tensors is made.

The inverse of the transformation matrix, $\beta_{\bar{j}i}$, can be obtained by replacing θ in Equation (1.4) by $-\theta$ as

$$B^{-1} = \begin{pmatrix} \cos\theta, & \sin\theta \\ -\sin\theta, & \cos\theta \end{pmatrix}^{-1}$$

$$= \begin{pmatrix} \cos(-\theta), & \sin(-\theta) \\ -\sin(-\theta), & \cos(-\theta) \end{pmatrix}$$

$$= \begin{pmatrix} \cos\theta, & -\sin\theta \\ \sin\theta, & \cos\theta \end{pmatrix},$$

---

[7] When the Kronecker delta, δ_{ij}, is multiplied by another tensor one of whose indices is contracted (summed) with one of the indices of δ_{ij}, it follows

$$\delta_{ij} a_{jklmn...} = a_{iklmn...}$$

that is, the Kronecker delta is absorbed.

[8] The matrix can be easily memorized if you realize that the second row is the differentiation of the first row.

or
$$B^{-1} = B^T,$$

where B^T is the transpose of B.

Here is how to enter the 2-D transformation matrix in *Mathematica*.

```
In[1]:= β = {{Cos[θ], Sin[θ]}, {-Sin[θ], Cos[θ]}}

Out[1]= {{Cos[θ], Sin[θ]}, {-Sin[θ], Cos[θ]}}

In[2]:= xnew = β.Table[x[i], {i, 1, 2}]

Out[2]= {Cos[θ] x[1] + Sin[θ] x[2], -Sin[θ] x[1] + Cos[θ] x[2]}
```

We can enter the Greek letter, β, from a keyboard by hitting Esc, b, and Esc keys. The transformation matrix, B, is entered as a 2×2 list. Note that, for the product between lists, a dot (·) must be used instead of a space or an asterisk.

1.3 Definition of Tensors

Tensors are geometrical quantities whose values are subject to the transformational rules. Tensor equations are invariant under any coordinate transformation, which is the reason why any physically significant equation must be written as a tensor equation.

In this book, only Cartesian tensors are used throughout. The Cartesian tensors are defined through the transformation matrix, $\beta_{ij}(\theta)$, of Equation (1.4) between two Cartesian coordinate systems. Zeroth- and first-rank Cartesian tensors have aliases, i.e., scalars and vectors, respectively, but there are no specific names assigned for tensors of rank 2 and higher.

Here are the definitions of Cartesian tensors of different ranks.

1.3.1 Tensor of Rank 0 (Scalar)

A single quantity, Φ, is called a zeroth-rank tensor (also called a tensor of rank 0 or a *scalar*) if Φ satisfies the following:

(a) The quantity, $\Phi(x_i)$, has a single component (obviously!).
(b) If $\Phi(x_{\bar{\imath}})$ denotes a quantity where x_i in $\Phi(x_i)$ is replaced by $x_{\bar{\imath}}$, the following is held:
$$\Phi(x_{\bar{\imath}}) = \Phi(x_i).$$

Examples
1. $\Phi(x_i) \equiv 1$.
   This is a scalar as Φ remains unchanged no matter what the coordinate system is.
2. $\Phi(x, y) \equiv x^2 + y^2$.
$$\begin{aligned}\Phi(\bar{x}, \bar{y}) &= \bar{x}^2 + \bar{y}^2 \\ &= (x\cos\theta + y\sin\theta)^2 + (-x\sin\theta + y\cos\theta)^2 \\ &= x^2 + y^2 \\ &= \Phi(x, y).\end{aligned}$$

Hence, $\Phi(x, y)$ is a scalar.

3. $\Phi(x, y) \equiv x^2 - y^2$.

$$\Phi(\bar{x}, \bar{y}) = \bar{x}^2 - \bar{y}^2$$
$$= (x\cos\theta + y\sin\theta)^2 - (-x\sin\theta + y\cos\theta)^2$$
$$\neq \Phi(x, y).$$

Hence, $\Phi(x, y)$ is not a scalar.

Here is a *Mathematica* code to show that $x_i x_i$ is a scalar.

```
In[1]:= β = {{Cos[θ], Sin[θ]}, {-Sin[θ], Cos[θ]}}

Out[1]= {{Cos[θ], Sin[θ]}, {-Sin[θ], Cos[θ]}}

In[2]:= xnew = β.Table[ xold[i], {i, 2}]

Out[2]= {Cos[θ] xold[1] + Sin[θ] xold[2], -Sin[θ] xold[1] + Cos[θ] xold[2]}

In[3]:= xnew[[1]]^2 + xnew[[2]]^2

Out[3]= (-Sin[θ] xold[1] + Cos[θ] xold[2])^2 + (Cos[θ] xold[1] + Sin[θ] xold[2])^2

In[4]:= Simplify[%]

Out[4]= xold[1]^2 + xold[2]^2
```

In the aforementioned example, β is the 2-D transformation matrix and xold represents (x, y) in the original coordinate system. The variables in xnew are the components in the rotated coordinate system, $(\bar{x}, \bar{y})$, which can be obtained by $\beta_{ij} x_j$. The quantity $x_{\bar{i}} x_{\bar{i}}$ is computed and is simplified by using the Simplify function to be reduced to $x_i x_i$.

1.3.2 Tensor of Rank 1 (Vector)

A tensor of rank 1 is defined as a quantity, Φ_j, a function of x_i, satisfying the following properties:

(a) Having two components in 2-D (3 in 3-D).
(b) When the coordinate system is changed from x_i to $x_{\bar{i}}$,

$$\Phi_{\bar{i}}(x_{\bar{1}}, x_{\bar{2}}) = \beta_{\bar{i}j} \Phi_j(x_1, x_2).$$

Examples

1. $\Phi_1 = x$, $\Phi_2 = y$.
    This is obviously a tensor (vector) as the components of Φ_i are identical to those of the coordinate components.
2. $\Phi_1 = y$, $\Phi_2 = x$.
    To check to see if this quantity is a tensor, one has to test the transformation rule for the first-rank tensor,

$$\Phi_{\bar{1}} = \bar{y}$$
$$= -x\sin\theta + y\cos\theta,$$

$$\beta_{\bar{1}j}\Phi_j = \beta_{\bar{1}1}\Phi_1 + \beta_{\bar{1}2}\Phi_2$$
$$= y\cos\theta + x\sin\theta,$$

Therefore,
$$\Phi_{\bar{1}} \neq \beta_{\bar{1}j}\Phi_j.$$

Hence, Φ_i is not a vector.

3. $\Phi_1 = -y$, $\Phi_2 = x$.
$$\Phi_{\bar{1}} = -\bar{y}$$
$$= x\sin\theta - y\cos\theta,$$
$$\beta_{\bar{1}j}\Phi_j = \beta_{\bar{1}1}\Phi_1 + \beta_{\bar{1}2}\Phi_2$$
$$= -y\cos\theta + x\sin\theta.$$

Therefore,
$$\Phi_{\bar{1}} = \beta_{\bar{1}j}\Phi_j.$$

Also,
$$\Phi_{\bar{2}} = \bar{x}$$
$$= x\cos\theta + y\sin\theta,$$
$$\beta_{\bar{2}j}\Phi_j = \beta_{\bar{2}1}\Phi_1 + \beta_{\bar{2}2}\Phi_2$$
$$= y\sin\theta + x\cos\theta.$$

Therefore,
$$\Phi_{\bar{2}} = \beta_{\bar{2}j}\Phi_j.$$

Hence, Φ_i is a vector.

Here is an example of a code to transform ϕ_i with $\phi_1 = y$, $\phi_2 = -x$.

```
In[5]:= β = {{Cos[θ], Sin[θ]}, {-Sin[θ], Cos[θ]}};
       ϕ = {-y, x};
       Table[Sum[β[[i, j]] ϕ[[j]], {j, 2}], {i, 2}]
Out[7]= {-y Cos[θ] + x Sin[θ], x Cos[θ] + y Sin[θ]}
```

1.3.3 Tensor of Rank 2

A tensor of rank 2 (a second-rank tensor) is defined as a quantity, Φ_{ij}, satisfying the following properties:

(a) Having 4 components (9 in 3-D).
(b) When the coordinate system is changed from x_i to $x_{\bar{i}}$, the following relation holds:
$$\Phi_{\bar{i}\bar{j}}(x_{\bar{k}}) = \beta_{\bar{i}i}\beta_{\bar{j}j}\Phi_{ij}(x_k).$$

Examples
1.
$$\Phi_{ij} = \begin{pmatrix} 1 & 0 \\ 0 & 1 \end{pmatrix},$$

$$\Phi_{\bar{i}\bar{j}} = \begin{pmatrix} 1 & 0 \\ 0 & 1 \end{pmatrix}.$$

To prove that this is a tensor, the tensorial transformation rule needs to be examined for every single index. For the indices (1 1),

$$\Phi_{\bar{1}\bar{1}} = 1,$$
$$\beta_{\bar{1}i}\beta_{\bar{1}j}\Phi_{ij} = \beta_{\bar{1}1}\beta_{\bar{1}1}\Phi_{11} + \beta_{\bar{1}1}\beta_{\bar{1}2}\Phi_{12} + \beta_{\bar{1}2}\beta_{\bar{1}1}\Phi_{21} + \beta_{\bar{1}2}\beta_{\bar{1}2}\Phi_{22}$$
$$= \beta_{\bar{1}1}\beta_{\bar{1}1} + \beta_{\bar{1}2}\beta_{\bar{1}2}$$
$$= \cos^2\theta + \sin^2\theta$$
$$= 1.$$

Hence
$$\Phi_{\bar{1}\bar{1}} = \beta_{\bar{1}i}\beta_{\bar{1}j}\Phi_{ij}.$$

This can be repeated for the rest of the indices; thus, Φ_{ij} is a second-rank tensor.

2. A constant matrix:
$$\Phi_{ij} = \begin{pmatrix} 1 & 1 \\ 1 & 1 \end{pmatrix},$$

$$\Phi_{\bar{i}\bar{j}} = \begin{pmatrix} 1 & 1 \\ 1 & 1 \end{pmatrix}.$$

This is not a second-rank tensor. For example, for the indices (1, 1),

$$\Phi_{\bar{1}\bar{1}} = 1,$$
$$\beta_{\bar{1}i}\beta_{\bar{1}j}\Phi_{ij} = \beta_{\bar{1}1}\beta_{\bar{1}1}\Phi_{11} + \beta_{\bar{1}1}\beta_{\bar{1}2}\Phi_{12} + \beta_{\bar{1}2}\beta_{\bar{1}1}\Phi_{21} + \beta_{\bar{1}2}\beta_{\bar{1}2}\Phi_{22}$$
$$= \cos^2\theta + 2\cos\theta\sin\theta + \sin^2\theta.$$

Hence
$$\Phi_{\bar{1}\bar{1}} \neq \beta_{\bar{1}i}\beta_{\bar{1}j}\Phi_{ij}.$$

It is not necessary to check the rest of the indices.

3.
$$\Phi_{ij} = \begin{pmatrix} x^2 & xy \\ xy & y^2 \end{pmatrix}.$$

It is possible to prove that this is a second-rank tensor (exercise).[9]

---

[9] This is known as the moment.

1.3.4 Tensor of Rank 3

A tensor of rank 3 (third-rank tensor) is defined as a quantity, Φ_{ijk}, satisfying the following:

(a) Having eight components (27 in 3-D).
(b) When the coordinate system is changed from x_i to $x_{\bar{i}}$, the following relationship holds:

$$\Phi_{\bar{i}\bar{j}\bar{k}} = \beta_{\bar{i}i}\beta_{\bar{j}j}\beta_{\bar{k}k}\Phi_{ijk}.$$

1.3.5 Tensor of Rank 4

A tensor of rank 4 (fourth-rank tensor) is defined as a quantity, Φ_{ijkl}, satisfying the following:

(a) Having 16 components in 2-D (81 in 3-D).
(b) When the coordinate system is changed from x_i to $x_{\bar{i}}$, the following relationship holds:

$$\Phi_{\bar{i}\bar{j}\bar{k}\bar{l}} = \beta_{\bar{i}i}\beta_{\bar{j}j}\beta_{\bar{k}k}\beta_{\bar{l}l}\Phi_{ijkl}.$$

Fourth-rank tensors are important in continuum mechanics, as elastic constants are an example of fourth-rank tensors.

Here is an example of a code to implement a fourth-rank tensor, C_{ijkl}, defined as

$$C_{ijkl} = \mu(\delta_{ik}\delta_{jl} + \delta_{il}\delta_{jk}) + \lambda\delta_{ij}\delta_{kl},$$

and transform the tensor into the coordinate system rotated from the original by θ.

```
In[9]:= delta[i_, j_] := If[i == j, 1, 0]
 β = {{Cos[θ], Sin[θ]}, {-Sin[θ], Cos[θ]}};
 c = Table[
     µ (delta[i, k] delta[j, l] + delta[i, l] delta[j, k]) +
      λ delta[i, j] delta[k, l],
     {i, 2}, {j, 2}, {k, 2}, {l, 2}];
 cnew = Table[Sum[β[[i, i1]] β[[j, j1]] β[[k, k1]]
       β[[1, 11]] c[[i1, j1, k1, 11]], {i1, 2}, {j1, 2},
      {k1, 2}, {11, 2}], {i, 2}, {j, 2}, {k, 2}, {l, 2}];
```

The quantity c is defined as a fourth-rank tensor. The quantity cnew is an expression of c in the coordinate system rotated from the original coordinate system by θ.

In[13]:= c

Out[13]= {{{{λ + 2 µ, 0}, {0, λ}}, {{0, µ}, {µ, 0}}},
 {{{0, µ}, {µ, 0}}, {{λ, 0}, {0, λ + 2 µ}}}}

It is noted that the fourth-rank tensor is stored as a nested list. For instance, the C_{1111} component can be referenced as

In[14]:= c[[1, 1, 1, 1]]
Out[14]= λ + 2 µ

The components of C_{ijkl} in the new coordinate system are expressed as

In[15]:= **cnew**

Out[15]= $\{\{\{\{(\lambda+2\mu)\text{Cos}[\theta]^4+2\lambda\text{Cos}[\theta]^2\text{Sin}[\theta]^2+4\mu\text{Cos}[\theta]^2\text{Sin}[\theta]^2+$
$(\lambda+2\mu)\text{Sin}[\theta]^4,\lambda\text{Cos}[\theta]^3\text{Sin}[\theta]+2\mu\text{Cos}[\theta]^3\text{Sin}[\theta]-$
$(\lambda+2\mu)\text{Cos}[\theta]^3\text{Sin}[\theta]-\lambda\text{Cos}[\theta]\text{Sin}[\theta]^3-$
$2\mu\text{Cos}[\theta]\text{Sin}[\theta]^3+(\lambda+2\mu)\text{Cos}[\theta]\text{Sin}[\theta]^3\},$
$\{\lambda\text{Cos}[\theta]^3\text{Sin}[\theta]+2\mu\text{Cos}[\theta]^3\text{Sin}[\theta]-$
$(\lambda+2\mu)\text{Cos}[\theta]^3\text{Sin}[\theta]-\lambda\text{Cos}[\theta]\text{Sin}[\theta]^3-$
$2\mu\text{Cos}[\theta]\text{Sin}[\theta]^3+(\lambda+2\mu)\text{Cos}[\theta]\text{Sin}[\theta]^3,$
$\lambda\text{Cos}[\theta]^4-4\mu\text{Cos}[\theta]^2\text{Sin}[\theta]^2+$
$2(\lambda+2\mu)\text{Cos}[\theta]^2\text{Sin}[\theta]^2+\lambda\text{Sin}[\theta]^4\}\},$
$\{\{\lambda\text{Cos}[\theta]^3\text{Sin}[\theta]+2\mu\text{Cos}[\theta]^3\text{Sin}[\theta]-$
$(\lambda+2\mu)\text{Cos}[\theta]^3\text{Sin}[\theta]-\lambda\text{Cos}[\theta]\text{Sin}[\theta]^3-$
$2\mu\text{Cos}[\theta]\text{Sin}[\theta]^3+(\lambda+2\mu)\text{Cos}[\theta]\text{Sin}[\theta]^3,$
$\mu\text{Cos}[\theta]^4-2\lambda\text{Cos}[\theta]^2\text{Sin}[\theta]^2-2\mu\text{Cos}[\theta]^2\text{Sin}[\theta]^2+$
$2(\lambda+2\mu)\text{Cos}[\theta]^2\text{Sin}[\theta]^2+\mu\text{Sin}[\theta]^4\},$
$\{\mu\text{Cos}[\theta]^4-2\lambda\text{Cos}[\theta]^2\text{Sin}[\theta]^2-2\mu\text{Cos}[\theta]^2\text{Sin}[\theta]^2+$
$2(\lambda+2\mu)\text{Cos}[\theta]^2\text{Sin}[\theta]^2+\mu\text{Sin}[\theta]^4,$
$-\lambda\text{Cos}[\theta]^3\text{Sin}[\theta]-2\mu\text{Cos}[\theta]^3\text{Sin}[\theta]+$
$(\lambda+2\mu)\text{Cos}[\theta]^3\text{Sin}[\theta]+\lambda\text{Cos}[\theta]\text{Sin}[\theta]^3+$
$2\mu\text{Cos}[\theta]\text{Sin}[\theta]^3-(\lambda+2\mu)\text{Cos}[\theta]\text{Sin}[\theta]^3\}\},$
$\{\{\{\lambda\text{Cos}[\theta]^3\text{Sin}[\theta]+2\mu\text{Cos}[\theta]^3\text{Sin}[\theta]-$
$(\lambda+2\mu)\text{Cos}[\theta]^3\text{Sin}[\theta]-\lambda\text{Cos}[\theta]\text{Sin}[\theta]^3-$
$2\mu\text{Cos}[\theta]\text{Sin}[\theta]^3+(\lambda+2\mu)\text{Cos}[\theta]\text{Sin}[\theta]^3,$
$\mu\text{Cos}[\theta]^4-2\lambda\text{Cos}[\theta]^2\text{Sin}[\theta]^2-2\mu\text{Cos}[\theta]^2\text{Sin}[\theta]^2+$
$2(\lambda+2\mu)\text{Cos}[\theta]^2\text{Sin}[\theta]^2+\mu\text{Sin}[\theta]^4\},$
$\{\mu\text{Cos}[\theta]^4-2\lambda\text{Cos}[\theta]^2\text{Sin}[\theta]^2-2\mu\text{Cos}[\theta]^2\text{Sin}[\theta]^2+$
$2(\lambda+2\mu)\text{Cos}[\theta]^2\text{Sin}[\theta]^2+\mu\text{Sin}[\theta]^4,$
$-\lambda\text{Cos}[\theta]^3\text{Sin}[\theta]-2\mu\text{Cos}[\theta]^3\text{Sin}[\theta]+$
$(\lambda+2\mu)\text{Cos}[\theta]^3\text{Sin}[\theta]+\lambda\text{Cos}[\theta]\text{Sin}[\theta]^3+$
$2\mu\text{Cos}[\theta]\text{Sin}[\theta]^3-(\lambda+2\mu)\text{Cos}[\theta]\text{Sin}[\theta]^3\}\},$
$\{\{\lambda\text{Cos}[\theta]^4-4\mu\text{Cos}[\theta]^2\text{Sin}[\theta]^2+2(\lambda+2\mu)\text{Cos}[\theta]^2\text{Sin}[\theta]^2+$
$\lambda\text{Sin}[\theta]^4,-\lambda\text{Cos}[\theta]^3\text{Sin}[\theta]-2\mu\text{Cos}[\theta]^3\text{Sin}[\theta]+$
$(\lambda+2\mu)\text{Cos}[\theta]^3\text{Sin}[\theta]+\lambda\text{Cos}[\theta]\text{Sin}[\theta]^3+$
$2\mu\text{Cos}[\theta]\text{Sin}[\theta]^3-(\lambda+2\mu)\text{Cos}[\theta]\text{Sin}[\theta]^3\},$
$\{-\lambda\text{Cos}[\theta]^3\text{Sin}[\theta]-2\mu\text{Cos}[\theta]^3\text{Sin}[\theta]+$
$(\lambda+2\mu)\text{Cos}[\theta]^3\text{Sin}[\theta]+\lambda\text{Cos}[\theta]\text{Sin}[\theta]^3+$
$2\mu\text{Cos}[\theta]\text{Sin}[\theta]^3-(\lambda+2\mu)\text{Cos}[\theta]\text{Sin}[\theta]^3,$
$(\lambda+2\mu)\text{Cos}[\theta]^4+2\lambda\text{Cos}[\theta]^2\text{Sin}[\theta]^2+$
$4\mu\text{Cos}[\theta]^2\text{Sin}[\theta]^2+(\lambda+2\mu)\text{Sin}[\theta]^4\}\}\}\}$

which can be simplified to

In[16]:= `Simplify[cnew]`

Out[16]= `{{{{λ + 2 μ, 0}, {0, λ}}, {{0, μ}, {μ, 0}}},`
`{{{0, μ}, {μ, 0}}, {{λ, 0}, {0, λ + 2 μ}}}}`

which shows that C_{ijkl} is isotropic.

1.3.6 Differentiation

In index notation, partial derivatives are abbreviated by a comma (,) as

$$\frac{\partial}{\partial x_i} \Rightarrow , i.$$

Examples

1.
$$x_{1,2} = \frac{\partial x}{\partial y} = 0.$$

2.
$$x_{i,i} = x_{1,1} + x_{2,2} + x_{3,3}$$
$$= \frac{\partial x}{\partial x} + \frac{\partial y}{\partial y} + \frac{\partial z}{\partial z}$$
$$= 3.$$

3.
$$x_{i,j} = \frac{\partial x_i}{\partial x_j}$$
$$= \delta_{ij}.$$

4.
$$f_{,12} = \frac{\partial^2 f}{\partial x_1 \partial x_2}$$
$$= \frac{\partial^2 f}{\partial x \partial y},$$

that is, higher order derivatives can be specified by additional indices after the comma.

5.
$$v_{,i} = \frac{\partial v}{\partial x_i}$$
$$\equiv \nabla v,$$

hence, this is the gradient operator (∇).

6.
$$v_{i,i} = v_{1,1} + v_{2,2} + v_{3,3}$$
$$= \frac{\partial v_x}{\partial x} + \frac{\partial v_y}{\partial y} + \frac{\partial v_z}{\partial z}$$
$$\equiv \nabla \cdot v,$$

hence, this is the divergence operator ($\nabla \cdot$).

Example Prove
$$\Delta(r^n) = n(n+1)r^{n-2},$$
where $r = \sqrt{x^2 + y^2 + z^2}$.
Solution: Note that $r_{,i} = x_i/r$.
$$(r^n)_{,i} = nr^{n-1}\frac{x_i}{r} = nr^{n-2}x_i.$$
Therefore,
$$(r^n)_{,ii} = n(n-2)r^{n-3}\frac{x_i}{r}x_i + nr^{n-2}x_{i,i}$$
$$= n(n-2)r^{n-3}x_i\frac{x_i}{r} + 3nr^{n-2}$$
$$= n(n-2)r^{n-2} + 3nr^{n-2}$$
$$= n(n+1)r^{n-2}.$$

Here is a *Mathematica* code to do this automatically.

```
In[1]:= rvector = Table[x[i], {i, 1, 3}]
Out[1]= {x[1], x[2], x[3]}

In[2]:= r = Sqrt[ rvector.rvector]
Out[2]= √(x[1]² + x[2]² + x[3]²)

In[3]:= laplace[f_] := Sum[D[f, {x[i], 2}], {i, 3}]

In[4]:= laplace[r^n] // Simplify
Out[4]= n (1 + n) (x[1]² + x[2]² + x[3]²)^(-1+n/2)

In[5]:= % /. {x[1]^2 + x[2]^2 + x[3]^2 → R^2}
Out[5]= n (1 + n) (R²)^(-1+n/2)
```

In the aforementioned code, the position vector, $\mathbf{r} = (x_1, x_2, x_3)$, is entered as a function, x[i]. The magnitude of $\mathbf{r}$ is stored in r. The Laplacian operator is denoted as laplace. The previous output can be represented by a % (a percent sign). The right arrow symbol ($\rightarrow$) is the substitution rule (replace) that replaces what is to the left of the arrow by what is to the right of the arrow.

1.3.7 Differentiation of Cartesian Tensors

If a Cartesian tensor, say, $v_{ij...}$, is differentiated, its derivative, $v_{ij...,kl...}$, is also a Cartesian tensor.[10]

---

[10] This is true for Cartesian tensors but not true for general tensors as will be discussed later.

Proof. For illustration purposes, assume v_i is a first-rank tensor. Hence, its transformation rule is
$$v_{\bar{i}} = \beta_{\bar{i}j} v_j. \tag{1.5}$$

Differentiating both sides of Equation (1.5) with respect to $x_{\bar{j}}$ yields
$$v_{\bar{i},\bar{j}} = \beta_{\bar{i}j} v_{j,\bar{j}} {}^{11}$$
$$= \beta_{\bar{i}j} v_{j,k} x_{k,\bar{j}} {}^{12}$$
$$= \beta_{\bar{i}j} v_{j,k} (\beta^{-1})_{k\bar{j}}$$
$$= \beta_{\bar{i}j} v_{j,k} (\beta^T)_{k\bar{j}}$$
$$= \beta_{\bar{i}j} \beta_{\bar{j}k} v_{j,k}. \tag{1.6}$$

Equation (1.6) implies that $v_{i,j}$ is a second-rank tensor. The rank of the resulting tensor depends on the operation performed. For example, if v is a scalar, $v_{,i}$ is a first-rank tensor, $v_{,ij}$ is a second-rank tensor, but $v_{,ii}$ is a scalar.

Example (curl, rotation) The ith component of rotation of a vector, **v**, is expressed as
$$(\nabla \times v)_i = \epsilon_{ijk} v_{j,k}.$$

The rotation operator, rot **v**, can be implemented as

In[4]:= `Table[ Sum[ Signature[{i, j, k}] Dt[v, x[k]], {j, 3}, {k, 3}], {i, 3}]`

Out[4]= `{-Dt[v, x[2]] + Dt[v, x[3]], Dt[v, x[1]] - Dt[v, x[3]], -Dt[v, x[1]] + Dt[v, x[2]]}`

1.4 Invariance of Tensor Equations

Theorem 1.1 *If $A_{ij...}$ and $B_{ij...}$ are both tensors, their linear combination, $\alpha A_{ij...} + \beta B_{ij...}$, is also a tensor.*

Proof. Let $C_{ij...} \equiv \alpha A_{ij...} + \beta B_{ij...}$, it follows
$$C_{\bar{ij}...} = \alpha A_{\bar{ij}...} + \beta B_{\bar{ij}...}$$
$$= \alpha \beta_{\bar{i}i} \beta_{\bar{j}j} \ldots A_{ij...} + \beta \beta_{\bar{i}i} \beta_{\bar{j}j} \ldots B_{ij...}$$
$$= \beta_{\bar{i}i} \beta_{\bar{j}j} \ldots (\alpha A_{ij...} + \beta B_{ij...})$$
$$= \beta_{\bar{i}i} \beta_{\bar{j}j} \ldots C_{ij...}.$$

---

[11] As it is not possible to differentiate v_j with respect to $x_{\bar{j}}$ (a different coordinate system), the chain differentiation rule,
$$\frac{\partial}{\partial x_{\bar{j}}} = \frac{\partial x_k}{\partial x_{\bar{j}}} \frac{\partial}{\partial x_k}$$
was used.

[12] Note that
$$\frac{\partial x_{\bar{i}}}{\partial x_j} \frac{\partial x_j}{\partial x_{\bar{k}}} = \delta_{\bar{i}\bar{k}},$$
So,
$$\frac{\partial x_j}{\partial x_{\bar{k}}} = \left(\frac{\partial x_{\bar{i}}}{\partial x_j} \right)^{-1}.$$

Theorem 1.2 *If $v_{ij...} = 0$ in one coordinate system, $v_{\bar{i}\bar{j}...}$ vanishes in any other coordinate system.*

Proof.
$$v_{\bar{i}\bar{j}...} = \beta_{\bar{i}i}\beta_{\bar{j}j}\ldots v_{ij...}$$
$$= \beta_{\bar{i}i}\beta_{\bar{j}j}\ldots 0$$
$$= 0.$$

Theorem 1.3 *If $a_{ij...} = b_{ij...}$ holds in one coordinate system, it holds in any other coordinate system.*[13]

Proof. Let $C_{ij...} \equiv a_{ij...} - b_{ij...}$.
It follows

(a) $C_{ij...}$ is a tensor (from Theorem 1).
(b) Since $C_{ij...} = 0$, $C_{\bar{i}\bar{j}...} = a_{\bar{i}\bar{j}...} - b_{\bar{i}\bar{j}...} = 0$ (from Theorem 2).
(c) Hence $a_{\bar{i}\bar{j}...} = b_{\bar{i}\bar{j}...}$.

By this statement, it was shown that a (Cartesian) tensor equation is invariant in any coordinate system.

1.5 Quotient Rule

If $AB = C$ holds and A and C are both tensors (indices not shown), B must be a tensor.

Examples
1. Force
   The force, f, is a tensor as it is defined by
   $$fu_i = W,$$
   where u_i is the displacement and W is the work done. As both the displacement, u_i, (a first-rank tensor) and the work done, W, (a zeroth-rank tensor) are tensors, f must be a tensor from the quotient rule. Moreover, f must have an index, i, for both sides to be consistent. Thus, the force must be a first-rank tensor, f_i.
2. Elastic constants
   The elastic constant, C, is defined as the proportionality factor between the stress, σ_{ij}, and the strain, ϵ_{ij}, as
   $$\sigma = C\epsilon.$$
   Since both the stress and the strain are second-rank tensors as will be shown in the next chapter, the above equation must be written as
   $$\sigma_{ij} = C_?\epsilon_{kl}.^{14}$$

---

[13] Because of this theorem, tensor equations are independent of the frame of reference.
[14] The direction of σ is not necessarily colinear with the direction of ϵ, hence, different indices (ij and kl).

From the quotient rule, C must be a tensor as both σ and ϵ are tensors. Furthermore, C must have indices of $ijkl$ for both sides of the aforementioned equation to be consistent. Thus, we have

$$\sigma_{ij} = C_{ijkl}\epsilon_{kl}.$$

The component, C_{ijkl}, can be interpreted as the ij component of the stress for the kl component of the applied strain.

3. Thermal conductivity
   The thermal conductivity, k, is defined as the proportionality factor between the heat flux, h_j, and the temperature gradient, $T_{,i}$, as

$$h_j = k_? T_{,i}.$$

As both h_j and $T_{,i}$ are first-rank tensors, k must be a second-rank tensor as

$$h_i = -k_{ij} T_{,j},$$

where the minus sign is introduced for a historical reason.

1.6 Exercises

1. Rewrite the following using the summation convention.
   (a) $a_1 x_1 + a_2 x_2 + a_3 x_3 = p.$
   (b) $(a_{11} + a_{22} + a_{33})(a_{11}^2 + a_{12}^2 + a_{21}^2 + a_{22}^2).$
   (c) $a_{11} x^2 + a_{12} xy + a_{21} xy + a_{22} y^2 = c.$
   (d) $\frac{\partial f}{\partial x} dx + \frac{\partial f}{\partial y} dy + \frac{\partial f}{\partial z} dz.$
   (e) $a_{11} a_{22} - a_{12} a_{21}.$

2. Given $\phi_1 = -y$, $\phi_2 = x$,
   (a) Is ϕ_i a vector ? Why ?
   (b) Is $\phi_{i,j}$ a tensor ? Why ?

3. Compute $(\frac{1}{r})_{,ii}$ where $r = \sqrt{x^2 + y^2}$.

4. Is $v_{ij} = \begin{pmatrix} xy & y^2 \\ x^2 & -xy \end{pmatrix}$ a tensor ? Why ?

5. Thermal conductivity for an orthotropic material is a second-rank tensor and is expressed as $k_{ij} = \begin{pmatrix} k_{11} & 0 \\ 0 & k_{22} \end{pmatrix}$.
   (a) Obtain the components of $k_{\bar{i}\bar{j}}$ in the coordinate system $(x_{\bar{i}})$ rotated from the original coordinate system by $\pi/4$.
   (b) Calculate $k_{\bar{i}\bar{j}} k_{\bar{i}\bar{j}}$ in the rotated coordinate system $(x_{\bar{i}})$.

6. Express

$$C_{ijkl} w_{,ijkl}$$

where

$$C_{ijkl} = \mu(\delta_{ik}\delta_{jl} + \delta_{il}\delta_{jk}) + \lambda \delta_{ij}\delta_{kl},$$

without using index notation (i.e., using x and y) in 2-D.

7. Express the following in full notation (in terms of x and y assuming 2-D).
$$C_{ijkl} w_{,ij} w_{,kl} = 0,$$
where $C_{ijkl} = \alpha(\delta_{ik}\delta_{jl} + \delta_{il}\delta_{jk}) + \beta\delta_{ij}\delta_{kl}$.

8. Compute $\Delta\Delta r$ where $r^2 = x^2 + y^2 + z^2$ and $\Delta = \frac{\partial^2}{\partial x^2} + \frac{\partial^2}{\partial x^2} + \frac{\partial^2}{\partial x^2}$.

9. Is
$$v_{ij} = \begin{pmatrix} 0 & -1 \\ 1 & 0 \end{pmatrix}$$
a tensor? Give your rationale.

10. Write down $(\mathbf{a} \times \mathbf{b}) \cdot \mathbf{a}$ in index notation and simplify the result.

11. (a) Prove the following identities using *Mathematica*.
$$\nabla \cdot (r^n \mathbf{r}) = (n+3)r^n,$$
$$\nabla \times (r^n \mathbf{r}) = 0,$$
$$\Delta(r^n) = n(n+1)r^{n-2}.$$

(b) Write a code in *Mathematica* to automatically generate all the equations of the following formula without index notation.
$$G\left(u_{i,kk} + \frac{1}{1-2v} u_{k,ki}\right) + X_i = \rho \frac{\partial^2 u_i}{\partial t^2}.$$
Hint: Use Dt rather than D. Note: Refer to the companion webpage for solutions for the Exercises section.

References

Einstein A et al. 1916 The foundation of the general theory of relativity. *Annalen der Physik* **49**(769–822), 31.
Fung Y 1965 *Foundations of Solid Mechanics*, vol. 351. Prentice-Hall, Englewood Cliffs, NJ.
Garvan F 2001 *The Maple Book*. Chapman & Hall/CRC.
Romano A, Lancellotta R and Marasco A 2006 Continuum mechanics using mathematica. *Continuum Mechanics using Mathematica: Fundamentals, Applications and Scientific Computing, Modeling and Simulation in Science, Engineering and Technology*. Birkhäuser, Boston, MA. ISBN: 978-0-8176-3240-3.
Wolfram S 1999 *The Mathematica Book*. Cambridge university press.

2

Field Equations

In this chapter, based on the index notation introduced in Chapter 1, the equations for physical and mechanical fields are derived. Implementation of the equations, transformations, and curvilinear coordinate systems in *Mathematica* is discussed in detail. Some of the worked-out examples in this chapter are taken from the celebrated textbook (Fung 1977) along with their *Mathematica* implementation.

2.1 Concept of Stress

The concept of stress is essential to the understanding of mechanics for deformable bodies. The difference between stress and force is that the stress is a combination of the force and the direction of the surface on which the traction force is acting. One way of understanding the stress is to think of it as a black box as shown in Figure 2.1 that accepts an input, n, (the normal to the plane), which is the orientation of the surface, and outputs the surface traction, t. This output is the traction force, t, on the surface as shown in Figure 2.2.

The simplest way to realize this mechanism between n and t shown in Figure 2.1 is to assume that the traction, t, is a product between the stress, σ, and the normal, n, as

$$\sigma n = t. \tag{2.1}$$

As both n and t are first-rank tensors (vectors), Equation (2.1) must be written as

$$\sigma n_j = t_i,$$

which implies that σ must be a second-rank tensor from the quotient rule with the index j as the dummy index and the index i as the free index (Figure 2.2). Thus, Equation (2.1) can be written as

$$\sigma_{ij} n_j = t_i.$$

The component, σ_{ij}, thus, can be interpreted as the ith component of the traction force, t, acting on the surface, j. Here, the surface j is defined as the surface whose normal is aligned with the j-axis. For example, σ_{11} is the normal stress component on the surface 1 while σ_{21} is the shear stress component on the same surface. It follows that if two indices are the same, it represents

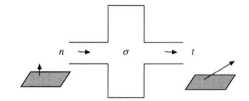

Figure 2.1 Stress as a black box

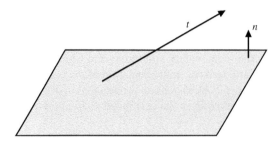

Figure 2.2 Traction acting on a plane

a normal stress, and if two indices are different, it represents a shear stress.[1]

Example: For the stress components

$$\sigma_{ij} = \begin{pmatrix} 1 & -3 & 2 \\ -3 & 1 & -7 \\ 2 & -7 & -2 \end{pmatrix},$$

shown in Figure 2.3, compute t_i (traction force), N_i (normal stress), S_i (shear stress), N (magnitude of N_i), and S (magnitude of S_i) at $(1, 1, 1)$ on the curved surface given by $xyz^3 = 1$.

Solution
The normal to the surface can be obtained by taking the total derivative of $xyz^3 = 1$ as

$$yz^3 dx + xz^3 dy + 3xyz^2 dz = 0,$$

which implies that $(yz^3, xz^3, 3xyz^2)_{(1,1,1)} = (1, 1, 3)$ is perpendicular to the surface element of (dx, dy, dz). By normalizing this vector, the components of **n** are

$$\mathbf{n} = \left(\frac{1}{\sqrt{11}}, \frac{1}{\sqrt{11}}, \frac{3}{\sqrt{11}} \right).$$

Hence, the traction force, t_i, can be obtained as

$$t_i = \sigma_{ij} n_j = \left(\frac{4}{\sqrt{11}}, \frac{-23}{\sqrt{11}}, -\sqrt{11} \right).$$

---

[1] The distinction between the normal and the shear components is lost once the coordinate system is changed.

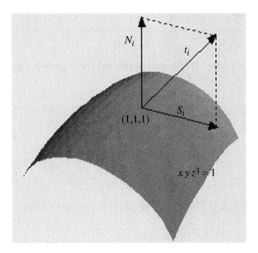

Figure 2.3 Surface traction on a curved surface

The magnitude of N_i, N, can be obtained as

$$N = t_i n_i = -\frac{52}{11}.$$

Note that the minus sign implies that the direction of N_i is opposite to n_i. The components of the normal stress, N_i, can be obtained as

$$N_i = N n_i = \left(\frac{-52}{11\sqrt{11}}, \frac{-52}{11\sqrt{11}}, \frac{-156}{11\sqrt{11}} \right).$$

The components of the shear stress, S_i, are the difference between t_i and N_i as

$$S_i = t_i - N_i = \left(\frac{96}{11\sqrt{11}}, \frac{-201}{11\sqrt{11}}, \frac{35}{11\sqrt{11}} \right).$$

The magnitude of S_i is computed as

$$S = \sqrt{S_i S_i} = \frac{\sqrt{4622}}{11} \sim 6.1804.$$

Mathematica code. In *Mathematica*, the stress components can be entered as a 3×3 matrix (list) as

```
In[1]:= sigma = {{1, -3, 2}, {-3, 1, -7}, {2, -7, -2}}
Out[1]= {{1, -3, 2}, {-3, 1, -7}, {2, -7, -2}}
```

To obtain the normal vector, the gradient of f at $(1, 1, 1)$ is computed as

In[2]:= **f = x y z^3**

Out[2]= $x y z^3$

In[3]:= **n = {D[f, x], D[f, y], D[f, z]} /. {x → 1, y → 1, z → 1}**

Out[3]= {1, 1, 3}

The Norm function can be used to normalize vectors.

In[4]:= **n = n / Norm[n]**

Out[4]= $\left\{\dfrac{1}{\sqrt{11}}, \dfrac{1}{\sqrt{11}}, \dfrac{3}{\sqrt{11}}\right\}$

The product, $\sigma_{ij}n_j$, can be computed as the product between the matrix and the vector as

In[7]:= **t = sigma.n**

Out[7]= $\left\{\dfrac{4}{\sqrt{11}}, -\dfrac{23}{\sqrt{11}}, -\sqrt{11}\right\}$

Note the usage of a dot (·). Ordinary products are either a space or an asterisk (*).
By using the N function, the result can be viewed as numerics.

In[8]:= **N[t]**

Out[8]= {1.20605, −6.93476, −3.31662}

The magnitude of the normal stress of N_i, N, can be computed by the dot product between t_i and n_i as

In[10]:= **nn = t.n**

Out[10]= $-\dfrac{52}{11}$

The components of the normal stress, N_i, can be computed as

In[11]:= **ni = nn n**

Out[11]= $\left\{-\dfrac{52}{11\sqrt{11}}, -\dfrac{52}{11\sqrt{11}}, -\dfrac{156}{11\sqrt{11}}\right\}$

The components of the shear stress, S_i, can be computed as

In[12]:= **si = t - ni**

Out[12]= $\left\{\dfrac{96}{11\sqrt{11}}, -\dfrac{201}{11\sqrt{11}}, \dfrac{156}{11\sqrt{11}} - \sqrt{11}\right\}$

The magnitude of the shear stress of S_i, S, can be computed as

In[14]:= **Norm[si] // Simplify**

Out[14]= $\dfrac{\sqrt{4622}}{11}$

Field Equations

2.1.1 Properties of Stress

As the stress is an extension of the force, it inherits the properties of the force, which includes the balance of forces and moments. The following are the three important equations for the stress:

1. Definition:
$$t_i = \sigma_{ij} n_j. \tag{2.2}$$

   This relationship is more of the definition than the property based on the quotient rule.

2. Balance of forces:
$$\sigma_{ij,j} + b_i = 0,$$

where b_i is the body force.

Proof. When a body is in static equilibrium, the sum of the surface traction force on the boundary of the body and the sum of the body force (such as gravitational force) inside the body must be balanced, which is expressed as

$$\oint_{\partial D} t_i \, dS + \int_D b_i \, dV = 0,$$

where D denotes the entire region and ∂D is its boundary. By using Equation (2.2), the definition of the stress, the first term in the given integral is rewritten as

$$\oint_{\partial D} t_i \, dS = \oint_{\partial D} \sigma_{ij} n_j \, dS$$
$$= \int_D \sigma_{ij,j} \, dV,$$

so that the balance of the force equation becomes[2]

$$\int_D (\sigma_{ij,j} + b_i) \, dV = 0,$$

which must be held for an arbitrary control volume. Therefore, it follows that

$$\sigma_{ij,j} + b_i = 0.$$

3. Balance of moments:
$$\sigma_{ij} = \sigma_{ji}.$$

Proof. The source of the moment acting on a continuum comes from the traction force and the body force. Therefore, the balance of moments is expressed as

$$\oint_{\partial D} \mathbf{r} \times \mathbf{t} \, dS + \int_D \mathbf{r} \times \mathbf{b} \, dV = 0,$$

---

[2] You are reminded of the Gauss theorem, i.e.,

$$\oint_{\partial D} n_i v \, dS = \int_D v_{,i} \, dV,$$

where the integral on the left-hand side is a contour integral defined along the boundary of the domain D and the integral in the right-hand side is a 2-D integral over D.

or in index notation,

$$\oint_{\partial D} \epsilon_{ijk} x_j t_k \, dS + \int_D \epsilon_{ijk} x_j b_k \, dV = 0. \tag{2.3}$$

By using Equation (2.2), the first integral can be written as

$$\oint_{\partial D} \epsilon_{ijk} x_j t_k \, dS = \oint_{\partial D} \epsilon_{ijk} x_j \sigma_{kl} n_l \, dS,$$

$$= \int_D (\epsilon_{ijk} x_j \sigma_{kl})_{,l} \, dV$$

$$= \int_D (\epsilon_{ijk} x_{j,l} \sigma_{kl} + \epsilon_{ijk} x_j \sigma_{kl,l}) \, dV$$

$$= \int_D (\epsilon_{ijk} \delta_{jl} \sigma_{kl} + \epsilon_{ijk} x_j \sigma_{kl,l}) \, dV$$

$$= \int_D (\epsilon_{ijk} \sigma_{kj} + \epsilon_{ijk} x_j \sigma_{kl,l}) \, dV,$$

where the Gauss theorem was used noting that $x_{j,l} = \delta_{jl}$. Therefore, Equation (2.3) is written as

$$\int_D (\epsilon_{ijk} \sigma_{kj} + \epsilon_{ijk} x_j (\sigma_{kl,l} + b_k)) \, dV = 0. \tag{2.4}$$

As Equation (2.4) has to be held for an arbitrary control volume noting that $\sigma_{kl,l} + b_k = 0$, it follows

$$\epsilon_{ijk} \sigma_{kj} = 0, \tag{2.5}$$

which leads to

$$\sigma_{kj} = \sigma_{jk}.$$

2.1.2 (Stress) Boundary Conditions

When two materials are in contact each other, the traction force must be continuous as otherwise the two pieces will not stay statically. The displacement must also be continuous across the material boundary as well for the two pieces not to be separated.

Example 2.1: Two materials are attached together as shown in Figure 2.4. Along the boundary, the normal vector is

$$\mathbf{n} = (0, 1).$$

The components of the traction force are given as

$$t_1 = \sigma_{11} n_1 + \sigma_{12} n_2 = \sigma_{12},$$
$$t_2 = \sigma_{21} n_1 + \sigma_{22} n_2 = \sigma_{22}.$$

Field Equations

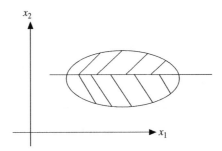

Figure 2.4 Boundary condition along the line parallel to the x_1 axis

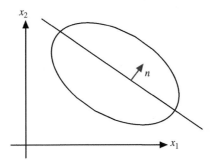

Figure 2.5 Boundary condition along the line perpendicular to **n**

Both t_1 and t_2 must be continuous. Hence, it follows that both σ_{12} and σ_{22} must be continuous across the boundary but σ_{11} does not have to be continuous.

Example 2.2: Same as the previous with the boundary given in Figure 2.5. This time, the normal, **n**, is $\mathbf{n} = (1/\sqrt{2}, 1/\sqrt{2})$. Therefore,

$$t_1 = \sigma_{11} n_1 + \sigma_{12} n_2 = (\sigma_{11} + \sigma_{12})/\sqrt{2},$$

$$t_2 = \sigma_{21} n_1 + \sigma_{22} n_2 = (\sigma_{21} + \sigma_{22})/\sqrt{2}.$$

Hence, $\sigma_{11} + \sigma_{12}$ and $\sigma_{21} + \sigma_{22}$ must be continuous but individual components of the stress tensor do not have to be continuous.

2.1.3 Principal Stresses

Stress tensors have nine components (4 in 2-D) of which 6 (3 in 2-D) are independent because of $\sigma_{ij} = \sigma_{ji}$. However, it is not necessary to keep track of all the nine components. With a proper coordinate transformation (rotation), the stress tensor can be completely described by three components called the *principal stresses*. An equivalent statement in linear algebra is that a symmetrical matrix can always be diagonalized (Greenberg 2001).

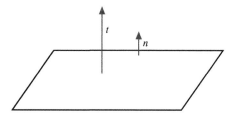

Figure 2.6 Principal stresses

When the surface traction is perpendicular to the surface on which the stress is defined as in Figure 2.6, the state of stress is in *principal stresses*.
When this happens, *t* and *n* must be colinear. Therefore,

$$t_i = \alpha n_i, \tag{2.6}$$

where α is a proportional factor and a scalar. Combining Equation (2.6) with Equation (2.2) yields

$$\sigma_{ij} n_j = \alpha n_i$$
$$= \alpha \delta_{ij} n_j. \tag{2.7}$$

Equation (2.7) is equivalent in matrix–vector form to

$$S\mathbf{n} = \alpha I \mathbf{n}, \tag{2.8}$$

where S is the matrix that represents σ_{ij} and I is the identity matrix. Equation (2.8) has a nontrivial solution if the matrix $S - \alpha I$ is singular, i.e., its determinant must vanish as

$$|S - \alpha I| = 0. \tag{2.9}$$

By solving Equation (2.9) for α, the principal stresses, α, can be found. The principal direction, n, can be obtained by back-substituting α into Equation (2.8). The principal stresses and the principal directions are known as the eigenvalues and the eigenvectors in linear algebra.

In 3D, Equation (2.9) is written as

$$\begin{vmatrix} \sigma_{11} - \alpha, & \sigma_{12}, & \sigma_{13} \\ \sigma_{21}, & \sigma_{22} - \alpha, & \sigma_{23} \\ \sigma_{31}, & \sigma_{32}, & \sigma_{33} - \alpha \end{vmatrix} = 0,$$

the expansion of which yields

$$\alpha^3 - I_1 \alpha^2 + I_2 \alpha - I_3 = 0,$$

where

$$I_1 = \sigma_{11} + \sigma_{22} + \sigma_{33},$$
$$I_2 = -\sigma_{12}^2 - \sigma_{23}^2 - \sigma_{31}^2 + \sigma_{11}\sigma_{22} + \sigma_{22}\sigma_{33} + \sigma_{33}\sigma_{11}$$
$$= \begin{vmatrix} \sigma_{22} & \sigma_{23} \\ \sigma_{32} & \sigma_{33} \end{vmatrix} + \begin{vmatrix} \sigma_{33} & \sigma_{31} \\ \sigma_{13} & \sigma_{11} \end{vmatrix} + \begin{vmatrix} \sigma_{11} & \sigma_{12} \\ \sigma_{21} & \sigma_{22} \end{vmatrix},$$

Field Equations

$$I_3 = -\sigma_{13}^2\sigma_{22} - \sigma_{23}^2\sigma_{11} - \sigma_{12}^2\sigma_{33} + 2\sigma_{12}\sigma_{13}\sigma_{23} + \sigma_{11}\sigma_{22}\sigma_{33}$$

$$= \begin{vmatrix} \sigma_{11} & \sigma_{12} & \sigma_{13} \\ \sigma_{21} & \sigma_{22} & \sigma_{23} \\ \sigma_{31} & \sigma_{32} & \sigma_{33} \end{vmatrix}.$$

It is clear that the quantities, I_1, I_2, and I_3, are all scalars as the cubic equation they define is invariant. From the standpoint of tensor algebra, it is also possible to show that the stress invariants, I_1, I_2, and I_3, are indeed scalars, i.e., the zeroth-rank tensors too. To show this, choose a coordinate system in which the principal stresses, σ_1, σ_2, and σ_3, are realized. In such a coordinate system, the stress invariants will have the forms:

$$I_1 = \sigma_1 + \sigma_2 + \sigma_3,$$

$$I_2 = \sigma_1\sigma_2 + \sigma_2\sigma_3 + \sigma_3\sigma_1,$$

$$I_3 = \sigma_1\sigma_2\sigma_3.$$

It can be shown that the aforementioned invariants are rewritten in index notation as

$$I_1 = \sigma_{ii},$$

$$I_2 = \frac{\sigma_{ii}\sigma_{jj} - \sigma_{ij}\sigma_{ij}}{2},$$

$$I_3 = \frac{\sigma_{ii}\sigma_{jj}\sigma_{kk} - \sigma_{ij}\sigma_{jk}\sigma_{ki} - 3(\sigma_{ii}\sigma_{kl}\sigma_{kl} - \sigma_{ij}\sigma_{jk}\sigma_{ki})}{6}.$$

The stress invariants, I_1–I_3, are expressed in terms of zeroth-rank tensors.

Example

$$\sigma_{ij} = \begin{pmatrix} -1 & 2 & 2 \\ 2 & -2 & -1 \\ 2 & -1 & -2 \end{pmatrix}.$$

The three eigenvalues can be found by solving

$$\begin{vmatrix} -1-\alpha, & 2 & 2 \\ 2, & -2-\alpha & -1 \\ 2 & -1 & -2-\alpha \end{vmatrix} = -\alpha^3 - 5\alpha^2 + \alpha + 5$$

$$= -(\alpha - 1)(\alpha + 1)(\alpha + 5)$$

$$= 0.$$

The three eigenvalues are

$$(\alpha_1, \alpha_2, \alpha_3) = (-1, 1, -5).$$

The corresponding eigenvectors can be found by solving for (n_1, n_2, n_3) for each α.

$$\begin{pmatrix} -1-\alpha, & 2 & 2 \\ 2, & -2-\alpha & -1 \\ 2 & -1 & -2-\alpha \end{pmatrix} \begin{pmatrix} n_1 \\ n_2 \\ n_3 \end{pmatrix} = \begin{pmatrix} 0 \\ 0 \\ 0 \end{pmatrix}.$$

For $\alpha_1 = -1$, the aforementioned equations are expressed as

$$0\, n_1 + 2n_2 + 2n_3 = 0,$$
$$2n_1 - n_2 - n_3 = 0,$$
$$2n_1 - n_2 - n_3 = 0,$$

among which only the first two equations are independent as

$$n_1 = 0, \quad n_2 + n_3 = 0.$$

Therefore, one can choose (after normalizing)

$$\mathbf{n} = \begin{pmatrix} 0 \\ -1/\sqrt{2} \\ 1/\sqrt{2} \end{pmatrix}.$$

Similarly, for $\alpha_2 = 1$, one can choose

$$\mathbf{n} = \begin{pmatrix} 2/\sqrt{6} \\ 1/\sqrt{6} \\ 1/\sqrt{6} \end{pmatrix},$$

and for $\alpha_3 = -5$, one can choose

$$\mathbf{n} = \begin{pmatrix} -1/\sqrt{3} \\ 1/\sqrt{3} \\ 1/\sqrt{3} \end{pmatrix}.$$

Mathematica implementation: In *Mathematica*, the `Eigensystem` function is available to compute both the eigenvalues (principal stresses) and the eigenvectors (principal directions).

In[3]:= `s = {{-1, 2, 2}, {2, -2, -1}, {2, -1, -2}}`

Out[3]= `{{-1, 2, 2}, {2, -2, -1}, {2, -1, -2}}`

In[5]:= `{eval, evec} = Eigensystem[s]`

Out[5]= `{{-5, -1, 1}, {{-1, 1, 1}, {0, -1, 1}, {2, 1, 1}}}`

In[8]:= `Table[evec[[i]] / Norm[evec[[i]]], {i, 3}]`

Out[8]= $\left\{\left\{-\dfrac{1}{\sqrt{3}},\, \dfrac{1}{\sqrt{3}},\, \dfrac{1}{\sqrt{3}}\right\},\, \left\{0,\, -\dfrac{1}{\sqrt{2}},\, \dfrac{1}{\sqrt{2}}\right\},\, \left\{\sqrt{\dfrac{2}{3}},\, \dfrac{1}{\sqrt{6}},\, \dfrac{1}{\sqrt{6}}\right\}\right\}$

The eigenvalues are stored in a variable, `eval`, as a list and the corresponding eigenvectors are stored in a variable, `evec`. However, as the eigenvectors returned by the `Eigensystem` function are not normalized, it is necessary to normalize the result using the `Norm` function.

2.1.4 Stress Deviator

Problems involving deformation can be often simplified if the source of deformation is split into the shear part (deviatoric part) and the volume expansion part (hydrostatic part) as the two are completely decoupled.

A second-rank tensor (3-D), v_{ij}, can be uniquely decomposed into its shear part (deviatoric part), v'_{ij}, and its volume expansion part (hydrostatic part), v_{kk}, as[3]

$$v_{ij} = v'_{ij} + \frac{1}{3}\delta_{ij}v_{kk}. \qquad (2.10)$$

As an example, consider the following matrix:

$$v_{ij} = \begin{pmatrix} 1 & 2 & 3 \\ 2 & 4 & 5 \\ 3 & 5 & 6 \end{pmatrix}.$$

Noting that $v_{ii} = 1 + 4 + 6 = 11$, the deviatoric part of v_{ij} can be obtained as

$$v'_{ij} = v_{ij} - \frac{1}{3}\delta_{ij}v_{kk}$$

$$= \begin{pmatrix} 1 & 2 & 3 \\ 2 & 4 & 5 \\ 3 & 5 & 6 \end{pmatrix} - \frac{1}{3}\begin{pmatrix} 11 & 0 & 0 \\ 0 & 11 & 0 \\ 0 & 0 & 11 \end{pmatrix}$$

$$= \begin{pmatrix} -8/3 & 2 & 3 \\ 2 & 1/3 & 5 \\ 3 & 5 & 7/3 \end{pmatrix}.$$

It should be noted that the trace[4] of the deviator tensor is identically 0 as

$$v'_{ii} = 0.$$

Proof. If we set $i = j$ in Equation (2.10),

$$v_{ii} = v'_{ii} + \frac{1}{3}\delta_{ii}v_{kk}$$

$$= v'_{ii} + v_{kk}.$$

Since $v_{ii} = v_{kk}$, it follows that

$$v'_{ii} = 0,$$

---

[3] For 2-D, the decomposition is

$$v_{ij} = v'_{ij} + \frac{1}{2}\delta_{ij}v_{kk}.$$

[4] The sum of diagonal elements.

i.e., the deviatoric part represents part of the deformation that does not accompany volume expansion.[5]

Examples:
1. Strain energy
   The elastic strain energy, U, is defined as
   $$U \equiv \frac{1}{2}\sigma_{ij}\epsilon_{ij},$$
   where ϵ_{ij} is the strain tensor. By using the decomposition of σ_{ij} and ϵ_{ij} as
   $$\sigma_{ij} = \sigma'_{ij} + \frac{1}{3}\delta_{ij}\sigma_{kk},$$
   $$\epsilon_{ij} = \epsilon'_{ij} + \frac{1}{3}\delta_{ij}\epsilon_{kk},$$
   the strain energy is expressed as
   $$U = \frac{1}{2}\left(\sigma'_{ij} + \frac{1}{3}\delta_{ij}\sigma_{kk}\right)\left(\epsilon'_{ij} + \frac{1}{3}\delta_{ij}\epsilon_{kk}\right)$$
   $$= \frac{1}{2}\left(\sigma'_{ij}\epsilon'_{ij} + \frac{1}{3}\delta_{ij}\sigma'_{ij}\epsilon_{ll} + \frac{1}{3}\delta_{ij}\epsilon'_{ij}\sigma_{kk} + \frac{1}{9}\delta_{ij}\delta_{ij}\sigma_{kk}\epsilon_{ll}\right)$$
   $$= \frac{1}{2}\left(\sigma'_{ij}\epsilon'_{ij} + \frac{1}{3}\sigma'_{jj}\epsilon_{ll} + \frac{1}{3}\sigma_{kk}\epsilon'_{jj} + \frac{1}{3}\sigma_{kk}\epsilon_{ll}\right)$$
   $$= \frac{1}{2}\left(\sigma'_{ij}\epsilon'_{ij} + \frac{1}{3}\sigma_{kk}\epsilon_{ll}\right).$$
   The strain energy expression is split into two parts: the first part from the deviatoric parts of σ_{ij} and ϵ_{ij} and the second part from the hydrostatic parts of σ_{ij} and ϵ_{ij}; thus, the result implies that the deviatoric part and the hydrostatic part of the strain energy do not couple with each other.

2. Shear deformation
   Shear deformation can be defined as the deviatoric part of either the stress or the strain tensor. Similar to the stress invariants, I_1, I_2, and I_3, of σ_{ij}, the stress invariants, J_1, J_2, and J_3, of the deviatoric part of the stress, σ'_{ij}, can be defined as
   $$J_1 = \sigma'_{ii}$$
   $$= 0,$$
   $$J_2 = \frac{1}{2}(\sigma'_{ii}\sigma'_{jj} - \sigma'_{ij}\sigma'_{ij})$$

---

[5]
$$\frac{\Delta V}{V} = \frac{(a+\Delta a)(b+\Delta b)(c+\Delta c) - abc}{abc}$$
$$\sim \frac{\Delta a}{a} + \frac{\Delta b}{b} + \frac{\Delta c}{c}$$
$$= \epsilon_x + \epsilon_y + \epsilon_z$$
$$= \epsilon_{ii}..$$

$$= -\frac{1}{2}\sigma'_{ij}\sigma'_{ij},$$

$$J_3 = (\sigma'_{ii}\sigma'_{jj}\sigma'_{kk} - \sigma'_{ij}\sigma'_{jk}\sigma'_{ki} - 3(\sigma'_{ii}\sigma'_{kl}\sigma'_{kl} - \sigma'_{ij}\sigma'_{jk}\sigma'_{ki}))/6$$

$$= -\sigma'_{ij}\sigma'_{jk}\sigma'_{ki}/6.$$

3. von Mises failure criterion
   Metal failure is defined as the yielding point that is caused by dislocation of atoms for which the shear force is required to cause dislocation. Hence, the deviatoric part of σ_{ij} is responsible for the yielding, and as yielding is a phenomenon independent of the frame of the reference, it must be a function of the deviatoric parts of the stress alone. As $J_1 = 0$, its simplest form is to assume that

$$J_2 = c,$$

or

$$\sigma'_{ij}\sigma'_{ij} = c, \tag{2.11}$$

where c is a constant. When the body is subject to the uniaxial yield strength, σ_Y, σ'_{ij} is expressed as

$$\sigma'_{ij} = \begin{pmatrix} \sigma_Y & 0 & 0 \\ 0 & 0 & 0 \\ 0 & 0 & 0 \end{pmatrix} - \frac{\sigma_Y}{3}\begin{pmatrix} 1 & 0 & 0 \\ 0 & 1 & 0 \\ 0 & 0 & 1 \end{pmatrix}$$

$$= \begin{pmatrix} \frac{2}{3}\sigma_Y & 0 & 0 \\ 0 & -\frac{1}{3}\sigma_Y & 0 \\ 0 & 0 & -\frac{1}{3}\sigma_Y \end{pmatrix}.$$

Therefore, it follows that

$$\sigma'_{ij}\sigma'_{ij} = \frac{2}{3}\sigma_Y^2. \tag{2.12}$$

For the state of 2-D principal stresses (biaxial stresses), σ_1 and σ_2, the stress deviator, σ'_{ij}, is expressed as

$$\sigma'_{ij} = \sigma_{ij} - \frac{1}{3}\delta_{ij}\sigma_{kk}$$

$$= \begin{pmatrix} \sigma_1 & 0 & 0 \\ 0 & \sigma_2 & 0 \\ 0 & 0 & 0 \end{pmatrix} - \begin{pmatrix} \frac{\sigma_1+\sigma_2}{3} & 0 & 0 \\ 0 & \frac{\sigma_1+\sigma_2}{3} & 0 \\ 0 & 0 & \frac{\sigma_1+\sigma_2}{3} \end{pmatrix}$$

$$= \begin{pmatrix} \frac{2\sigma_1-\sigma_2}{3} & 0 & 0 \\ 0 & \frac{2\sigma_2-\sigma_1}{3} & 0 \\ 0 & 0 & \frac{-\sigma_1-\sigma_2}{3} \end{pmatrix}.$$

Therefore, it follows that

$$\sigma'_{ij}\sigma'_{ij} = \left(\frac{2\sigma_1 - \sigma_2}{3}\right)^2 + \left(\frac{2\sigma_2 - \sigma_1}{3}\right)^2 + \left(\frac{-\sigma_1 - \sigma_2}{3}\right)^2$$

$$= \frac{2}{3}(\sigma_1^2 - \sigma_1\sigma_2 + \sigma_2^2). \tag{2.13}$$

Hence, by equating Equation (2.12) with Equation (2.13), the yield condition under biaxial stresses is expressed as

$$\sigma_1^2 - \sigma_1\sigma_2 + \sigma_2^2 = \sigma_Y^2. \tag{2.14}$$

Equation (2.14) is widely used to predict metal failure under biaxial stresses.

2.1.5 Mohr's Circle

The transformation rule for the 2-D stress tensor, $\sigma_{i'j'} = \beta_{i'i}\beta_{j'j}\sigma_{ij}$, in the Cartesian coordinate system can be explicitly written as

$$\sigma_{\bar{1}\bar{1}} = \cos^2\theta\, \sigma_{11} + 2\cos\theta\sigma_{12} + \sin^2\theta\, \sigma_{12},$$

$$\sigma_{\bar{1}\bar{2}} = -\sigma_{12}\sin^2\theta + \sigma_{12}\cos^2\theta - \sigma_{11}\sin\theta\cos\theta + \sigma_{22}\sin\theta\cos\theta,$$

$$\sigma_{\bar{2}\bar{2}} = \sigma_{11}\sin^2\theta + \sigma_{22}\cos^2\theta - 2\sigma_{12}\sin\theta\cos\theta,$$

or equivalently

$$\sigma_{\bar{1}\bar{1}} = \frac{1}{2}(2\sigma_{12}\sin 2\theta + \sigma_{11}\cos 2\theta - \sigma_{22}\cos 2\theta + \sigma_{11} + \sigma_{22}), \tag{2.15}$$

$$\sigma_{\bar{1}\bar{2}} = \frac{1}{2}(-\sigma_{11}\sin 2\theta + \sigma_{22}\sin 2\theta + 2\sigma_{12}\cos 2\theta), \tag{2.16}$$

$$\sigma_{\bar{2}\bar{2}} = \frac{1}{2}(-2\sigma_{12}\sin 2\theta + \sigma_{11}(-\cos 2\theta) + \sigma_{22}\cos 2\theta + \sigma_{11} + \sigma_{22}).$$

Here is a *Mathematica* implementation of the derivation of the aforementioned formulas. First, we represent stress tensor, σ_{ij}, by a function sigma[i, j]. As it is a symmetrical second-rank tensor, the SetAttributes[sigma, Orderless] statement can be used so that the arguments of sigma[i,j] are always put in order, symmetrized. With this setting of the attributes, sigma[2,1] is automatically converted to sigma[1,2].

In[1]:= **SetAttributes[sigma, Orderless]**

The 2-D transformation matrix, β_{ij}, is entered as

In[2]:= **beta = {{Cos[th], Sin[th]}, {-Sin[th], Cos[th]}}**

Out[2]= {{Cos[th], Sin[th]}, {-Sin[th], Cos[th]}}

The transformation rule for second-rank tensors can be entered as

Field Equations

```
In[4]:= Table[Sum[beta[[ii, i]] beta[[jj, j]] sigma[i, j], {i, 2}, {j, 2}], {ii, 2}, {jj, 2}]

Out[4]= {{Cos[th]² sigma[1, 1] + 2 Cos[th] sigma[1, 2] Sin[th] + sigma[2, 2] Sin[th]²,
         Cos[th]² sigma[1, 2] - Cos[th] sigma[1, 1] Sin[th] +
         Cos[th] sigma[2, 2] Sin[th] - sigma[1, 2] Sin[th]²}, {Cos[th]² sigma[1, 2] -
         Cos[th] sigma[1, 1] Sin[th] + Cos[th] sigma[2, 2] Sin[th] - sigma[1, 2] Sin[th]²,
         Cos[th]² sigma[2, 2] - 2 Cos[th] sigma[1, 2] Sin[th] + sigma[1, 1] Sin[th]²}}
```

By using the `TrigReduce` function, products and powers of trigonometric functions can be reduced to first-order terms as

```
In[5]:= TrigReduce[%]

Out[5]= {{1/2 (sigma[1, 1] + Cos[2 th] sigma[1, 1] +
          sigma[2, 2] - Cos[2 th] sigma[2, 2] + 2 sigma[1, 2] Sin[2 th]),
          1/2 (2 Cos[2 th] sigma[1, 2] - sigma[1, 1] Sin[2 th] + sigma[2, 2] Sin[2 th])},
         {1/2 (2 Cos[2 th] sigma[1, 2] - sigma[1, 1] Sin[2 th] + sigma[2, 2] Sin[2 th]),
          1/2 (sigma[1, 1] - Cos[2 th] sigma[1, 1] + sigma[2, 2] +
          Cos[2 th] sigma[2, 2] - 2 sigma[1, 2] Sin[2 th])}}
```

Mohr's circle is a graphical visualization of evaluating Equations (2.15) and (2.16), commonly written as

$$\sigma = \frac{\sigma_x + \sigma_y}{2} + \frac{\sigma_x - \sigma_y}{2} \cos 2\theta + \tau_{xy} \sin 2\theta,$$

$$\tau = -\frac{\sigma_x - \sigma_y}{2} \sin 2\theta + \tau_{xy} \cos 2\theta.$$

Note that the angle of rotation, θ, is defined by comparing the two normals on the surfaces as shown in Figure 2.7.

2.1.5.1 How to Draw Mohr's Circle?

Each point on Mohr's circle represents a surface of the given orientation. Therefore, one can start by mapping each side of the element to a corresponding point on Mohr's circle, i.e.,

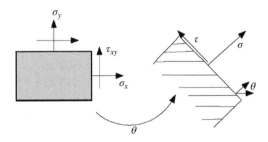

Figure 2.7 Mohr's circle

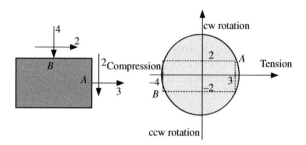

Figure 2.8 Mohr's circle example

1. Normal stress ⇒ Mapped to the horizontal axis
   (a) Tensile stress ⇒ Mapped to the right side
   (b) Compressive stress ⇒ Mapped to the left side
2. Shear stress ⇒ Mapped to the vertical axis
   (a) Clockwise rotation ⇒ Mapped to the upper side
   (b) Counterclockwise rotation ⇒ Mapped to the lower side
3. Draw a circle by connecting the four corners.
4. Rotation of θ in the physical elements is translated into 2θ on the Mohr circle space.

Note that terms such as "positive rotation" were avoided on purpose.

Examples:
1. Draw Mohr's circle for $(\sigma_x, \sigma_y, \tau_{xy}) = (3, -4, -2)$ as shown in Figure 2.8.
2. By using Mohr's circle, one can explain why the angle of the failure surface of a metal bar is 45°. The maximum shear stress theory in failure of metals postulates that *failure (yielding) of metals is predicted when the maximum shear stress reaches a critical value*. When a metal bar is subject to uniaxial tension, the corresponding Mohr's circle is the one centered at $(\sigma/2, 0)$ with a radius of $\sigma/2$. So, the maximum shear stress is at the top of the circle $(\sigma/2, \sigma/2)$, which is rotation from the surface subject to the uniaxial tension by 90°, which is translated into rotation of the physical element by 45°.
3. Drawing Mohr's circle can also explain why hydrostatic stresses never cause failure in metals. If an element is subject to hydrostatic stresses $(\sigma_x = \sigma_y = \sigma)$, the corresponding Mohr's circle is degenerated to a single point whose maximum shear stress remains 0.

2.2 Strain

When a deformable body is subjected to stress, it undergoes deformation. Strain is defined as the rate of deformation and is related to the stress. While the concept of the stress is based on physics (force), the concept of the strain is based on geometry and is not a physical quantity; thus, one can arbitrarily define the strain at will. For instance, in Figure 2.9, if one wants to measure the rate of deformation based on the original length, one can define the strain as

$$E \equiv \frac{\ell - \ell_o}{\ell_o}.$$

Figure 2.9 Strains in one dimension

On the other hand, the rate of deformation can also be defined based on the current length as

$$e \equiv \frac{\ell - \ell_o}{\ell}.$$

Since the definitions are different, their values are different naturally. However, if one assumes that the deformation is infinitesimal, i.e.,

$$\ell = \ell_o(1 + \epsilon),$$

where ϵ is an infinitesimal quantity,

$$E = \frac{\ell_o(1+\epsilon) - \ell_o}{\ell_o}$$
$$= \epsilon,$$
$$e = \frac{\ell_o(1+\epsilon) - \ell_o}{\ell}$$
$$= \frac{\epsilon}{1+\epsilon}$$
$$= \epsilon(1 - \epsilon + \epsilon^2 - \epsilon^3 + \cdots)$$
$$\sim \epsilon,$$

the two definitions coincide.

When a point in a deformable continua moves from a to x as shown in Figure 2.10, the rate of deformation cannot be measured by the distance between a and x (displacement) alone as a mere transformation or rotation makes the displacement nonzero. Instead, the rate change of the distance must be considered as a measurement of the deformation. Therefore, instead of

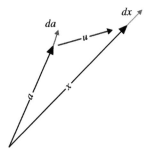

Figure 2.10 Displacements

a and x, their derivatives, da and dx, must be compared as

$$ds^2 \equiv |dx|^2 - |da|^2$$
$$= dx_i dx_i - da_i da_i. \qquad (2.17)$$

If Equation (2.17) is to be expressed based on the original point, a, the position, x, can be regarded as a function of a as

$$x_i = x_i(a_j),$$

so that

$$dx_i = \frac{\partial x_i}{\partial a_j} da_j. \qquad (2.18)$$

Substituting Equation (2.18) to Equation (2.17) yields

$$ds^2 = dx_i dx_i - da_i da_i$$
$$= \frac{\partial x_i}{\partial a_j}\frac{\partial x_i}{\partial a_k} da_j da_k - \delta_{jk} da_j da_k$$
$$= \left(\frac{\partial x_i}{\partial a_j}\frac{\partial x_i}{\partial a_k} - \delta_{jk}\right) da_j da_k$$
$$\equiv 2E_{jk} da_j da_k,$$

where

$$E_{jk} \equiv \frac{1}{2}\left(\frac{\partial x_i}{\partial a_j}\frac{\partial x_i}{\partial a_k} - \delta_{jk}\right).$$

This is called the Green strain (the Lagrangian strain in fluid mechanics).

On the other hand, if one wants to measure the rate of deformation based on the current position, the point a can be regarded as a function of x as

$$a_i = a_i(x_j),$$

so that

$$da_i = \frac{\partial a_i}{\partial x_j} dx_j.$$

By substituting this to Equation (2.17), one obtains

$$ds^2 = dx_i dx_i - da_i da_i$$
$$= \delta_{jk} dx_j dx_k - \frac{\partial a_i}{\partial x_j} dx_j \frac{\partial a_i}{\partial x_k} dx_k$$
$$= \left(\delta_{jk} - \frac{\partial a_i}{\partial x_j}\frac{\partial a_i}{\partial x_k}\right) dx_j dx_k$$
$$\equiv 2e_{jk} dx_j dx_k,$$

Field Equations

where
$$e_{jk} \equiv \frac{1}{2}\left(\delta_{jk} - \frac{\partial a_i}{\partial x_j}\frac{\partial a_i}{\partial x_k}\right).$$

This is called the Cauchy strain (the Euler strain in fluid mechanics).

It should be noted that E_{ij} and e_{ij} are different in values as the definitions are different. Nevertheless, their combinations with the infinitesimal elements are the same, i.e.,

$$ds^2 = 2E_{ij}da_i da_j$$
$$= 2e_{ij}dx_i dx_j.$$

It is possible to express the Green strains and the Cauchy strains in terms of the displacement, i.e.,
$$u_i = x_i - a_i.$$

For the Green strains,
$$x_i(a_j) = u_i(a_j) + a_i,$$

so that
$$\frac{\partial x_i}{\partial a_j} = \frac{\partial u_i}{\partial a_j} + \delta_{ij}.$$

Therefore,
$$\begin{aligned}
E_{jk} &= \frac{1}{2}\left(\frac{\partial x_i}{\partial a_j}\frac{\partial x_i}{\partial a_k} - \delta_{jk}\right) \\
&= \frac{1}{2}\left(\left(\frac{\partial u_i}{\partial a_j} + \delta_{ij}\right)\left(\frac{\partial u_i}{\partial a_k} + \delta_{ik}\right) - \delta_{jk}\right) \\
&= \frac{1}{2}\left(\left(\frac{\partial u_i}{\partial a_j}\frac{\partial u_i}{\partial a_k} + \frac{\partial u_k}{\partial a_j} + \frac{\partial u_j}{\partial a_k} + \delta_{jk}\right) - \delta_{jk}\right) \\
&= \frac{1}{2}\left(\frac{\partial u_k}{\partial a_j} + \frac{\partial u_j}{\partial a_k} + \frac{\partial u_i}{\partial a_j}\frac{\partial u_i}{\partial a_k}\right).
\end{aligned} \quad (2.19)$$

For the Cauchy strains,
$$a_k(x_i) = x_k - u_k(x_i),$$

so that
$$\frac{\partial a_k}{\partial x_i} = \delta_{ki} - \frac{\partial u_k}{\partial x_i}.$$

Therefore,
$$\begin{aligned}
e_{ij} &= \frac{1}{2}\left(\delta_{ij} - \left(\delta_{ki} - \frac{\partial u_k}{\partial x_i}\right)\left(\delta_{kj} - \frac{\partial u_k}{\partial x_j}\right)\right) \\
&= \frac{1}{2}\left(\delta_{ij} - \left(\delta_{ij} - \frac{\partial u_i}{\partial x_j} - \frac{\partial u_j}{\partial x_i} + \frac{\partial u_k}{\partial x_i}\frac{\partial u_k}{\partial x_j}\right)\right) \\
&= \frac{1}{2}\left(\frac{\partial u_i}{\partial x_j} + \frac{\partial u_j}{\partial x_i} - \frac{\partial u_k}{\partial x_i}\frac{\partial u_k}{\partial x_j}\right).
\end{aligned} \quad (2.20)$$

If the deformation is small such that the second-order terms in Equations (2.19) and (2.20) are dropped, Equations (2.19) and (2.20) become the same and both can be expressed as

$$\epsilon_{ij} \equiv \frac{1}{2}\left(\frac{\partial u_i}{\partial x_j} + \frac{\partial u_j}{\partial x_i}\right). \qquad (2.21)$$

Equation (2.21) is known as the infinitesimal strain. Note that this definition differs from the definition of engineering strain by 1/2 for the shear components.

Examples
1. A square plate is rotated from the first quarter to the third quarter as shown in Figure 2.11. Compute the strains.

   Answer

   Since

   $$x = -a, \quad y = -b,$$

   it follows that

   $$u = -x - (-x) = 2x, \quad v = -y - (-y) = 2y.$$

   Therefore,

   $$\epsilon_{11} = \frac{\partial u}{\partial x} = 2, \quad \epsilon_{12} = \frac{1}{2}\left(\frac{\partial u}{\partial y} + \frac{\partial v}{\partial x}\right) = 0, \quad \epsilon_{22} = \frac{\partial v}{\partial y} = 2.$$

   There is something wrong with this answer as the rotation should not accompany any deformation (rigid body rotation).

   Correction:

   The displacement in the rotation is not considered to be small. Hence, one has to use the formula for finite deformation (either E_{ij} (Green strains) or e_{ij} (Cauchy strain)). The Green strain, E_{ij}, is computed as

   $$u = -2a, \quad v = -2b,$$

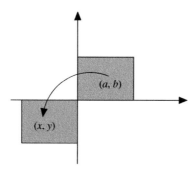

Figure 2.11 Rigid rotation by 90°

so
$$E_{11} = \frac{1}{2}\left(\frac{\partial u_1}{\partial a_1} + \frac{\partial u_1}{\partial a_1} + \frac{\partial u_k}{\partial a_1}\frac{\partial u_k}{\partial a_1}\right)$$
$$= \frac{1}{2}\left(\frac{\partial u}{\partial a} + \frac{\partial u}{\partial a} + \left(\frac{\partial u}{\partial a}\right)^2 + \left(\frac{\partial v}{\partial a}\right)^2\right)$$
$$= \frac{1}{2}((-2) + (-2) + (-2)^2)$$
$$= 0.$$

Other values of E_{ij} can be shown to be 0.

2. Compute the Green and Euler strains for Figure 2.12.

Solution
$$u_1 = x_1 - a_1$$
$$= \cos\theta a_1 - \sin\theta a_2 - a_1$$
$$= (\cos\theta - 1)a_1 - \sin\theta a_2,$$
$$u_2 = x_2 - a_2$$
$$= \sin\theta a_1 + \cos\theta a_2 - a_2$$
$$= -\sin\theta a_1 + (\cos\theta - 1)a_2.$$

Therefore, the Green strains are computed as
$$E_{11} = (\cos\theta - 1) + \frac{1}{2}\left((\cos\theta - 1)^2 + \sin^2\theta\right)$$
$$= 0,$$
$$E_{22} = (\cos\theta - 1) + \frac{1}{2}\left(\sin^2\theta_+ (\cos\theta - 1)^2\right)$$
$$= 0,$$
$$E_{12} = -\sin\theta - \frac{1}{2}((\cos\theta - 1)\sin\theta + \sin\theta(\cos\theta - 1))$$
$$= -\cos\theta\sin\theta.$$

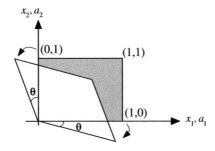

Figure 2.12 Computations of strains

For infinitesimal deformation,
$$\epsilon_{ij} = \begin{pmatrix} 0 & \theta \\ \theta & 0 \end{pmatrix}.$$

The Euler strains can be computed as
$$e_{ij} = \begin{pmatrix} -\dfrac{\cos^2\theta}{2(\cos^2\theta-\sin^2\theta)^2} - \dfrac{\sin^2\theta}{2(\cos^2\theta-\sin^2\theta)^2} + \dfrac{1}{2} & \dfrac{\cos\theta\sin\theta}{(\cos^2\theta-\sin^2\theta)^2} \\ \dfrac{\cos\theta\sin\theta}{(\cos^2\theta-\sin^2\theta)^2} & -\dfrac{\cos^2\theta}{2(\cos^2\theta-\sin^2\theta)^2} - \dfrac{\sin^2\theta}{2(\cos^2\theta-\sin^2\theta)^2} + \dfrac{1}{2} \end{pmatrix}.$$

For infinitesimal deformation, the Euler strains are also reduced to
$$\epsilon_{ij} = \begin{pmatrix} 0 & \theta \\ \theta & 0 \end{pmatrix}.$$

Mathematica code:
The aforementioned computation can be implemented by *Mathematica*. The Green strain can be defined as a function as

```
In[3]:= Eij[i_, j_] := 1 / 2
        (Sum[D[x[alpha], a[i]] D[x[alpha], a[j]],
        {alpha, 1, 2}] -
        IdentityMatrix[2][[i, j]])
```

The relationship between (x_1, x_2) and (a_1, a_2) is defined as

```
In[4]:= Clear[a, x];
        {x[1], x[2]} =
        {{Cos[th], Sin[th]}, {Sin[th], Cos[th]}}.
        {a[1], a[2]}
Out[5]= {a[1] Cos[th] + a[2] Sin[th],
        a[2] Cos[th] + a[1] Sin[th]}
```

The Green strain for the aforementioned problem can be computed as

```
In[6]:= Table[Eij[i, j], {i, 2}, {j, 2}] // Simplify
Out[6]= {{0, Cos[th] Sin[th]}, {Cos[th] Sin[th], 0}}
```

The Cauchy strain is defined as

```
In[7]:= eij[i_, j_] :=
        1 / 2 (IdentityMatrix[2][[i, j]] -
        Sum[D[a[alpha], x[i]] D[a[alpha], x[j]],
        {alpha, 1, 2}])
```

The relationship between (x_1, x_2) and (a_1, a_2) is entered as

Field Equations

In[12]:= `Clear[a, x]`
`{a[1], a[2]} =`
`Inverse[{{Cos[th], Sin[th]}, {Sin[th], Cos[th]}}].{x[1], x[2]}`

Out[13]= $\left\{ \dfrac{\text{Cos}[\text{th}]\, x[1]}{\text{Cos}[\text{th}]^2 - \text{Sin}[\text{th}]^2} - \dfrac{\text{Sin}[\text{th}]\, x[2]}{\text{Cos}[\text{th}]^2 - \text{Sin}[\text{th}]^2}, \right.$
$\left. -\dfrac{\text{Sin}[\text{th}]\, x[1]}{\text{Cos}[\text{th}]^2 - \text{Sin}[\text{th}]^2} + \dfrac{\text{Cos}[\text{th}]\, x[2]}{\text{Cos}[\text{th}]^2 - \text{Sin}[\text{th}]^2} \right\}$

The Cauchy strain thus can be computed as

In[14]:= `Table[eij[i, j], {i, 2}, {j, 2}] // Simplify`

Out[14]= $\left\{ \left\{ -\dfrac{1}{2} \text{Tan}[2\,\text{th}]^2,\; \text{Cos}[\text{th}]\,\text{Sec}[2\,\text{th}]^2\,\text{Sin}[\text{th}] \right\}, \right.$
$\left. \left\{ \text{Cos}[\text{th}]\,\text{Sec}[2\,\text{th}]^2\,\text{Sin}[\text{th}],\; -\dfrac{1}{2}\text{Tan}[2\,\text{th}]^2 \right\} \right\}$

The infinitesimal strain can be computed by expanding the aforementioned result by Taylor's series as

In[9]:= `Series[%, {th, 0, 1}]`

Out[9]= $\left\{ \left\{ O[\text{th}]^2,\; \text{th} + O[\text{th}]^2 \right\},\; \left\{ \text{th} + O[\text{th}]^2,\; O[\text{th}]^2 \right\} \right\}$

2.2.1 Shear Deformation

Shear deformation is defined as part of the deformation that does not accompany volume expansion, which is equivalent to the statement that the trace of the strain tensor is 0.[6] Consider the two types of deformation shown in Figure 2.13. Figure 2.13a is generally called *pure shear* while Figure 2.13b is called *simple shear*, and they seem to represent different types of

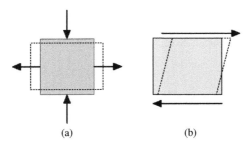

(a)   (b)

Figure 2.13 Shear deformation. (a) Pure shear; (b) simple shear

---

[6] $\epsilon_{ii} = 0$.

deformation. However, it can be shown that they indeed represent the same type of deformation if the strain components are computed.

1. Pure shear:
   The relationship of the positions before and after the deformation is expressed as
   $$X' = (1+\epsilon)x, \quad y' = \frac{y}{1+\epsilon}.$$
   Therefore, the displacement components are
   $$u = \epsilon x, \quad v = \left(1 - \frac{1}{1+\epsilon}\right)y,$$
   from which one can compute the infinitesimal strain components as
   $$\epsilon_{ij} = \begin{pmatrix} \epsilon & 0 \\ 0 & -\frac{\epsilon}{1+\epsilon} \end{pmatrix}$$
   $$\sim \begin{pmatrix} \epsilon & 0 \\ 0 & -\epsilon \end{pmatrix}.$$

2. Simple shear
   The relationship of the positions before and after the deformation is expressed as
   $$X' = x + 2\epsilon y, \quad y' = y.$$
   The displacement components are
   $$u = 2\epsilon y, \quad v = 0.$$
   Therefore, the infinitesimal strain components are
   $$\epsilon_{ij} = \begin{pmatrix} 0 & \epsilon \\ \epsilon & 0 \end{pmatrix}.$$

Both types of the deformation can be characterized by the fact that the trace is 0, i.e.,
$$\epsilon_{ii} = 0.$$
In fact, the strain components in simple shear become pure shear if the coordinate system is rotated by $-45°$ as
$$\epsilon_{\bar{i}\bar{j}} = \beta_{\bar{i}i}\beta_{\bar{j}j}\epsilon_{ij},$$
or
$$\begin{pmatrix} \epsilon_{\bar{1}\bar{1}} & \epsilon_{\bar{1}\bar{2}} \\ \epsilon_{\bar{2}\bar{1}} & \epsilon_{\bar{2}\bar{2}} \end{pmatrix} = \begin{pmatrix} \cos\theta & \sin\theta \\ -\sin\theta & \cos\theta \end{pmatrix}_{\theta=\pi/4} \begin{pmatrix} 0 & \epsilon \\ \epsilon & 0 \end{pmatrix} \begin{pmatrix} \cos\theta & \sin\theta \\ -\sin\theta & \cos\theta \end{pmatrix}^T_{\theta=\pi/4}$$
$$= \begin{pmatrix} \epsilon & 0 \\ 0 & -\epsilon \end{pmatrix}.$$

It is seen that what was pure shear earlier is now simple shear if observed from a coordinate system rotated by 45° from the original.

2.3 Compatibility Condition

The strain–displacement relationship expressed as

$$u_{i,j} + u_{j,i} = 2\epsilon_{ij}, \qquad (2.22)$$

is thought as a set of differential equations for u_i provided ϵ_{ij} is known. Equation (2.22) is overspecified as there are six equations for three unknowns (u_i) in 3-D and there are three equations for two unknowns (u_i) in 2-D.[7] This implies that not all the components of ϵ_{ij} are independent and there must be three constraints for ϵ_{ij} in 3-D and one in 2-D to match the number of unknowns with the number of equations. Such constraints are called the *compatibility conditions*.

Two new operators are introduced in order to derive the compatibility conditions:

Definitions 2.1
1. *Symmetrization operator.*
   When indices are enclosed by parentheses, symmetrization on the indices is performed, i.e.,

$$a_{(ij)} \equiv \tfrac{1}{2}(a_{ij} + a_{ji}).$$

2. *Anti-symmetrization operator*
   When indices are enclosed by square brackets, antisymmetrization on the indices is performed, i.e.,

$$a_{[ij]} \equiv \tfrac{1}{2}(a_{ij} - a_{ji}).$$

Note that

$$a_{(ij)} = a_{(ji)}, \qquad a_{[ij]} = -a_{[ji]}.$$

Any tensor with two indices can be uniquely decomposed into its symmetrical part and antisymmetrical part as

$$\begin{aligned} v_{ij} &= \tfrac{1}{2}(v_{ij} + v_{ji}) + \tfrac{1}{2}(v_{ij} - v_{ji}) \\ &= v_{(i,j)} + v_{[ij]}. \end{aligned}$$

For example, the displacement gradient, $u_{i,j}$, can be decomposed as

$$\begin{aligned} u_{i,j} &= \tfrac{1}{2}(u_{i,j} + u_{j,i}) + \tfrac{1}{2}(u_{i,j} - u_{j,i}) \\ &= \epsilon_{ij} + \omega_{ij}, \end{aligned}$$

where ϵ_{ij} is the strain tensor and ω_{ij} is the rotation. The strain is the symmetrical part of the displacement gradient and the rotation is the antisymmetrical part of the displacement gradient. If u_i is the velocity field in fluid mechanics, ϵ_{ij} defines the strain rate and ω_{ij} defines the vorticity.

---
[7] Because of the symmetry, $\epsilon_{ij} = \epsilon_{ji}$, the number of independent components for ϵ_{ij} is 6 in 3-D and 3 in 2-D.

In general, the compatibility condition for a system of partial differential equations,

$$u_{,i} = g_i,$$

can be written as

$$g_{[i,j]} = 0. \qquad (2.23)$$

It is clear that this is a necessary condition for the existence of the solution, u_i. It can be shown that this condition is indeed a sufficient condition (Malvern 1969) as well. Hence, Equation (2.23) is an equivalent condition to the existence of the solution, u_i. Extending this idea to the strain–displacement relation implies that the compatibility condition for the relation,

$$u_{(i,j)} = \epsilon_{ij},$$

can be written as

$$\epsilon_{[i[j,k]l]} = 0,$$

which is explicitly written as

$$\epsilon_{[i[j,k]l]} = \frac{1}{4}(\epsilon_{ij,kl} + \epsilon_{kl,ij} - \epsilon_{lj,ki} - \epsilon_{ik,jl})$$
$$= 0.$$

Note that this condition yields only one independent equation in 2-D as

$$\epsilon_{11,22} + \epsilon_{22,11} - 2\epsilon_{12,12} = 0,$$

and six conditions in 3-D as

$$\epsilon_{11,23} + \epsilon_{23,11} - \epsilon_{13,12} - \epsilon_{12,13} = 0,$$
$$\epsilon_{22,13} + \epsilon_{13,22} - \epsilon_{12,23} - \epsilon_{23,12} = 0,$$
$$\epsilon_{33,12} + \epsilon_{12,33} - \epsilon_{12,13} - \epsilon_{13,23} = 0,$$
$$2\epsilon_{12,12} - \epsilon_{11,22} - \epsilon_{22,11} = 0,$$
$$2\epsilon_{23,23} - \epsilon_{22,33} - \epsilon_{33,22} = 0,$$
$$2\epsilon_{13,13} - \epsilon_{33,11} - \epsilon_{11,33} = 0.$$

However, the aforementioned six equations are not independent of one another. There are three constraints among the six equations known as the Bianchi formulas in Riemannian geometry (Malvern 1969), which makes the number of independent compatibility conditions in 3-D as 3. Thus, the number of independent equations for the stress components and the number of stress components match in both 2-D and 3-D.

2.4 Constitutive Relation, Isotropy, Anisotropy

The simplest relationship between the stress, σ_{ij}, and the strain, ϵ_{ij}, is that they are proportional to each other as

$$\sigma = C\epsilon.$$

Field Equations

By the quotient rule, the proportionality constant, C, must be a fourth-rank tensor and has four indices as

$$\sigma_{ij} = C_{ijkl}\epsilon_{kl}.$$

The component, C_{ijkl}, represents the (ij) component of the stress subject to the (kl) component of the strain. It is called the elastic modulus.[8]

Without considering any symmetrical properties, the number of independent components of C_{ijkl} is 81. However, C_{ijkl} possesses the following three types of symmetrical properties:

1. $C_{ijkl} = C_{jikl}$.
   This symmetrical property comes from the symmetrical property of σ_{ij}. The component, C_{ijkl}, is the (ij) component of the stress corresponding to the (kl) component of the strain while the component, C_{jikl}, is the (ji) component of the stress corresponding to the (kl) component of the strain. As $\sigma_{ij} = \sigma_{ji}$, it follows that $C_{ijkl} = C_{jikl}$.
2. $C_{ijkl} = C_{ijlk}$.
   This symmetrical property comes from the symmetrical property of ϵ_{ij}. The component, C_{ijkl}, is the (ij) component of the stress corresponding to the (kl) component of the strain while the component, C_{ijlk}, is the (ij) component of the stress corresponding to the (lk) component of the strain. As $\epsilon_{kl} = \epsilon_{lk}$, it follows that $C_{ijkl} = C_{ijlk}$.
3. $C_{ijkl} = C_{jikl}$.
   This symmetrical property can be proven by assuming the existence of the strain energy, U. From the stress–strain relationship of $\sigma_{ij} = C_{ijkl}\epsilon_{kl}$, it follows

$$C_{ijkl} = \frac{\partial \sigma_{ij}}{\partial \epsilon_{kl}}.$$

On the other hand, the definition of the strain energy, U, is

$$dU = \sigma_{ij} d\epsilon_{ij},$$

so

$$\frac{\partial U}{\partial \epsilon_{ij}} = \sigma_{ij}.$$

Therefore,

$$C_{ijkl} = \frac{\partial}{\partial \epsilon_{kl}}\left(\frac{\partial U}{\partial \epsilon_{ij}}\right)$$

$$= \frac{\partial^2 U}{\partial \epsilon_{kl} \partial \epsilon_{ij}}$$

$$= C_{klij}.$$

hence,

$$C_{ijkl} = C_{klij}.$$

By using these symmetrical properties, the number of independent components of C_{ijkl} can be reduced from 81 to a mere 21.

---
[8] The plural is moduli.

2.4.1 Isotropy

Isotropy refers to the property invariant under rotation and reflection while *homogeneity* refers to the property invariant under translation. Note that the negation of isotropy is *anisotropy*. Anisotropic materials include woods and liquid crystals besides composite materials.

It is possible to show that if a tensor is isotropic, its order must be even, i.e., there is no odd-rank tensor that is isotropic.

1. Zeroth-rank tensors.
   It is clear that all zeroth-rank tensors (scalars) are isotropic because they remain unchanged in any coordinate system.
2. First-rank tensors.
   If v_i is a first-rank tensor and isotropic, it must be invariant under the reflection of

$$\beta_{ij} = \begin{pmatrix} -1 & 0 \\ 0 & 1 \end{pmatrix}. \tag{2.24}$$

For example,

$$\begin{aligned} v_{\bar{1}} &= \beta_{1j} v_j \\ &= \beta_{11} v_1 + \beta_{12} v_2 \\ &= -v_1 \quad \text{(tensor transformation)} \\ &= v_1, \quad \text{(isotropy)} \end{aligned}$$

which results in

$$v_1 = 0.$$

Similarly, it can be shown that

$$v_2 = 0.$$

Therefore, if v_i is isotropic, it must be a zero vector.

3. Second-rank tensors.
   If v_{ij} is a second-rank tensor and isotropic, it must be a scalar multiplication of the Kronecker delta, i.e.,

$$v_{ij} = \alpha \delta_{ij},$$

where α is a scalar.

Proof. First, it can be shown that if v_{ij} is isotropic, it must be diagonal, i.e.,

$$v_{12} = v_{21} = 0.$$

For the transformation matrix, β_{ij}, of Equation (2.24),

$$\begin{aligned} v_{\bar{1}\bar{2}} &= \beta_{1i} \beta_{2j} v_{ij} \\ &= \beta_{11} \beta_{22} v_{12} \\ &= -v_{12} \quad \text{(tensor transformation)} \\ &= v_{12}. \quad \text{(isotropy)} \end{aligned}$$

Therefore,
$$v_{12} = 0.$$
Similarly, it can be shown that
$$v_{21} = 0,$$
Thus, v_{ij} must be diagonal, i.e.,
$$v_{ij} = \begin{pmatrix} v_{11} & 0 \\ 0 & v_{22} \end{pmatrix}.$$
If one imposes a rotation on v_{ij} with
$$\beta_{ij} = \begin{pmatrix} \cos\theta & \sin\theta \\ -\sin\theta & \cos\theta \end{pmatrix},$$
then,
$$\begin{aligned} v_{\bar{1}\bar{1}} &= \beta_{1i}\beta_{1j}v_{ij} \\ &= \beta_{11}\beta_{11}v_{11} + \beta_{12}\beta_{12}v_{22} \\ &= \cos^2\theta v_{11} + \sin^2\theta v_{22} \quad \text{(tensor transformation)} \\ &= v_{11}, \quad \text{(isotropy)} \end{aligned}$$
which is equivalent to
$$v_{11} = \cos^2\theta v_{11} + \sin^2\theta v_{22},$$
or
$$\sin^2\theta(v_{22} - v_{11}) = 0.$$
Therefore,
$$\begin{aligned} v_{ij} &= \begin{pmatrix} v_{11} & 0 \\ 0 & v_{11} \end{pmatrix} \\ &= v_{11}\begin{pmatrix} 1 & 0 \\ 0 & 1 \end{pmatrix} \\ &= v_{11}\delta_{ij}. \end{aligned}$$

4. Third-rank tensors.
   Similar to first-rank tensors, no third-rank tensor is isotropic.
5. Fourth-rank tensors.
   If v_{ijkl} is a fourth-rank tensor, it must be a combination of Kronecker's deltas, i.e.,
$$v_{ijkl} = A\delta_{ij}\delta_{kl} + B\delta_{ik}\delta_{jl} + C\delta_{il}\delta_{jk}. \tag{2.25}$$
6. Higher-rank tensors.
   Only even-rank tensors can be isotropic, and they are expressed as a combination of the Kronecker deltas.

2.4.2 Elastic Modulus

In general, isotropic fourth-rank tensors have three independent constants as shown in Equation (2.25). However, for elastic moduli, C_{ijkl}, the number of independent components is reduced to 2 because of the symmetry, $C_{ijkl} = C_{ijlk}$, as

$$C_{ijkl} = A\delta_{ij}\delta_{kl} + B\delta_{ik}\delta_{jl} + C\delta_{il}\delta_{jk}, \tag{2.26}$$

$$C_{ijlk} = A\delta_{ij}\delta_{lk} + B\delta_{il}\delta_{jk} + C\delta_{ik}\delta_{lj}. \tag{2.27}$$

Subtracting Equation (2.27) from Equation (2.26) yields

$$(B - C)(\delta_{ik}\delta_{jl} - \delta_{il}\delta_{ik}) = 0,$$

or

$$B = C,$$

Thus,

$$C_{ijkl} = A\delta_{ij}\delta_{kl} + B(\delta_{ik}\delta_{jl} + \delta_{il}\delta_{jk}).$$

Conventionally, λ and μ are used in place of A and B and they are called the Lamé constants. The elastic modulus thus can be written as

$$C_{ijkl} = \lambda\delta_{ij}\delta_{kl} + \mu(\delta_{ik}\delta_{jl} + \delta_{il}\delta_{jk}). \tag{2.28}$$

The constant, μ, is identical to the shear modulus (the modulus of rigidity commonly denoted by G) and λ is related to the Poisson ratio.

By using Equation (2.28), the stress–strain relationship (the constitutive equation) for linear isotropic elastic materials can be written as

$$\begin{aligned}\sigma_{ij} &= C_{ijkl}\epsilon_{kl} \\ &= \lambda\delta_{ij}\delta_{kl}\epsilon_{kl} + \mu\delta_{ik}\delta_{jl}\epsilon_{kl} + \mu\delta_{il}\delta_{jk}\epsilon_{kl} \\ &= \mu\epsilon_{ij} + \mu\epsilon_{ji} + \lambda\delta_{ij}\epsilon_{ll} \\ &= 2\mu\epsilon_{ij} + \lambda\delta_{ij}\epsilon_{kk}.\end{aligned} \tag{2.29}$$

The inversion of Equation (2.29) is

$$\epsilon_{ij} = \frac{1}{2\mu}\sigma_{ij} - \frac{\lambda}{2\mu(2\mu + 3\lambda)}\delta_{ij}\sigma_{kk}. \tag{2.30}$$

Proof. Decompose both σ_{ij} and ϵ_{ij} into the deviatoric and hydrostatic parts as

$$\sigma_{ij} = \sigma'_{ij} + \frac{1}{3}\delta_{ij}\sigma_{kk}, \tag{2.31}$$

$$\epsilon_{ij} = \epsilon'_{ij} + \frac{1}{3}\delta_{ij}\epsilon_{kk}. \tag{2.32}$$

By setting $i = j$ in Equation (2.29), it follows that

$$\sigma_{ii} = (2\mu + 3\lambda)\epsilon_{ii},$$

or
$$\sigma_{ii} = 3K\epsilon_{ii},$$
where
$$K \equiv \frac{2\mu + 3\lambda}{3},$$
which defines the bulk modulus. Substitution of Equations (2.31) and (2.32) into Equation (2.29) yields
$$\sigma'_{ij} + \frac{1}{3}\delta_{ij}\sigma_{kk} = 2\mu\left(\epsilon'_{ij} + \frac{1}{3}\delta_{ij}\epsilon_{kk}\right) + \lambda\delta_{ij}\epsilon_{kk}.$$
That is,
$$\sigma'_{ij} + \frac{2\mu + 3\lambda}{3}\delta_{ij}\epsilon_{kk} = 2\mu\epsilon'_{ij} + \frac{2\mu + 3\lambda}{3}\delta_{ij}\epsilon_{kk},$$
which is reduced to
$$\sigma'_{ij} = 2\mu\epsilon'_{ij}.$$
Therefore,
$$\epsilon'_{ij} = \frac{1}{2\mu}\sigma'_{ij}, \quad \epsilon_{kk} = \frac{1}{3K}\sigma_{kk}.$$
which is equivalent to Equation (2.30).

If one sets $(i, j) = (1, 1)$, Equation (2.30) becomes
$$\epsilon_{11} = \frac{1}{2\mu}\sigma_{11} - \frac{\lambda}{2\mu(2\mu + 3\lambda)}\delta_{ij}\sigma_{kk}$$
$$= \frac{\mu + \lambda}{\mu(2\mu + 3\lambda)}\sigma_{11} - \frac{\lambda}{2\mu(2\mu + 3\lambda)}(\sigma_{22} + \sigma_{33}),$$
which can be compared with one of the engineering strain–stress relations,
$$\epsilon_{11} = \frac{1}{E}\sigma_{11} - \frac{\nu}{E}(\sigma_{22} + \sigma_{33}),$$
where E is the Young modulus and ν is the Poisson ratio. Therefore, one can derive one of the relationships between the engineering constants and the Lamé constants as
$$\frac{1}{E} = \frac{\mu + \lambda}{\mu(2\mu + 3\lambda)},$$
$$\frac{\nu}{E} = \frac{\lambda}{2\mu(2\mu + 3\lambda)}.$$
The inverse of the elastic modulus, C_{ijkl}, is called the elastic compliance and is defined as
$$\epsilon_{ij} = S_{ijkl}\sigma_{kl}.$$
The reciprocal relationship is
$$C_{ijkl}S_{klmn} = I_{ijmn},$$
where
$$I_{ijmn} = \frac{1}{2}(\delta_{im}\delta_{jn} + \delta_{in}\delta_{jm}).$$

It can be shown that the elastic compliance for isotropic materials is expressed as

$$S_{ijkl} = \frac{1}{4\mu}(\delta_{ik}\delta_{jl} + \delta_{il}\delta_{jk}) - \frac{\lambda}{2\mu(3\lambda + 2\mu)}\delta_{ij}\delta_{kl}.$$

2.4.3 Orthotropy

The term *orthotropy* refers to the property that is invariant under reflection with respect to the x, y, and z axes. It can be shown that, for the components of orthotropic elastic modulus to be nonzero, the number of repeated indices must be an even number. For example, for the reflection transformation matrix with respect to the x_2–x_3 plane,

$$\beta_{ij} = \begin{pmatrix} -1 & 0 & 0 \\ 0 & 1 & 0 \\ 0 & 0 & 1 \end{pmatrix},$$

$$C_{\bar{1}3\bar{2}3} = \beta_{1i}\beta_{3j}\beta_{2k}\beta_{3l}C_{ijkl}$$

$$= \cdots$$

$$= \beta_{11}\beta_{33}\beta_{22}\beta_{33}C_{1323}$$

$$= -C_{1323} \quad \text{(tensor transformation)}$$

$$= C_{1323} \quad \text{(orthotropy)}.$$

Therefore, it follows that

$$C_{1323} = 0.$$

This result can be generalized, i.e., for C_{ijkl} to be nonzero, the number of indices must be balanced. Using this property, after eliminating redundant components, the independent components for orthotropic materials are

$$C_{1111}, C_{2222}, C_{3333}, C_{1122}, C_{2233}, C_{3311}, C_{1212}, C_{2323}, C_{3131}.$$

By using the engineering constants, the strain–stress relationship can be expressed as

$$\begin{pmatrix} \epsilon_x \\ \epsilon_y \\ \epsilon_y \end{pmatrix} = \begin{pmatrix} \frac{1}{E_x} & -\frac{\nu_{xy}}{E_x} & -\frac{\nu_{xz}}{E_x} \\ -\frac{\nu_{yx}}{E_y} & \frac{1}{E_y} & -\frac{\nu_{yz}}{E_y} \\ -\frac{\nu_{zx}}{E_z} & -\frac{\nu_{zy}}{E_z} & \frac{1}{E_z} \end{pmatrix} \begin{pmatrix} \sigma_x \\ \sigma_y \\ \sigma_z \end{pmatrix},$$

$$\epsilon_{xy} = \frac{1}{2G_{xy}}\sigma_{xy}, \quad \epsilon_{yz} = \frac{1}{2G_{yz}}\sigma_{yz}, \quad \epsilon_{zx} = \frac{1}{2G_{zx}}\sigma_{zx}.$$

However, because of the symmetry,

$$S_{ijkl} = S_{klij},$$

the following relationships must hold:

$$\frac{\nu_{xy}}{E_x} = \frac{\nu_{yx}}{E_y}, \quad \frac{\nu_{xz}}{E_x} = \frac{\nu_{zx}}{E_z}, \quad \frac{\nu_{yz}}{E_y} = \frac{\nu_{zy}}{E_z}.$$

Therefore, the number of independent components is 9 (= 12 − 3) for 3-D orthotropic materials.

2.4.4 2-D Orthotropic Materials

The constitutive relation for 2-D orthotropic materials as shown in Figure 2.14 is reduced to

$$\begin{pmatrix} \epsilon_{11} \\ \epsilon_{22} \end{pmatrix} = \begin{pmatrix} \frac{1}{E_1} & -\frac{\nu_{12}}{E_1} \\ -\frac{\nu_{21}}{E_2} & \frac{1}{E_2} \end{pmatrix} \begin{pmatrix} \sigma_{11} \\ \sigma_{22} \end{pmatrix},$$

$$\epsilon_{12} = \frac{1}{2G_{12}} \sigma_{12}.$$

Therefore, in term of the compliance components, it follows that

$$S_{1111} = \frac{1}{E_1}, \quad S_{2222} = \frac{1}{E_2},$$

$$S_{1122} = S_{2211} = -\frac{\nu_{12}}{E_1} = -\frac{\nu_{21}}{E_2},$$

$$S_{1212} = S_{2112} = S_{2121} = S_{1221} = \frac{1}{4G_{12}}.$$

All other values of S_{ijkl} are 0. The number of independent components is 4 for 2-D orthotropic materials (Figure 2.14).

2.4.5 Transverse Isotropy

Transverse isotropy is a subset of orthotropy in which the material is isotropic on one of the three planes. If the *x–y* plane is the plane of isotropy, for instance, the directional preference in the *x* and *y* directions becomes identical and thus

$$C_{1111} = C_{2222}, \quad C_{1133} = C_{2233}, \quad C_{1313} = C_{2323},$$

so that the number of independent components for transversely isotropic materials is further reduced to 5.

An example of transversely isotropic materials is a unidirectionally reinforced composite.

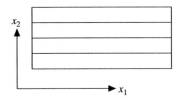

Figure 2.14 2-D orthotropic material.

2.5 Constitutive Relation for Fluids

The following are the major differences between fluids and solids in constitutive relation:

1. Strains (ϵ_{ij}) in solids must be replaced by strain rates (V_{ij}).
2. There always exists hydrostatic pressure in the absence of V_{ij} in fluids.

This leads to the following constitutive relation for fluids as

$$\sigma_{ij} = D_{ijkl} V_{kl} - p\delta_{ij}. \tag{2.33}$$

For isotropic fluids,

$$D_{ijkl} = \lambda \delta_{ij} \delta_{kl} + \mu(\delta_{ik}\delta_{jl} + \delta_{il}\delta_{jk}),$$

so that Equation (2.33) becomes

$$\sigma_{ij} = 2\mu V_{ij} + \lambda \delta_{ij} V_{kk} - p\delta_{ij}. \tag{2.34}$$

The following is called *Stokes' hypothesis*, which is generally believed to be true (Gad-el Hak 1995).

$$\sigma_{kk} = -3P.$$

By setting $i = j$ in Equation (2.34), one obtains

$$\sigma_{ii} = 2\mu V_{ii} + 3\lambda V_{kk} - 3P.$$

Therefore, it follows that

$$2\mu + 3\lambda = 0,$$

and Equation (2.34) becomes

$$\sigma_{ij} = 2\mu V_{ij} - \frac{2\mu}{3}\delta_{ij} V_{kk} - p\delta_{ij}.$$

The fluid that this constitutive relation holds is called the *Stokes fluid* where μ is the viscosity.

If the fluid is incompressible, $V_{kk} = 0$, the constitutive relation becomes

$$\sigma_{ij} = 2\mu V_{ij} - p\delta_{ij}. \tag{2.35}$$

For a nonviscous fluid, $\mu = 0$, Equation (2.35) becomes

$$\sigma_{ij} = -p\delta_{ij}.$$

2.5.1 Thermal Effect

In the absence of applied stress, a temperature rise causes deformation. The simplest relationship between the deformation, ϵ_{ij}, and the temperature rise, ΔT, is to assume that they are proportional to each other as

$$\epsilon_{ij} = \alpha_{ij} \Delta T,$$

where α_{ij} is the proportionality factor and a second-rank tensor called the thermal expansion coefficient. If the material is isotropic, α_{ij} has to be a scalar multiplication of the Kronecker delta, δ_{ij}, as

$$\alpha_{ij} = \alpha \delta_{ij}.$$

Field Equations

As the total deformation is the sum of the elastic deformation and the thermal deformation, it follows that

$$\epsilon_{ij}^{total} = \epsilon_{ij}^{elastic} + \epsilon_{ij}^{thermal},$$

where $\epsilon_{ij}^{elastic}$ is a part of the strain that is directly proportional to the applied stress as

$$\begin{aligned}\sigma_{ij} &= C_{ijkl}\epsilon_{kl}^{elastic} \\ &= C_{ijkl}(\epsilon_{kl}^{total} - \epsilon_{kl}^{thermal}) \\ &= C_{ijkl}(\epsilon_{kl}^{total} - \alpha_{kl}\Delta T). \end{aligned} \quad (2.36)$$

Equation (2.36) is the modified constitutive relation when thermal stresses must be considered.

2.6 Derivation of Field Equations

2.6.1 Divergence Theorem (Gauss Theorem)

In order to derive field equations, it is necessary to use the Gauss divergence theorem extensively. The Gauss divergence theorem is a multidimensional version of the fundamental theorem of calculus that states that integration and differentiation are reciprocal to each other, i.e.,

$$\int_a^b f'(x)dx = [f]_a^b. \quad (2.37)$$

Referring to Figure 2.15, the right-hand side of Equation (2.37) can be written as

$$\begin{aligned}[f]_a^b &= -f|_{x=a} + f|_{x=b} \\ &= (-1) \times f|_{x=a} + (+1) \times f|_{x=b} \\ &= n_a f|_{x=a} + n_b f|_{x=b} \\ &= (nf)|_{x=a} + (nf)|_{x=b} \\ &= \sum_{boundary} nf. \end{aligned}$$

Therefore, Equation (2.37) can be written as

$$\int_a^b f'(x)dx = \sum_{boundary} nf. \quad (2.38)$$

Based on Equation (2.38) for 1-D, it is possible to rewrite this in 2-D and 3-D. For 2-D, the integral range needs to be changed to a 2-D region. The derivative with respect to x can be

Figure 2.15 1-D divergence theorem

rewritten as the derivative with respect to x_i and the normal n needs to be rewritten as n_i. Thus, Equation (2.38) can be extended to 2-D as

$$\int\int_S f_{,i}\,dS = \oint_{\partial S} n_i f\,d\ell, \qquad (2.39)$$

where dS is the area element and $d\ell$ is the line element. This is the Gauss divergence theorem in 2-D. It is straightforward to express the Gauss divergence theorem in 3-D from Equation (2.39) as

$$\int\int\int_V f_{,i}\,dV = \int\int_{\partial V} n_i f\,dS,$$

where dV is the volume element and dS is the area element over the boundary of the volume. It should be noted that the majority of the usage of the Gauss divergence theorem is to convert a boundary integral that contains the normal vector into a volume integral (i.e., from the right to the left).

2.6.2 Material Derivative

If a function, $f(x_j, t)$, is defined on a point in a continuum that flows, its variation at $t + \Delta t$, $f(x_j + \Delta x_j, t + \Delta t)$, can be expanded by Taylor's series as

$$f(x_j + \Delta x_j, t + \Delta t) = f(x_j, t) + \left(f_{,j}\Delta x_j + \frac{\partial f}{\partial t}\Delta t\right) + \cdots$$

Thus, the rate of time change is expressed as

$$\frac{f(x_j + \Delta x_j, t + \Delta t) - f(x_j, t)}{\Delta t} \sim f_{,j}\frac{\Delta x_j}{\Delta t} + \frac{\partial f}{\partial t}. \qquad (2.40)$$

By taking a limit, $\Delta t \to 0$, Equation (2.40) becomes

$$\frac{Df}{Dt} \equiv \frac{\partial f}{\partial t} + f_{,i}v_i, \qquad (2.41)$$

where v_i is the velocity of the continua. The quantity, $\frac{Df}{Dt}$, in Equation (2.41) is called *the material derivative* or *the Lagrangian derivative* and represents the substantial rate of time change for a quantity associated with a moving particle. The second term on the right-hand side of Equation (2.41) is the convective term that is carried with the flow.

2.6.2.1 Material Derivative for Functions Defined by Integrals

If a quantity is defined over the entire continua, it is possible to compute its material derivative. Let the integration of $f(x, t)$ over the entire material points be denoted as $I(t)$ as

$$I(t) \equiv \int_a^b f(x, t)\,dx.$$

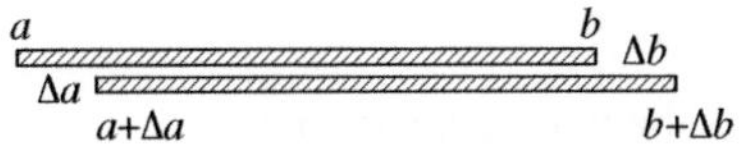

Figure 2.16 1-D material derivative for a function defined by an integral

Referring to Figure 2.16, the quantity, $I(t + \Delta t)$, is expressed in terms of the reference integral from a to b and the quantity between b and $b + \Delta b$ subtracting the quantity from a to $a + \Delta a$ as

$$\begin{aligned} I(t+\Delta t) &= \int_{a+\Delta a}^{b+\Delta b} f(x, t+\Delta t)dx \\ &= \int_a^b f(x,t+\Delta t)dx + \int_b^{b+\Delta b} f(x,t+\Delta t)dx - \int_a^{a+\Delta} f(x,t+\Delta t)dx \\ &\sim \int_a^b f(x,t+\Delta t)dx + f|_b \Delta b - f|_a \Delta a, \end{aligned}$$

where $f|_a$ and $f|_b$ are the evaluations of f at a and b (Figure 2.16). Therefore, the rate of time change is

$$\frac{I(t+\Delta t) - I(t)}{\Delta t} \sim \int_a^b \frac{f(x,t+\Delta t) - f(x,t)}{\Delta t} dx + f|_b \frac{\Delta b}{\Delta t} - f|_a \frac{\Delta a}{\Delta t}. \tag{2.42}$$

By taking a limit by letting Δt tend to 0, Equation (2.42) becomes

$$\begin{aligned} \frac{DI}{Dt} &= \lim_{\Delta t \to 0} \frac{I(t+\Delta t) - I(t)}{\Delta t} \\ &= \int_a^b \frac{\partial f}{\partial t} dx + (-1) f v|_a + (+1) f v|_b \\ &= \int_a^b \frac{\partial f}{\partial t} dx + \sum_{boundary} n f v. \end{aligned} \tag{2.43}$$

Equation (2.43) is known as the Leibniz integral rule. This result can be extended to 2-D as. Define $I(t)$ as

$$I(t) \equiv \int\int f(x_j, t) dS.$$

The material derivative for $I(t)$ is

$$\begin{aligned} \frac{DI}{Dt} &= \int\int \frac{\partial f}{\partial t} dS + \oint n_i f v_i d\ell \\ &= \int\int \frac{\partial f}{\partial t} dS + \int\int (f v_i)_{,i} dS \\ &= \int\int \left(\frac{\partial f}{\partial t} + (f v_i)_{,i} \right) dS. \end{aligned} \tag{2.44}$$

Equation (2.44) is used to derive various field equations.

2.6.3 Equation of Continuity

In Equation (2.44), choose f as the mass density, ρ, as

$$f(x_j, t) = \rho(x_j, t).$$

Then, its integral is the total mass defined as

$$M(t) = \int\int \rho(x_j, t) dS.$$

The material derivative of the total mass is expressed as

$$\frac{DM}{Dt} = \int\int \left(\frac{\partial \rho}{\partial t} + (\rho v_i)_{,i}\right) dS.$$

On the other hand, the total mass is conserved in classical mechanics. Therefore, one has

$$\frac{DM}{Dt} = 0.$$

By equating the aforementioned two equations, one obtains

$$\int\int \left(\frac{\partial \rho}{\partial t} + (\rho v_i)_{,i}\right) dS = 0,$$

or

$$\frac{\partial \rho}{\partial t} + (\rho v_i)_{,i} = 0, \tag{2.45}$$

which is known as the *equation of continuity*. If ρ is a constant (incompressible), Equation (2.45) becomes

$$v_{i,i} = 0.$$

2.6.4 Equation of Motion

By choosing $f = \rho v_i$ (momentum), its integral defines the total momentum as

$$I(t) \equiv \int\int \rho v_i dS.$$

The material derivative of the total momentum, therefore, is

$$\frac{DI}{Dt} = \int\int \left(\frac{\partial}{\partial t}(\rho v_i) + (\rho v_i v_j)_{,j}\right) dS. \tag{2.46}$$

According to Newton's second law, the rate change of the total momentum is due to the total force applied to the body as

$$\frac{DI}{Dt} = \oint t_i d\ell + \int\int b_i dS$$

$$= \oint \sigma_{ij} n_j d\ell + \int\int b_i dS$$

$$= \int\int (\sigma_{ij,j} + b_i) dS. \tag{2.47}$$

Field Equations

By equating Equation (2.46) with Equation (2.47), one obtains

$$\sigma_{ij,j} + b_i = \frac{\partial}{\partial t}(\rho v_i) + (\rho v_i v_j)_{,j}$$

$$= \frac{\partial \rho}{\partial t} v_i + \rho \frac{\partial v_i}{\partial t} + (\rho v_j)_{,j} v_i + \rho v_j v_{i,j}$$

$$= v_i \left(\frac{\partial \rho}{\partial t} + (\rho v_j)_{,j} \right) + \rho \left(\frac{\partial v_i}{\partial t} + v_j v_{i,j} \right)$$

$$= \rho \left(\frac{\partial v_i}{\partial t} + v_j v_{i,j} \right)$$

$$= \rho \frac{Dv_i}{Dt}.$$

where Equation (2.45) was used. Thus, the equation of motion is expressed as

$$\rho \frac{Dv_i}{Dt} = \sigma_{ij,j} + b_i. \tag{2.48}$$

Equation (2.48) is universal for any deformable bodies.

2.6.5 Equation of Energy

The balance of energy states that

$$\frac{D}{Dt} \int \int \left(\frac{1}{2} \rho v_i v_i + \rho E \right) dS = \int \int b_i v_i dS + \oint t_i v_i d\ell - \oint h_i n_i d\ell, \tag{2.49}$$

where v_i is the velocity, E is the internal energy, and h_i is the heat flux. Note that $1/2 \rho v_i v_i$ is the kinetic energy. The material derivative of the first term in Equation (2.49) is evaluated as

$$\frac{D}{Dt} \int \int \frac{1}{2} \rho v_i v_i dS = \int \int \left\{ \frac{\partial}{\partial t} \left(\frac{1}{2} \rho v_i v_i \right) + \left(\frac{1}{2} \rho v_i v_i v_j \right)_{,j} \right\} dS$$

$$= \int \int \left\{ \frac{1}{2} \frac{\partial}{\partial t}(\rho v_i) v_i + \frac{1}{2} \rho v_i \frac{\partial v_i}{\partial t} + \frac{1}{2}(\rho v_i v_j)_{,j} v_i + \frac{1}{2} \rho v_i v_j v_{i,j} \right\} dS$$

$$= \int \int \left\{ \frac{1}{2} v_i \left(\frac{\partial}{\partial t}(\rho v_i) + (\rho v_i v_j)_{,j} \right) + \frac{1}{2} \rho v_i \left(\frac{\partial v_i}{\partial t} + v_{i,j} v_j \right) \right\} dS$$

$$= \int \int \left\{ \frac{1}{2} v_i \left(v_i \frac{\partial \rho}{\partial t} + \rho \frac{\partial v_i}{\partial t} + (\rho v_j)_{,j} v_i + \rho v_j v_{i,j} \right) + \frac{1}{2} \rho v_i \frac{Dv_i}{Dt} \right\} dS$$

$$= \int \int \left\{ \left\{ \frac{1}{2} \rho v_i \left(\frac{\partial v_i}{\partial t} + v_{i,j} v_j \right) + \frac{1}{2} v_i v_i \left(\frac{\partial \rho}{\partial t} + (\rho v_j)_{,j} \right) \right\} + \frac{1}{2} \rho v_i \frac{Dv_i}{Dt} \right\} dS$$

$$= \int \int \left\{ v_i \rho \frac{Dv_i}{Dt} + \frac{1}{2} v_i v_i \left(\frac{\partial \rho}{\partial t} + (\rho v_j)_{,j} \right) \right\} dS$$

$$= \int \int v_i \rho \frac{Dv_i}{Dt} dS$$

$$= \int \int v_i (\sigma_{ij,j} + b_i) dS,$$

where both Equations (2.45) and (2.48) were used. The material derivative of the second term to the left of the equal sign of Equation (2.49) is evaluated as

$$\frac{D}{Dt}\int\int \rho E dS = \int\int \left\{\frac{\partial}{\partial t}(\rho E) + (\rho E v_i)_{,i}\right\} dS$$

$$= \int\int \left\{\rho\frac{\partial E}{\partial t} + \frac{\partial \rho}{\partial t}E + (\rho v_i)_{,i}E + \rho v_i E_{,i}\right\} dS$$

$$= \int\int \left\{\rho\left(\frac{\partial E}{\partial t} + E_{,i}v_i\right) + E\left(\frac{\partial \rho}{\partial t} + (\rho v_i)_{,i}\right)\right\} dS$$

$$= \int\int \rho\frac{DE}{Dt} dS,$$

where Equation (2.45) was used. The second and third terms to the right of the equal sign of Equation (2.49) can be modified as

$$\oint t_i v_i d\ell = \oint \sigma_{ij} n_j v_i d\ell$$

$$= \int\int (\sigma_{ij} v_i)_{,j} dS$$

$$= \int\int (\sigma_{ij,j} v_i + \sigma_{ij} v_{i,j}) dS,$$

and

$$\oint -h_i n_i d\ell = -\int\int h_{i,i} dS.$$

Therefore, Equation (2.49) becomes

$$\int\int \left\{v_i \sigma_{ij,j} + v_i b_i + \rho\frac{DE}{Dt}\right\} dS = \int\int \{v_i b_i + \sigma_{ij,j} v_i + \sigma_{ij} v_{i,j} - h_{i,i}\} dS,$$

or

$$\rho\frac{DE}{Dt} = -h_{i,i} + \sigma_{ij} v_{i,j}, \qquad (2.50)$$

which is called *the equation of energy*.

Example If a body is not in motion, it follows that $v_i = 0$ and $E = C_p T$ where C_p is the specific heat. According to Fourier's law, the heat flux is directly proportional to the negative gradient of the temperature field with the proportionality k_{ij} known as the thermal conductivity as

$$h_i = -k_{ij} T_{,j}.$$

Therefore, Equation (2.50) becomes

$$\rho C_p \frac{\partial T}{\partial t} = (k_{ij} T_{,j})_{,i}. \qquad (2.51)$$

Equation (2.51) is known as the transient heat conduction equation.

Field Equations

2.6.6 Isotropic Solids

For an isotropic solid, a combination of

$$\sigma_{ij} = 2\mu\epsilon_{ij} + \lambda\delta_{ij}\epsilon_{kk},$$

$$\epsilon_{ij} = \frac{1}{2}(u_{i,j} + u_{j,i})$$

yields

$$\sigma_{ij} = \mu u_{i,j} + \mu u_{j,i} + \lambda\delta_{ij}u_{k,k}.$$

Therefore,

$$\sigma_{ij,j} = \mu u_{i,jj} + \mu u_{j,ij} + \lambda\delta_{ij}u_{k,kj}$$
$$= \mu u_{i,jj} + \mu u_{j,ji} + \lambda u_{k,ki}$$
$$= \mu u_{i,jj} + (\mu + \lambda)u_{j,ji}.$$

Hence, the equation of motion for an isotropic solid becomes

$$\rho\frac{Dv_i}{Dt} = \mu u_{i,jj} + (\mu + \lambda)u_{k,ki} + b_i, \qquad (2.52)$$

or

$$\rho\frac{D\mathbf{v}}{Dt} = \mu\Delta\mathbf{u} + (\mu + \lambda)\nabla\nabla\cdot\mathbf{u} + \mathbf{b}. \qquad (2.53)$$

Equation (2.52) or Equation (2.53) is called the *Navier equation*.

2.6.7 Isotropic Fluids

For an isotropic fluid, a combination of

$$\sigma_{ij} = -p\delta_{ij} + 2\mu V_{ij} - \frac{2}{3}\mu V_{kk}\delta_{ij},$$

$$V_{ij} = \frac{1}{2}(v_{i,j} + v_{j,i})$$

yields

$$\sigma_{ij} = -p\delta_{ij} + \mu(v_{i,j} + v_{j,i}) - \frac{2}{3}\mu\delta_{ij}v_{k,k}.$$

Therefore,

$$\sigma_{ij,j} = -p_{,j}\delta_{ij} + \mu(v_{i,jj} + v_{j,ij}) - \frac{2}{3}\mu\delta_{ij}v_{k,kj}$$
$$= -p_{,i} + \mu v_{i,jj} + \mu v_{j,ij} - \frac{2}{3}\mu v_{k,ki}$$
$$= -p_{,i} + \mu v_{i,jj} + \frac{\mu}{3}v_{j,ji}.$$

Hence, the equation of motion becomes

$$\rho \frac{Dv_i}{Dt} = \mu v_{i,jj} + \frac{\mu}{3} v_{j,ji} - P_{,i} + b_i, \quad (2.54)$$

or

$$\rho \frac{D\mathbf{v}}{Dt} = \mu \Delta \mathbf{v} + \frac{\mu}{3} \nabla \nabla \cdot \mathbf{v} - \nabla P + \mathbf{b}. \quad (2.55)$$

Equation (2.54) or Equation (2.55) is called the *Navier–Stokes* equation.

2.6.8 Thermal Effects

A combination of Equation (2.36) with Equation (2.52) yields the Navier equation when thermal effects need to be considered as

$$\rho \frac{Dv_i}{Dt} = (C_{ijkl} u_{k,l})_{,j} + b_i - (C_{ijkl} \alpha_{kl} \Delta T)_{,j}. \quad (2.56)$$

It is seen from Equation (2.56) that the presence of the thermal stress term, $-(C_{ijkl}\alpha_{kl}\Delta T)_{,j}$, can be identified as a part of the body force.

2.7 General Coordinate System

2.7.1 Introduction to Tensor Analysis

All the quantities introduced so far were for the Cartesian coordinate systems where the length of the base vectors remains unchanged (unit length) and constant. However, in general curvilinear coordinate systems[9] such as the spherical coordinate system, the length of the base vectors is not necessarily constant and is a function of the position in general. For instance, the length of the base vector in the tangential direction in the polar coordinate system increases as the position moves away from the origin.

In order to write tensor equations valid for general curvilinear coordinate systems, it is important to distinguish two different quantities, i.e., the contravariant quantities indicated by superscripts (i.e., a^i) and the covariant quantities indicated by subscripts (i.e., b_i). The contravariant quantity typically represents the component of a physical quantity while the covariant quantity typically represents the unit of the physical quantity. For example, "1" in the length 1 cm is the component while "cm" is the unit. When the same length is expressed as $\frac{1}{100}$ m, the component becomes 1/100 while the unit is magnified by 100 times. It is seen, therefore, that the transformation rules for component and the unit are reciprocal to each other. As any physical quantity is a combination of the component (contravariant quantity) and the unit (covariant quantity), this trivial example can be generalized to tensorial quantities.

To formally define the contravariant and covariant quantities, the following convention is adopted, i.e.,

The indices of coordinate components are expressed by superscripts, i.e., x^i.

---

[9] Although some of the notations and symbols introduced in this chapter are often used in differential geometry, it should be noted that the nature of the space is always the Euclidean space, i.e., no curvature exists.

Field Equations

If the coordinate system is changed from x^j to $x^{\bar{i}}$, the following relationship is assumed:

$$x^{\bar{i}} = x^{\bar{i}}(x^j). \qquad (2.57)$$

Taking the total derivative of Equation (2.57) yields

$$dx^{\bar{i}} = \frac{\partial x^{\bar{i}}}{\partial x^j} dx^j$$

$$= \beta_j^{\bar{i}} dx^j, \qquad (2.58)$$

where

$$\beta_j^{\bar{i}} \equiv \frac{\partial x^{\bar{i}}}{\partial x^j}. \qquad (2.59)$$

On the other hand, consider the base vector, $\mathbf{e}_i$, defined as

$$\mathbf{e}_i \equiv \frac{\partial \mathbf{R}}{\partial x^i},$$

where $\mathbf{R}$ is the position vector and expressed in the Cartesian coordinate system as

$$\mathbf{R} = (x_1, x_2, x_3).$$

When the coordinate system is changed from x^i to $x^{\bar{j}}$, the new base vector, $\mathbf{e}_{\bar{i}}$, is subject to the following transformation:

$$\mathbf{e}_{\bar{i}} = \frac{\partial \mathbf{R}}{\partial x^{\bar{i}}}$$

$$= \frac{\partial \mathbf{R}}{\partial x^j} \frac{\partial x^j}{\partial x^{\bar{i}}}$$

$$= \beta_{\bar{i}}^j \mathbf{e}_j, \qquad (2.60)$$

where

$$\beta_{\bar{i}}^j \equiv \frac{\partial x^j}{\partial x^{\bar{i}}}. \qquad (2.61)$$

It is seen from Equations (2.58) and (2.60) that the transformation rules for dx^i and $\mathbf{e}_i$ are reciprocal to each other. Note that

$$\beta_j^{\bar{i}} \beta_{\bar{k}}^j = \delta_k^i.$$

Based on the aforementioned transformation rules, the following conventions are adopted:

- The indices of coordinate components are expressed by superscripts, i.e., x^i.
- Any quantity whose transformation is similar to Equation (2.58) is called a "contravariant" quantity indicated by a superscript.
- Any quantity whose transformation is similar to Equation (2.60) is called a "covariant" quantity indicated by a subscript.
- The summation convention is used between the contravariant and covariant quantities only.

The major reason for this convention is that in a general curvilinear coordinate system as shown in Figure 2.17, the length of the base vectors is not necessarily normalized so that the transformation rules for the components and the base vectors (unit) are different (inverse each other) which leads to the distinction between contravariant quantities (components) and covariant quantities (units).

The quantity, $\beta_i^{\bar{j}}$, is called the (contravariant) transformation matrix.

2.7.2 Definition of Tensors in Curvilinear Systems

By using the transformation matrices of Equations (2.59) and (2.61), tensors in general curvilinear coordinate systems (Figure 2.17) can be defined as follows:

1. Scalar (tensor of rank 0, zeroth-rank tensor).
   If a single quantity, $\Phi(x^i)$, is invariant under the coordinate transformation, i.e.,
   $$\Phi(x^{\bar{i}}) = \Phi(x^i),$$
   the quantity, Φ, is called a scalar.
2. Vector (tensor of rank 1, first-rank tensor)
   (a) Contravariant vector.
       A quantity, v^i, is called a contravariant vector if its transformation rule is
       $$v^{\bar{i}} = \beta_j^{\bar{i}} v^j.$$
   (b) Covariant vector.
       A quantity, v_i, is called a covariant vector if its transformation rule is
       $$v_{\bar{i}} = \beta_{\bar{i}}^j v_j.$$
   Note that $\beta_j^{\bar{i}}$ and $\beta_{\bar{j}}^i$ are reciprocal each other.
3. Second-rank tensors (tensors of rank 2).
   (a) Second-rank contravariant tensor.
       A quantity, v^{ij}, is called a second-rank contravariant vector if its transformation rule is
       $$v^{\bar{i}\bar{j}} = \beta_i^{\bar{i}} \beta_j^{\bar{j}} v^{ij}.$$

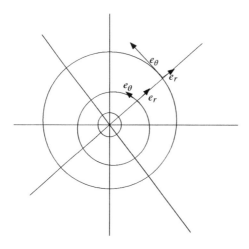

Figure 2.17 Curvilinear coordinate system

Field Equations

(b) Second-rank covariant tensor.
A quantity, v_{ij}, is called a second-rank covariant vector if its transformation rule is

$$v_{\bar{i}\bar{j}} = \beta_{\bar{i}}^{i}\beta_{\bar{j}}^{j}v_{ij}.$$

(c) Second-rank mixed tensor.
A quantity, v_j^i, is called a second-rank mixed tensor if its transformation rule is

$$v_{\bar{j}}^{\bar{i}} = \beta_i^{\bar{i}}\beta_{\bar{j}}^{j}v_j^i.$$

4. Other higher rank tensors can be defined in a similar manner.

2.7.3 Metric Tensor[10], g_{ij}

The metric tensor, g_{ij}, is defined through the inner product between the two base vectors as

$$g_{ij} = \mathbf{e}_i \cdot \mathbf{e}_j,$$

where $\mathbf{e}_i$ is the base vector defined as

$$\mathbf{e}_i \equiv \frac{\partial \mathbf{R}}{\partial x^i},$$

and $\mathbf{R}$ is the position vector. It can be shown that the metric tensor, g_{ij}, is a second-rank covariant tensor.

Examples:
1. Cartesian coordinate system (x, y, z):
As $\mathbf{R} = (x, y, z)$ and $\partial \mathbf{R}/\partial x = (1, 0, 0)$, $\partial \mathbf{R}/\partial y = (0, 1, 0)$ and $\partial \mathbf{R}/\partial z = (0, 0, 1)$, it follows that

$$g_{ij} = \begin{pmatrix} 1 & 0 \\ 0 & 1 \end{pmatrix}.$$

2. Polar coordinate system (r, θ):
As $\mathbf{R} = (r\cos\theta, r\sin\theta)$, $\partial \mathbf{R}/\partial r = (\cos\theta, \sin\theta)$ and $\partial \mathbf{R}/\partial \theta = (-r\sin\theta, r\cos\theta)$, it follows that

$$g_{ij} = \begin{pmatrix} 1 & 0 \\ 0 & r^2 \end{pmatrix}.$$

By using the metric tensor, g_{ij}, and its inverse, $g^{ij} (\equiv (g_{ij})^{-1})$, it is possible to convert a tensor format between the contravariant components and covariant components. For instance, a contravariant tensor, v^i, can be converted into a covariant tensor, v_j, by

$$v_j = v^i g_{ij}.$$

---

[10] Although the metric tensor, g_{ij}, plays an important role in differential geometry, the usage of g_{ij} in our application is limited and for the purpose of convenience.

A second-rank covariant tensor, v_{ij}, can be converted to a second-rank contravariant tensor, v^{kl}, through

$$v^{kl} = g^{ki} g^{lj} v_{ij}.$$

Conversion of higher rank tensors can be carried out in a similar manner. It is important to note that the difference between the covariant components and the contravariant components is not essential as they are interchangeable through g_{ij} or g^{ij}.

2.7.4 Covariant Derivatives

Unlike the Cartesian tensors in which the derivative of a tensor automatically yields yet another tensor, the partial derivative of a tensor in the general curvilinear coordinate system is not necessarily a tensor. For example, if v^i is a first-rank contravariant tensor subject to the transformation

$$v^{\bar{i}} = \beta^{\bar{i}}_i v^i,$$

where $\beta^{\bar{i}}_i$ is a function of the position, its partial derivative with respect to $x^{\bar{j}}$ is

$$v^{\bar{i}}_{,\bar{j}} = \beta^{\bar{i}}_{i,\bar{j}} v^i + \beta^{\bar{i}}_i v^i_{,\bar{j}}. \tag{2.62}$$

Because of the presence of the first term on the right-hand side of Equation (2.62), $v^{\bar{i}}_{,\bar{j}}$ is not a tensor. This makes it harder to write field equations in a general curvilinear coordinate system as the derivative of a tensorial quantity is no longer a tensor.

This problem can be circumvented with the modification of the partial differentiation called the *covariant derivative*. In the covariant derivative, the Christoffel symbol,[11] denoted as Γ^i_{jk}, is introduced as a correction factor so that the combination of the partial derivative and the correction term proportional to the given tensor becomes a tensor. In the aforementioned example, the correction term consists of the proportionality factor, Γ^i_{jk}, and the vector v^k as

$$v^i_{,j} + \Gamma^i_{jk} v^k.$$

The components of the Christoffel symbol, Γ^i_{jk}, are expressed as

$$\Gamma^i_{jk} \equiv \frac{1}{2} g^{il} (g_{lj,k} + g_{lk,j} - g_{jk,l}). \tag{2.63}$$

Examples of the Christoffel symbol
1. Polar coordinate system (r, θ).
   In the polar coordinate system, (r, θ), the position vector, **R**, and the base vectors are expressed as

   $$\mathbf{R} = (r \cos \theta, r \sin \theta),$$

   $$\mathbf{e}_1 = \frac{\partial \mathbf{R}}{\partial r} = (\cos \theta, \sin \theta),$$

   $$\mathbf{e}_2 = \frac{\partial \mathbf{R}}{\partial \theta} = (-r \sin \theta, r \cos \theta).$$

---
[11] The Christoffel symbol is not a tensor.

Field Equations

The metric tensor, g_{ij}, and its inverse are expressed as

$$g_{ij} = \begin{pmatrix} 1 & 0 \\ 0 & r^2 \end{pmatrix}, g^{ij} = \begin{pmatrix} 1 & 0 \\ 0 & \frac{1}{r^2} \end{pmatrix}.$$

Using Equation (2.63), the nonzero components of Γ^i_{jk} are

$$\Gamma^1_{22} = -r, \Gamma^2_{21} = \Gamma^2_{12} = 1/r.$$

All other components of Γ^i_{jk} are 0.

2. Spherical coordinate system.

The spherical coordinate system is defined by

$$\mathbf{R} = (r \sin \theta \cos \phi, r \sin \theta \sin \phi, r \cos \theta).$$

The base vectors are expressed as

$$\mathbf{e}_1 = (\cos \phi \sin \theta, \sin \phi \sin \theta, \cos \theta),$$
$$\mathbf{e}_2 = (r \cos \phi \cos \theta, r \cos \theta \sin \phi, -r \sin \theta),$$
$$\mathbf{e}_3 = (-r \sin \phi \sin \theta, r \cos \phi \sin \theta, 0).$$

The metric tensor, g_{ij}, and its inverse, g^{ij}, are expressed as

$$g_{ij} = \begin{pmatrix} 1 & 0 & 0 \\ 0 & r^2 & 0 \\ 0 & 0 & r^2 \sin^2 \theta \end{pmatrix}, g^{ij} = \begin{pmatrix} 1 & 0 & 0 \\ 0 & \frac{1}{r^2} & 0 \\ 0 & 0 & \frac{1}{r^2 \sin^2 \theta} \end{pmatrix}.$$

The nonzero components of the Christoffel symbol, Γ^i_{jk}, are

$$\Gamma^1_{22} = -r, \quad \Gamma^1_{33} = -r \sin^2 \theta,$$
$$\Gamma^2_{12} = \Gamma^2_{21} = \frac{1}{r}, \quad \Gamma^2_{33} = -\cos \theta \sin \theta,$$
$$\Gamma^3_{13} = \Gamma^3_{31} = \frac{1}{r}, \quad \Gamma^3_{23} = \Gamma^3_{32} = \cot \theta.$$

A *Mathematica* code is given as follows:
First, the spherical coordinates are defined.

```
In[15]:= Clear[x]
x = {r, theta, phi};
R = {r Sin[theta] Cos[phi], r Sin[theta] Sin[phi], r Cos[theta]};
```

Next, the base vectors and the metric tensor are computed as

```
In[6]:= e = Table[D[R, x[[i]]], {i, 3}]
        g = Table[e[[i]].e[[j]] // Simplify, {i, 3}, {j, 3}]
Out[6]= {{Cos[phi] Sin[theta], Sin[phi] Sin[theta], Cos[theta]},
         {r Cos[phi] Cos[theta], r Cos[theta] Sin[phi], -r Sin[theta]},
         {-r Sin[phi] Sin[theta], r Cos[phi] Sin[theta], 0}}
Out[7]= {{1, 0, 0}, {0, r², 0}, {0, 0, r² Sin[theta]²}}
```

The inverse of the metric tensor and the Christoffel symbols are computed as

```
In[37]:= ginv = Inverse[g]
         gamma = Table[Sum[1/2 ginv[[i, l]]
           (D[g[[l, j]], x[[k]]] + D[g[[l, k]], x[[j]]] -
            D[g[[j, k]], x[[l]]]), {l, 3}], {i, 3}, {j, 3}, {k, 3}]
```

$$Out[37]= \left\{\{1, 0, 0\}, \left\{0, \frac{1}{r^2}, 0\right\}, \left\{0, 0, \frac{\text{Csc[theta]}^2}{r^2}\right\}\right\}$$

$$Out[38]= \left\{\{\{0, 0, 0\}, \{0, -r, 0\}, \{0, 0, -r \text{ Sin[theta]}^2\}\},\right.$$
$$\left\{\left\{0, \frac{1}{r}, 0\right\}, \left\{\frac{1}{r}, 0, 0\right\}, \{0, 0, -\text{Cos[theta] Sin[theta]}\}\right\},$$
$$\left.\left\{\left\{0, 0, \frac{1}{r}\right\}, \{0, 0, \text{Cot[theta]}\}, \left\{\frac{1}{r}, \text{Cot[theta]}, 0\right\}\right\}\right\}$$

The covariant derivative is denoted by a semicolon (;) instead of the regular comma (,) used for partial derivatives. Correction terms in the covariant derivatives are added for the superscripted indices (contravariant component) and are subtracted for the subscripted indices (covariant components). The following are the formulas for the covariant derivatives for different ranks of tensors.

1. Covariant derivative of a scalar, v:

$$v_{;i} \equiv v_{,i}. \quad \text{(same as partial derivative)}$$

2. Covariant derivative of v^i:

$$v^i_{;j} \equiv v^i_{,j} + \Gamma^i_{jk} v^k.$$

3. Covariant derivative of v_i:

$$v_{i;j} \equiv v_{i,j} - \Gamma^k_{ij} v_k.$$

4. Covariant derivative of v^{ij}:

$$v^{ij}_{;k} \equiv v^{ij}_{,k} + \Gamma^i_{kl} v^{lj} + \Gamma^j_{kl} v^{il}.$$

5. Covariant derivative of v_{ij}:

$$v_{ij;k} \equiv v_{ij,k} - \Gamma^l_{ik} v_{lj} - \Gamma^l_{jk} v_{il}.$$

6. Covariant derivative of v^i_j:

$$v^i_{j;k} \equiv v^i_{j,k} + \Gamma^i_{lk} v^l_j - \Gamma^l_{jk} v^i_l.$$

Field Equations

2.7.4.1 Physical and Tensor Components

Every physical quantity consists of the component and the unit. Since the length of the base vector, ($\mathbf{e}_i$), in a curvilinear coordinate system is not necessarily normalized, the magnitude of the components does not necessarily represent the physical length to be actually measured. Noting that $\sqrt{g_{ii}}$ (i not summed) represents the length of $\mathbf{e}_i$, the base vector, $\mathbf{e}_i$, can be normalized by dividing it with $\sqrt{g_{ii}}$ (i not summed). Thereby, for instance, a vector, $\mathbf{v}$, is expressed as

$$\mathbf{v} = v^1 \mathbf{e}_1 + v^2 \mathbf{e}_2$$
$$= v^1 \sqrt{g_{11}} \left(\frac{\mathbf{e}_1}{\sqrt{g_{11}}} \right) + v^2 \sqrt{g_{22}} \left(\frac{\mathbf{e}_2}{\sqrt{g_{22}}} \right),$$

where $v^1 \sqrt{g_{11}}$ and $v^2 \sqrt{g_{22}}$ do represent the physical length. Quantities such as $v^1 \sqrt{g_{11}}$ and $v^2 \sqrt{g_{22}}$ are called "physical components" while v^1 and v^2 are called "tensor components." Physical components represent the actual physical values (those that can be measured in the lab), but they are not subject to the tensor transformation. Table 2.1 shows the relationship between the physical components and tensor components.

The following steps are necessary to convert tensor equations in the Cartesian coordinate system into a curvilinear coordinate system:

1. Change commas (,) to semicolons (;).
2. Change δ_{ij} to g_{ij}.
3. Adjust dummy indices so that summation occurs between superscripts and subscripts.
4. Use physical components.

2.7.5 Examples

2.7.5.1 Stress Equilibrium Equation

In this example, the stress equilibrium equation expressed in the Cartesian coordinate system as

$$\sigma_{ij,j} = 0, \tag{2.64}$$

Table 2.1 Relation between physical and tensor components.

Tensor components	Physical components
v^1	$v^1 \sqrt{g_{11}}$
v^{11}	$v^{11} \sqrt{g_{11}} \sqrt{g_{11}}$
v^{12}	$v^{12} \sqrt{g_{11}} \sqrt{g_{22}}$
v_1	$v_1 \sqrt{g^{11}}$
v_{11}	$v_{11} \sqrt{g^{11}} \sqrt{g^{11}}$
v_{12}	$v_{12} \sqrt{g^{11}} \sqrt{g^{22}}$
v_1^2	$v_1^2 \sqrt{g^{11}} \sqrt{g_{22}}$
...	...

is shown to be rewritten in the curvilinear coordinate system. Equation (2.64) can be written in the general curvilinear coordinate system as

$$\sigma^j_{i;j} = 0.$$

For the polar coordinate system (r, θ),

$$g_{ij} = \begin{pmatrix} 1 & 0 \\ 0 & r^2 \end{pmatrix}, g^{ij} = \begin{pmatrix} 1 & 0 \\ 0 & \frac{1}{r^2} \end{pmatrix},$$

and

$$\Gamma^1_{22} = -r \quad \Gamma^2_{12} = \Gamma^2_{21} = \frac{1}{r}, \quad \text{all other values of } \Gamma \text{ are zero.}$$

Hence, for $i = 1$,

$$\begin{aligned}
\sigma^j_{1;j} &= \sigma^1_{1;1} + \sigma^2_{1;2} \\
&= (\sigma^1_{1,1} + \Gamma^1_{1i}\sigma^i_1 - \Gamma^i_{11}\sigma^1_i) + (\sigma^2_{1,2} + \Gamma^2_{2i}\sigma^i_1 - \Gamma^i_{12}\sigma^2_i) \\
&= \sigma^1_{1,1} + \sigma^2_{1,2} + \Gamma^2_{21}\sigma^1_1 - \Gamma^2_{12}\sigma^2_2 \\
&= \sigma^1_{1,1} + \sigma^2_{1,2} + \frac{1}{r}\sigma^1_1 - \frac{1}{r}\sigma^2_2.
\end{aligned}$$

The final step is to convert the tensor components to the corresponding physical components using the following table.

Tensor components	Physical components
σ^1_1	$\sigma^1_1 \sqrt{g_{11}} \sqrt{g^{11}} \equiv \tilde{\sigma}_{rr}$
σ^2_1	$\sigma^2_1 \sqrt{g_{22}} \sqrt{g^{11}} \equiv \tilde{\sigma}_{\theta r}$
σ^2_2	$\sigma^2_2 \sqrt{g^{22}} \sqrt{g_{22}} \equiv \tilde{\sigma}_{\theta\theta}$

With the relationship,

$$\sigma^1_1 = \tilde{\sigma}_{rr},$$

$$\sigma^2_1 = \frac{1}{r}\tilde{\sigma}_{\theta r},$$

$$\sigma^2_2 = \tilde{\sigma}_{\theta\theta},$$

the first of the three equilibrium equations can be written as

$$\frac{\partial}{\partial r}\tilde{\sigma}_{rr} + \frac{\partial}{\partial \theta}\left(\frac{1}{r}\tilde{\sigma}_{\theta r}\right) + \frac{1}{r}(\tilde{\sigma}_{rr} - \tilde{\sigma}_{\theta\theta}) = 0.$$

2.7.5.2 Strain–Displacement Relation

The infinitesimal strain–displacement relation in the Cartesian coordinate system,

$$\epsilon_{ij} = \frac{1}{2}(u_{i,j} + u_{j,i}),$$

can be written in the general curvilinear coordinate system as

$$\epsilon_{ij} = \frac{1}{2}(u_{i;j} + u_{j;i}).$$

For example, the (12) component in (r, θ) can be expressed as

$$\epsilon_{12} = \frac{1}{2}(u_{1;2} + u_{2;1})$$
$$= \frac{1}{2}(u_{1,2} - \Gamma^i_{12} u_i + u_{2,1} - \Gamma^i_{21} u_i)$$
$$= \frac{1}{2}\left(u_{1,2} - \frac{1}{r} u_2 + u_{2,1} - \frac{1}{r} u_2\right)$$
$$= \frac{1}{2}(u_{1,2} + u_{2,1}) - \frac{u_2}{r}.$$

By using the following table,

Tensor component	Physical components
ϵ_{12}	$\epsilon_{12}\sqrt{g^{11}}\sqrt{g^{22}} \equiv \tilde{\epsilon}_{r\theta}$
u_1	$u_1\sqrt{g^{11}} \equiv \tilde{u}_r$
u_2	$u_2\sqrt{g^{22}} \equiv \tilde{u}_\theta$

i.e.,

$$\epsilon_{12} = r\tilde{\epsilon}_{r\theta}, \quad u_1 = \tilde{u}_r, \quad u_2 = r\tilde{u}_\theta,$$

the aforementioned strain–displacement equation can be written as

$$r\tilde{\epsilon}_{r\theta} = \frac{1}{2}\left(\frac{\partial \tilde{u}_r}{\partial \theta} + \frac{\partial}{\partial r}(r\tilde{u}_\theta)\right) - r\frac{\tilde{u}_\theta}{r}.$$

2.7.6 Vector Analysis
2.7.6.1 Divergence

The divergence of a vector in the Cartesian coordinate system is expressed as

$$v_{i,i} = 0,$$

which can be rewritten in the general curvilinear coordinate system as

$$v^i_{;i} = 0, \tag{2.65}$$

In the polar coordinate system, (r, θ), Equation (2.65) can be expressed as

$$\begin{aligned} v^i_{;i} &= v^i_{,i} + \Gamma^i_{ij} v^j \\ &= (v^1_{,1} + v^2_{,2}) + \Gamma^1_{1j} v^j + \Gamma^2_{2j} v^j \\ &= (v^1_{,1} + v^2_{,2}) + (\Gamma^1_{11} v^1 + \Gamma^1_{12} v^2) + (\Gamma^2_{21} v^1 + \Gamma^2_{22} v^2) \\ &= v^1_{,1} + v^2_{,2} + \frac{1}{r} v^1. \end{aligned} \qquad (2.66)$$

By using the following table,

Tensor component	Physical components
v^1	$v^1 \sqrt{g_{11}} \equiv V_r$
v^2	$v^2 \sqrt{g_{22}} \equiv V_\theta$

Equation (2.66) can be written as

$$\begin{aligned} \nabla \cdot v &= \frac{\partial}{\partial r} \left(\frac{1}{\sqrt{g_{11}}} V_r \right) + \frac{\partial}{\partial \theta} \left(\frac{1}{\sqrt{g_{22}}} V_\theta \right) \\ &= \frac{\partial V_r}{\partial r} + \frac{\partial}{\partial \theta} \left(\frac{1}{r} V_\theta \right) + \frac{V_r}{r}. \end{aligned}$$

2.7.6.2 Laplacian

The Laplacian of a in the Cartesian coordinate system is expressed as

$$a_{,ii} = 0.$$

It is noted that covariant derivative of a scalar is the same as the partial derivative, i.e.,

$$a_{;i} = a_{,i}.$$

Taking another derivative with respect to x^i on this result and taking summation over i are a violation of the summation convention in tensor analysis. It is necessary to raise the index i by g^{ij} as

$$b^j \equiv a_{,i} g^{ij},$$

and covariant-differentiate it as

$$b^j_{;j} = (a_{,i} g^{ij})_{;j}.$$

Thus, the Laplacian in the general curvilinear coordinate system is expressed as

$$(a_{,i} g^{ij})_{,j} = 0.$$

Field Equations

It is noted that
$$b^j_{;j} = b^j_{,j} + \Gamma^j_{jl}b^l$$
$$= b^1_{,1} + b^2_{,2} + \Gamma^2_{21}b^1.$$

Also, it is noted that
$$b^1 = a_{,i}g^{i1}, = a_{,1}, \quad b^2 = a_{,i}g^{i2} = g^{22}a_{,2} = \frac{1}{r^2}a_{,2}.$$

Therefore, in the polar coordinate system, it follows that
$$(a_{,1})_{,1} + \left(\frac{1}{r^2}a_{,2}\right)_{,2} + \frac{1}{r}a_{,1} = \frac{\partial^2 a}{\partial r^2} + \frac{1}{r}\frac{\partial a}{\partial r} + \frac{1}{r^2}\frac{\partial^2 a}{\partial \theta^2}.$$

2.8 Exercises

1. A state of deformation in 2-D is described as
$$x = a, \quad y = 2\epsilon a + b.$$

   (a) Compute the Green strains.
   (b) By assuming ϵ is infinitesimal, obtain the simplified expression of (a).
   (c) Obtain the deviatoric part of the strains in (b).

2. The stress components are given as $\sigma_{ij} = \begin{pmatrix} 0 & 0 & -1 \\ 0 & 0 & -2 \\ -1 & -2 & 1 \end{pmatrix}$. Compute the magnitude (N) of the normal components (N_i) of the traction force vector (t_i) at $(1, 1, 1)$ on the surface of $2xy - z^2 = 1$.

3. A 2-D displacement field is given as $u_i(x, y) = -z\,\phi(x, y)_{,i}$ where $i = 1, 2$ and $\phi(x, y)$ is a function of x and y.
   (a) Obtain the stress field, σ_{ij}, by assuming small deformation.
   (b) Write down the equilibrium equation without body force (2-D) for σ_{ij} in (a) and derive the equation that $\phi(x, y)$ must satisfy.

4. What does C_{1322} of an elastic modulus tensor signify?

5. If the interface between two materials is along $x^2 + y^2 = 1$, what kind of continuity condition must be held for σ_{ij}?

6. (a) Obtain the convection term in the material derivative of $\phi = v_i v_i$, i.e., $\frac{D\phi}{Dt}$ minus $\frac{\partial \phi}{\partial t}$.
   (b) Express the above result in universal coordinate systems.
   (c) For $\mathbf{v} = (x, -y)$ in the 2-D Cartesian coordinate system, compute the convection term in (a).

7. For an incompressible continuum, show that
$$\frac{D}{Dt}\int\int ABdS = \int\int \left(\frac{DA}{Dt}B + A\frac{DB}{Dt}\right)dS,$$
where A and B are arbitrary quantities.

8. If the body force, b_i, in 2-D is given as

$$b_1 = -\rho g, \quad b_2 = 0, \quad \rho, g: \text{constants}$$

compute

$$\oint t_i d\ell,$$

where t_i is the surface traction force and the integral range is along the unit circle $(x^2 + y^2 = 1)$.

9. The generalized stress–strain and strain–displacement relationships are expressed as

$$\sigma_{ij} = 2\mu\epsilon_{ij} + \lambda g_{ij}\epsilon_k^k, \quad \epsilon_{ij} = \frac{1}{2}(u_{i;j} + u_{j;i}),$$

where g_{ij} is the metric tensor. Express $\tilde{\sigma}_{r\theta}$ in terms of $\tilde{u}_r$ and $\tilde{u}_\theta$ in the polar coordinate system. The quantities, $\tilde{\sigma}_{r\theta}, \tilde{u}_r, \tilde{u}_\theta$, are the physical components of σ_{12}, u_1, u_2, respectively.

10. The bulk modulus, K, is defined as $\sigma_{kk} = 3K\epsilon_{kk}$ in 3-D for isotropic materials.
    (a) Express K in terms of μ and λ.
    (b) By using $\mu = \frac{E}{2(1+\nu)}$ and $\lambda = \frac{\nu E}{(1+\nu)(1-2\nu)}$, show that the Poisson ratio $\nu < 0.5$.

11. The strain energy in 3-D due to the deviatoric stress and strain is defined as $W = \frac{1}{2}\sigma'_{ij}\epsilon'_{ij}$ where $a'_{ij} \equiv a_{ij} - \frac{1}{3}a_{kk}\delta_{ij}$.
    (a) Calculate W for

    $$\epsilon_{ij} = \begin{pmatrix} 0 & \tau & 0 \\ \tau & 0 & 0 \\ 0 & 0 & 0 \end{pmatrix}.$$

    (b) Calculate W for

    $$\epsilon_{ij} = \begin{pmatrix} \epsilon & 0 & 0 \\ 0 & -\epsilon & 0 \\ 0 & 0 & 0 \end{pmatrix},$$

    both assuming isotropy.

12. Consider a 2-D orthotropic body as shown in Figure 2.18 with $C_{1111}, C_{2222}, C_{1122}$, and C_{1212} as the four independent components.

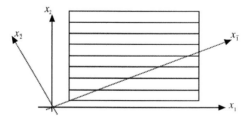

Figure 2.18 Two-dimensional orthotropic body

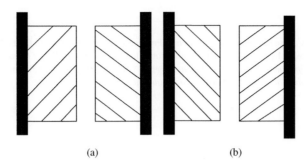

(a) (b)

Figure 2.19 Gate door

(a) Express $C_{ijkl}\epsilon_{ij}\epsilon_{kl}$ where

$$\epsilon_{ij} = \begin{pmatrix} -\epsilon & 0 \\ 0 & \epsilon \end{pmatrix},$$

with the aforementioned components and ϵ.

(b) If the coordinate system is rotated by $\pi/4$ from the original coordinate system, find the new components of $\epsilon_{\bar{i}\bar{j}}$ in the new coordinate system.

(c) Compute

$$C_{\overline{ijkl}}\epsilon_{\bar{i}\bar{j}}\epsilon_{\overline{kl}},$$

in the new coordinate system.

13. When you make a gate door by 2-D orthotropic materials such as wood as shown in Figure 2.19, which option should you choose?

14. For the parabolic cylindrical coordinate system defined as

$$x = \frac{1}{2}(\xi^2 - \eta^2),$$

$$y = \xi\,\eta,$$

$$z = z.$$

(a) Compute the metric tensor, g_{ij}.
(b) Compute the Christoffel symbols, Γ^i_{jk} and show all nonzero components.
(c) Express

$$u_{i,i},$$

in the parabolic cylindrical coordinate system. The final form must be written in terms of physical components.

Note: Refer www.wiley.com/go/nomura0615 for solutions for the Exercise section.

References

Fung YC 1977 *A First Course in Continuum Mechanics*. Prentice-Hall, Inc., Englewood Cliffs, NJ, 351 p.
Gad-el Hak M 1995 Stokes' hypothesis for a newtonin, isotropic fluid. *Journal of Fluids Engineering* **117**(1), 3–5.
Greenberg MD 2001 *Differential Equations & Linear Algebra*, vol. **1**. Prentice Hall, Englewood Cliffs, NJ.
Malvern LE 1969 *Introduction to the Mechanics of a Continuous Medium*. Prentice-Hall.

3

Inclusions in Infinite Media

In this chapter, three analytical methods are presented to derive the elastic fields in an infinitely extended body that contains an inclusion along with the corresponding *Mathematica* code and its implementations.

In Section 3.1, the celebrated work of Eshelby for a single ellipsoidal inclusion problem is introduced. There is a general consensus that the history of micromechanics began in 1957 with the pioneering work of Eshelby 1957 that demonstrated that the elastic field inside an ellipsoidal inclusion embedded in an isotropic and infinite medium subject to uniform stress at infinity is uniform. This paper (Eshelby 1957) is arguably the most cited scientific paper in applied mechanics (Bilby 1990). While Eshelby's work can be applied to many different types of material defects such as cracks and dislocations (Eshelby 1959), the most significant application of Eshelby's work is in composite materials where multiple inclusions are embedded in an infinite matrix (e.g., Mura 1987). Much of the important analytical work on composite materials available today owes Eshelby's inclusion paper. As Eshelby's method for a single inclusion is already well documented in detail in many references (e.g., Markenscoff and Gupta (2006)), this section only overviews the core of Eshelby's method instead of lengthy elucidation of the details and elaborates more on the implementation of Eshelby's method along with programming in *Mathematica*.

In Section 3.2, not directly related to Eshelby's method in Section 3.1, a direct analytical method to obtain the elastic field where a layer of concentric spherical inclusions are embedded in an infinite matrix is introduced. One of the limitations of Eshelby's method is that it is difficult to apply this approach to the elastic field outside an inclusion in which the elastic field fluctuates as a function of the position. Although Eshelby showed a method for the elastic field outside the inclusion, it is necessary to carry out numerical integrations. In addition, Eshelby's approach works for a single inclusion only and cannot be used when concentric (coated, layered) inclusions are present. The elastic field inside an inclusion that is coated by another inclusion is no longer uniform and requires a different approach. Christensen and Lo (1979) derived a solution for the elastic field in an infinite medium that has two concentric inclusions in both 2-D (circular inclusions) and 3-D (spherical inclusions). Although it was a mathematically refined approach, the obtained result was not completely analytical as it was still necessary to solve a lengthy algebraic equation. In Section 3.2, a unified analytical method that can derive the elastic field for any number of concentric inclusions is introduced. This is an exemplary problem of how *Mathematica* can be used to derive a new result, which would otherwise have been impossible to be realized.

In Section 3.3, thermal stress analysis is presented using the analytical approach developed in Section 3.2. When a mismatch of the thermal expansion coefficients and/or the thermal conductivities between an inclusion and a matrix exists, thermal stress is generated in the body even though there is no external stress applied. Although thermal stress problems are reduced to special cases of the general elasticity problem by choosing the body force to represent the thermal effect, understanding the thermal stress in composites is practically important as many of such materials are used in high-temperature environments (Vinson and Sierakowski 2006). Two cases of importance are considered in this section. First, the thermal stress for a medium with a spherical inclusion when a heat source exists inside the inclusion is derived. The temperature field is obtained first by solving the steady-state heat conduction equation, and thereafter, the displacement equilibrium equation is solved based on the temperature distribution. Next, the thermal stress when an inclusion and a matrix are both subject to a heat flux is solved. Both problems are based on the analytical method developed in Section 3.2, and *Mathematica* enables to derive the exact solution symbolically.

In Section 3.4, a different approach from the analytical methods developed in Sections 3.2 and 3.3 is introduced to solve for inclusion problems using the complex Airy stress function (Sokolnikoff and Specht 1956). An old fashioned classical approach is given new blood with the help of *Mathematica*, and it is shown that the concentric inclusion problems similar to those in Section 3.2 can be also solved by using the Airy stress function. Complex-valued analytic functions from which the Airy stress functions are constructed are set up for each phase expanded by the Taylor/Laurent series, and the coefficients of each function are determined to satisfy the continuity conditions of the displacement and traction across the phases as well as the boundary condition. The method demonstrated can be used to derive the elastic stress field of a composite comprised of an inclusion coated by arbitrary number of layers, and thus, it is a novel approach with many potential applications.

In Section 3.5, the effective properties of composites that contain multiple inclusions are discussed. The inclusion problems introduced in the previous sections are for an infinite medium that has a single inclusion. Practical composites used in such fields as aerospace and automotive industries contain multiple inclusions (fibers) in a matrix. Obtaining the effective properties of such composites has a long history dating back to the last century known as the Reuss and Voigt bounds (Reuss 1929). However, extensive research on the effective properties did not exist before Eshelby's work, and many papers on the effective properties of composites are based on Eshelby's result (Nemat-Nasser and Hori 2013). In this section, the self-consistent method and the variational method are presented along with their implementation by *Mathematica*.

3.1 Eshelby's Solution for an Ellipsoidal Inclusion Problem

The method Eshelby employed for the inclusion problem (Eshelby 1957) consists of two parts in its derivation. First, the elastic field in a body that contains eigenstrains (the strain induced by nonstress such as thermal stress and dislocation) is solved. Second, it is shown that the eigenstrain in an elastic body can simulate a second phase (inhomogeneity) that exhibits the same elastic field (equivalent problems). Hence, by solving the elastic field for a body with eigenstrains, the elastic field for inclusion problems can be obtained.

As both steps are based on the Green's function, it is appropriate to begin with the introduction of the concept of the Green's function and show how it is implemented in the analysis of inclusion problems.

It is noted first that many equations for physical and mechanical fields including static elasticity and steady-state heat conduction problems are expressed symbolically as

$$Lu + b = 0, \tag{3.1}$$

where L is a Hermitian (symmetric)[1] differential operator and b represents a generalized source term. Equation (3.1) along with a proper boundary condition constitutes a boundary value problem. For instance, the steady-state heat conduction is expressed as

$$LT + b = 0, \quad LT \equiv (k_{ij}T_{,i})_{,j},$$

where T is the temperature field, k_{ij} is the thermal conductivity, and b is the heat source. For static linear elasticity, Navier's equation for the displacement, u_i, is expressed as

$$(Lu)_i + b_i = 0, \quad (Lu)_i \equiv (C_{ijkl}u_{k,l})_{,j},$$

where C_{ijkl} is the elastic modulus and b_i is the body force. The index notation (also known as Einstein's summation convention) is used throughout; i.e., a repeated index denotes summation.

In general, for the boundary value problem of Equation (3.1), the Green's function, g, is defined and expressed symbolically as (Stakgold and Holst 2011)

$$L^*g + \delta = 0,$$

where L^* is the adjoint operator[2] of L and δ is the Dirac delta function. The Green's function, g, must satisfy the homogeneous boundary condition.

The Green's function to a differential equation is what the inverse matrix is to a matrix in linear algebra (Greenberg 1971). To illustrate this, consider a linear equation,

$$L\mathbf{u} + \mathbf{b} = \mathbf{0},$$

where L is a matrix and $\mathbf{b}$ is a known vector. The solution, $\mathbf{u}$, is

$$\mathbf{u} = -L^{-1}\mathbf{b} = G\mathbf{b}, \quad G \equiv -L^{-1}. \tag{3.2}$$

Similarly, in a system of differential equations, Equation (3.1), with a homogeneous boundary condition, the solution, u, is expressed as the convolution integral between the Green's function, $g(\mathbf{x}, \mathbf{x}')$, and the source term, $b(\mathbf{x})$, as

$$u(\mathbf{x}) = \int_\Omega g(\mathbf{x}, \mathbf{x}')b(\mathbf{x}')d\mathbf{x}'. \tag{3.3}$$

---

[1] If $(Lu, v) = (u, Lv)$, u and v are functions in a function space and $(., .)$ is an inner product, the linear operator, L, is called Hermitian or symmetrical.
[2] If $(Lu, v) = (u, L^*v)$, L^* is called the adjoint operator of L. If $L = L^*$, L is called a self-adjoint (symmetrical) operator.

Equation (3.3) is what is equivalent to Equation (3.2) as the convolution integral in Equation (3.3) can be thought of as a product between two functions, g and b, similar to G and $\mathbf{b}$ in Equation (3.2).

In linear elasticity, the displacement, u_i, must satisfy Navier's equation as

$$(C_{ijkl}u_{k,l})_{,j} + b_i = 0. \tag{3.4}$$

The Green's function for Equation (3.4) is defined as

$$\left((C_{ijkl}(\mathbf{x})g_{km}(\mathbf{x},\mathbf{x}'))_{,l}\right)_{,j} + \delta_{im}\delta(\mathbf{x}-\mathbf{x}') = 0, \tag{3.5}$$

where δ_{im} is the Kronecker delta and $\delta(\mathbf{x}-\mathbf{x}')$ is the Dirac delta function. The Green's function, $g_{km}(\mathbf{x},\mathbf{x}')$, must satisfy the homogeneous boundary condition that corresponds to the imposed boundary condition on u_i.

Not encouragingly, the analytical solutions to Equation (3.5) are available for only a few cases. First, the medium must be infinitely extended (no boundary) under which the Green's function, $g(\mathbf{x},\mathbf{x}')$, is translationally invariant, i.e., $g(\mathbf{x},\mathbf{x}') = g(\mathbf{x}-\mathbf{x}')$. Second, the elastic modulus, C_{ijkl}, must be isotropic or transversely isotropic (Pan and Chou 1979). No analytical solution exists if the aforementioned conditions are not met.

When the material is isotropic, it is possible to obtain the Green's function by employing the Fourier transform. The isotropic elastic modulus, C_{ijkl}, is expressed as

$$C_{ijkl} = \lambda \delta_{ij}\delta_{kl} + \mu(\delta_{ik}\delta_{jl} + \delta_{il}\delta_{jk}),$$

where μ and λ are the Lamé constants. With the isotropic C_{ijkl}, Equation (3.5) is reduced to

$$\mu g_{im,jj} + (\mu+\lambda)g_{mj,ji} + \delta_{im}\delta(\mathbf{x}-\mathbf{x}') = 0. \tag{3.6}$$

As the medium is extended to infinity, the Fourier transform[3] can be used to convert the partial differential equations of (3.6) into a system of algebraic equations. Noting that

$$g_{im}(\mathbf{x}) = \frac{1}{(2\pi)^3}\int\int\int \hat{G}_{im}(\mathbf{k})e^{i\mathbf{k}\cdot\mathbf{x}}d\mathbf{k},$$

$$g_{im,j}(\mathbf{x}) = \frac{1}{(2\pi)^3}\int\int\int ik_j\hat{G}_{im}(\mathbf{k})e^{i\mathbf{k}\cdot\mathbf{x}}d\mathbf{k},$$

where $\hat{G}_{im}$ is the Fourier transform of g_{im} and k_m is the variable in the Fourier transformed domain. The Fourier transform of Equation (3.6) is expressed as

$$-\mu\hat{G}_{im}k^2 - (\mu+\lambda)\hat{G}_{mj}k_jk_i + \delta_{im} = 0, \tag{3.7}$$

---

[3] The 3-D Fourier transform of $g(\mathbf{x})$ is defined as

$$\hat{G}(\mathbf{k}) \equiv \int\int\int g(x)e^{-i\mathbf{k}\cdot\mathbf{x}}d\mathbf{x}.$$

The inverse Fourier transform of $\hat{G}(\mathbf{k})$ is defined as

$$g(\mathbf{x}) = \hat{F}^{-1}(\hat{G}) = \frac{1}{(2\pi)^3}\int\int\int \hat{G}(\mathbf{k})e^{i\mathbf{k}\cdot\mathbf{x}}d\mathbf{k}.$$

where $k^2 = k_\ell k_\ell$. Equation (3.7) can be solved for $\hat{G}_{im}$ by multiplying k_i on both sides as

$$\mu \hat{G}_{im} k_i k^2 + (\mu + \lambda) \hat{G}_{mj} k_j k^2 = k_m,$$

or

$$\hat{G}_{im} k_i k^2 (2\mu + \lambda) = k_m,$$

from which it follows

$$\hat{G}_{im} k_i = \frac{1}{2\mu + \lambda} \frac{k_m}{k^2}. \tag{3.8}$$

Substituting Equation (3.8) into Equation (3.7) yields

$$\mu \hat{G}_{im} k^2 + \frac{\mu + \lambda}{2\mu + \lambda} \frac{k_m k_i}{k^2} = \delta_{im}.$$

Therefore, $\hat{G}_{im}$ can be solved as

$$\hat{G}_{im} = \frac{1}{\mu} \frac{\delta_{im}}{k^2} - \frac{\mu + \lambda}{\mu(2\mu + \lambda)} \frac{k_i k_m}{k^4}. \tag{3.9}$$

Applying the inverse Fourier transform on Equation (3.9) yields

$$g_{im}(\mathbf{x}) = \frac{1}{8\pi\mu(2\mu + \lambda)} \left\{ (3\mu + \lambda) \frac{\delta_{ij}}{r} + \frac{x_i x_j}{r^3} \right\}.$$

Therefore, the solution, $u_i(\mathbf{x})$, in Equation (3.4) in an infinite medium is expressed as

$$u_i(\mathbf{x}) = \int_\Omega g_{ij}(\mathbf{x} - \mathbf{x}') b_j(\mathbf{x}') d\mathbf{x}'. \tag{3.10}$$

3.1.1 Eigenstrain Problem

As an intermediate step for obtaining the elastic field for the inclusion problem, an elastic body that has a region in which an eigenstrain exists is considered. Eigenstrains are defined as the strains that are not related to the applied stresses exemplified by thermal strains and strains due to dislocations and other types of defects.

Consider an elastic body where an eigenstrain (inelastic strain source) denoted by ϵ_{ij}^* exists in the domain Ω as shown in Figure 3.1. The elastic moduli for both $D - \Omega$ and Ω remain unchanged throughout the body denoted as C_{ijkl}. As the elastic part of the strain, which is proportional to the applied stress, is the total strain less the eigenstrain, the stress–strain relationship is expressed as

$$\sigma_{ij} = C_{ijkl}(\epsilon_{kl} - \epsilon_{kl}^*), \tag{3.11}$$

where ϵ_{ij} is the compatible total strain $(= u_{(i,j)})$ induced by ϵ_{ij}^*. The stress self-equilibrium equation and the strain–displacement relation are expressed as

$$\sigma_{ij,j} = 0, \tag{3.12}$$

$$\epsilon_{ij} = u_{(j,i)}. \tag{3.13}$$

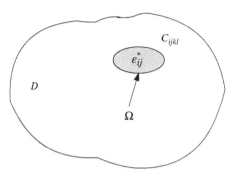

Figure 3.1 Inelastic strain

Combining Equations (3.11)–(3.13) yields

$$C_{ijkl}u_{l,jk} - C_{ijkl}\epsilon^*_{kl,j} = 0. \tag{3.14}$$

The $-C_{ijkl}\epsilon^*_{kl,j}$ term in Equation (3.14) is thought of as an equivalent body force term, b_i, in Equation (3.4). Equation (3.14) can be solved by the elastic Green's function, $g_{ij}(x, \xi)$, which is defined in Equation (3.5). Comparison of Equation (3.14) with Equation (3.5) yields the solution of Equation (3.14) for u_m as

$$u_m(\mathbf{x}) = -\int_\Omega C_{ijkl} g_{km}(\mathbf{x} - \mathbf{x}')\epsilon^*_{ji,l}(\mathbf{x}')d\mathbf{x}'. \tag{3.15}$$

Differentiating Equation (3.15) with respect to x_n, symmetrizing the indices, m and n, and integrating the result by parts yield

$$\epsilon_{mn}(\mathbf{x}) = -\int_\Omega C_{ijkl} g_{km,ln}(\mathbf{x} - \mathbf{x}')\epsilon^*_{ij}(\mathbf{x}')d\mathbf{x}'. \tag{3.16}$$

Equation (3.16) can be rewritten symbolically as

$$\epsilon_{mn} = S_{mnij}\epsilon^*_{ij}, \tag{3.17}$$

where S_{mnij} is an integral operator acting on an arbitrary second-rank tensor function, $f_{ij}(\mathbf{x})$, defined as

$$S_{mnij} f_{ij}(\mathbf{x}) \equiv -\int_\Omega C_{ijkl} g_{km,ln}(\mathbf{x} - \mathbf{x}')f_{ij}(\mathbf{x}')d\mathbf{x}'. \tag{3.18}$$

Eshelby showed that under the conditions that (1) ϵ^*_{ij} is uniform inside Ω, (2) the shape of Ω is ellipsoidal, and (3) C_{ijkl} is isotropic, the strain, ϵ_{mn}, inside Ω is uniform and the operator, S_{mnij}, in Equation (3.17) becomes a constant fourth-rank tensor known as the Eshelby tensor. The proof is found in his paper (Eshelby 1957) and omitted here with a note that the second-order derivative of the Green's function multiplied by C_{ijkl} behaves as a skewed Dirac delta function. The explicit formulas of the Eshelby tensor are shown

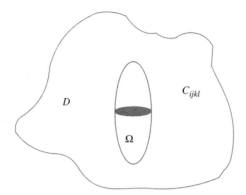

Figure 3.2 Ellipsoidal inclusion

in the next subsection. It is noted from Equation (3.18) that the symmetrical properties of S_{ijkl} are not the same as those of C_{ijkl}, i.e.,

$$S_{ijkl} = S_{jikl} = S_{ijlk}, \text{ but } S_{ijkl} \neq S_{klij}.$$

3.1.2 Eshelby Tensors for an Ellipsoidal Inclusion

When the shape of Ω is ellipsoidal expressed as

$$\left(\frac{x_1}{a_1}\right)^2 + \left(\frac{x_2}{a_2}\right)^2 + \left(\frac{x_3}{a_3}\right)^2 \leq 1,$$

and the eigenstrain, ϵ_{ij}^*, inside Ω is uniform, the Eshelby tensor in Equation (3.18) is known to be constant when the elastic modulus, C_{ijkl}, is isotropic, transversely isotropic, or orthotropic (Eshelby 1957, Lin and Mura 1973) (Figure 3.2).

If the medium is isotropic, the Eshelby tensor is expressed in closed form as shown in the next subsection. However, if the matrix medium is other than isotropic, deriving the Eshelby tensor for an ellipsoidal inclusion involves a considerable amount of algebra, and the previously available work on this subject left out numerical integrations in their final formula. With *Mathematica*, it is possible to express the Eshelby tensor analytically in closed form for a medium other than isotropic medium. The results for transversely isotropic media are shown in the subsequent section.

3.1.2.1 Isotropic Medium

When the elastic modulus, C_{ijkl}, is isotropic, the Eshelby tensor, S_{ijkl}, is expressed in closed form (Figure 3.3). The most common situation is when Ω is either a prolate spheroid ($a_1 =$

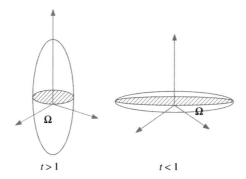

Figure 3.3 Prolate and oblate inclusions

$a_2 \leq a_3$) or an oblate spheroid ($a_1 = a_2 \geq a_3$) where x_3 is the axis of symmetry. The shape of Ω is expressed as

$$(x_1)^2 + (x_1)^2 + \left(\frac{x_3}{t}\right)^2 \leq a_1^2,$$

where $t = a_3/a_1$ is the aspect ratio of the ellipsoid. When $t > 1$, the ellipsoid is prolate, and when $t < 1$, the ellipsoid is oblate. When $t \to \infty$, the ellipsoid becomes cylindrical. Eshelby (1957) showed that S_{ijkl} is expressed as

$$S_{iiii} = \frac{3}{8\pi(1-v)} a_i^2 I_{ii} + \frac{1-2v}{8\pi(1-v)} I_i,$$

$$S_{iijj} = \frac{1}{8\pi(1-v)} a_j^2 I_{ii} - \frac{1-2v}{8\pi(1-v)} I_i,$$

$$S_{ijij} = \frac{3}{16\pi(1-v)} (a_i^2 + a_j^2) I_{ij} + \frac{1-2v}{16\pi(1-v)} (I_i + I_j),$$

($i \neq j$, no summation over i and j)

where v is the Poisson ratio and I_{ij} and I_i are defined as

$$I_{ij} = 2\pi a_1 a_2 a_3 \int_0^\infty \frac{du}{(a_i^2 + u)(a_j^2 + u)\Delta},$$

$$I_i = 2\pi a_1 a_2 a_3 \int_0^\infty \frac{du}{(a_i^2 + u)\Delta},$$

where

$$\Delta \equiv \sqrt{a_1^2 + u}\sqrt{a_2^2 + u}\sqrt{a_3^2 + u}.$$

The evaluation of I_{ij} and I_i for a general ellipsoid ($a_1 \neq a_2 \neq a_3$) requires elliptic integrals. However, if $a_1 = a_2$ (either prolate or oblate) is held, it is possible to express I_{ij} and I_i explicitly as

$$I_1 = I_2 = \frac{2\pi t \left(t - \frac{\cosh^{-1}(t)}{\sqrt{t^2-1}}\right)}{t^2 - 1},$$

$$I_3 = \frac{4\pi t\left(\sqrt{\frac{1}{t^2}-1}-\cos^{-1}(t)\right)}{(1-t^2)^{3/2}},$$

$$I_{11} = I_{22} = -\frac{\pi t\left(-2t^3 - \frac{3\cosh^{-1}(t)}{\sqrt{t^2-1}} + 5t\right)}{2a_1^2(t^2-1)^2},$$

$$I_{33} = -\frac{4\pi t(4t^4 - 5t^2 - 3\sqrt{t^2-1}t^3\cosh^{-1}(t) + 1)}{3a_1^2(t^3-t)^3},$$

$$I_{12} = -\frac{\pi t\left(-2t^3 - \frac{3\cosh^{-1}(t)}{\sqrt{t^2-1}} + 5t\right)}{2a_1^2(t^2-1)^2},$$

$$I_{13} = I_{23} = \frac{2\pi\left(t^2 - \frac{3t\cosh^{-1}(t)}{\sqrt{t^2-1}} + 2\right)}{a_1^2(t^2-1)^2},$$

where t is the aspect ratio ($= a_3/a_1$) of the ellipsoid. It should be noted that the aforementioned expressions are valid for both a prolate ellipsoid ($t > 1$) and an oblate ellipsoid ($t < 1$) even though the term $\sqrt{t^2-1}$ may become complex as one can always take the principal value exemplified by

$$\frac{\cosh^{-1}t}{\sqrt{t^2-1}} = \frac{\tan^{-1}\sqrt{\frac{1-t^2}{t^2}}}{\sqrt{1-t^2}}.$$

The left-hand side can be used for $t > 1$, while the right-hand side can be used for $t < 1$.

Here is *Mathematica* implementation of the Eshelby tensor, S_{ijkl}, when the medium is isotropic. The components of the Eshelby tensor are stored in a nested list, Sijkl[[i,j,k,l]].

```
In[1]:= Δ = Product[ Sqrt[a[i]^2 + u], {i, 1, 3}];
   Q = 3/8/Pi/(1-v); R = (1-2v)/8/Pi/(1-v);
   a[2] = a[1]; a[3] = t a[1];
   tmp1 = Table[2 Pi a[1] a[2] a[3] / (a[i]^2 + u) / Δ, {i, 3}];
   tmp2 =
      Table[2 Pi a[1] a[2] a[3] / (a[i]^2 + u) / (a[j]^2 + u) / Δ, {i, 3}, {j, 3}];
   iI = Assuming[ a[1] > 0 && t > 0, Integrate[tmp1, {u, 0, ∞}]];
   iIij = Assuming[ a[1] > 0 && t > 1, Integrate[tmp2, {u, 0, ∞}]];
   Sijkl = Table[0, {i, 3}, {j, 3}, {k, 3}, {l, 3}];
   Do[Sijkl[[i, i, i, i]] = Q a[i]^2 iIij[[i, i]] + R iI[[i]] // Simplify, {i, 3}]

In[10]:= Do[If[i ≠ j, Sijkl[[i, i, j, j]] = Q/3 a[j]^2 iIij[[i, j]] -
      R iI[[i]]], {i, 3}, {j, 3}]
   Do[If[i ≠ j, Sijkl[[i, j, i, j]] = Q/6 (a[i]^2 + a[j]^2) iIij[[i, j]] +
      R/2 (iI[[i]] + iI[[j]])], {i, 3}, {j, 3}]
   Do[If[i ≠ j, Sijkl[[i, j, j, i]] = Sijkl[[i, j, i, j]]],
      {i, 3}, {j, 3}, {k, 3}, {l, 3}]
```

In[13]:= **Sijkl[[1, 1, 1, 1]]**

Out[13]= $\left\{\left\{\left(t\left(3i\left(t\sqrt{-1+t^2}(-5+2t^2)+3\text{ArcCosh}[t]\right)\right)\Big/(1-t^2)^{5/2}+\dfrac{4(1-2v)\left(t-\dfrac{\text{ArcCosh}[t]}{\sqrt{-1+t^2}}\right)}{-1+t^2}\right)\right\}\Big/(16(1-v))\right\}$

For a spherical inclusion ($t = 1$), it can be shown that

$$S_{1111} = S_{2222} = S_{3333} = \dfrac{7 - 5\nu}{15(1 - \nu)},$$

$$S_{1122} = S_{2211} = S_{1133} = S_{3311} = S_{2233} = S_{3322} = \dfrac{5\nu - 1}{15(1 - \nu)},$$

$$S_{1212} = S_{2323} = S_{1313} = \dfrac{4 - 5\nu}{15(1 - \nu)},$$

All other $S_{ijkl} = 0$.

This can be verified by *Mathematica* as

In[15]:= **Limit[Sijkl[[3, 3, 3, 3]], t → 1]**

Out[15]= $\dfrac{7 - 5\nu}{15 - 15\nu}$

In[16]:= **Limit[Sijkl[[1, 2, 1, 2]], t → 1]**

Out[16]= $\dfrac{4 - 5\nu}{15 - 15\nu}$

For a cylindrical inclusion ($t \to \infty$), it can be shown that

$$S_{1111} = S_{2222} = \dfrac{5 - 4\nu}{8(1 - \nu)}, \quad S_{3333} = 0,$$

$$S_{1122} = S_{2211} = \dfrac{1 - 4\nu}{8(-1 + \nu)},$$

$$S_{2233} = S_{1133} = \dfrac{\nu}{2(1 - \nu)}, \quad S_{3322} = S_{3311} = 0,$$

$$S_{1212} = \dfrac{3 - 4\nu}{8(1 - \nu)}, \quad S_{2323} = S_{1313} = \dfrac{1}{4},$$

All other $S_{ijkl} = 0$.

This can be verified by *Mathematica* as

In[17]:= **Limit[Sijkl[[1, 1, 1, 1]], t → ∞]**

Out[17]= $\dfrac{5 - 4\nu}{8 - 8\nu}$

In[18]:= **Limit[Sijkl[[3, 3, 3, 3]], t → ∞]**

Out[18]= 0

```
In[19]:= Limit[Sijkl[[1, 1, 2, 2]], t -> ∞]
```

$$\text{Out[19]} = \frac{1 - 4\nu}{8(-1 + \nu)}$$

```
In[20]:= Limit[Sijkl[[2, 2, 3, 3]], t -> ∞]
```

$$\text{Out[20]} = \frac{\nu}{2 - 2\nu}$$

```
In[21]:= Limit[Sijkl[[3, 3, 2, 2]], t -> ∞]
```

Out[21]= 0

```
In[22]:= Limit[Sijkl[[1, 2, 1, 2]], t -> ∞]
```

$$\text{Out[22]} = \frac{3 - 4\nu}{8 - 8\nu}$$

For a penny-shaped inclusion ($a_3 \to 0$), it can be shown that

$$S_{3333} = 1, S_{3311} = S_{2211} = \frac{\nu}{1-\nu}, S_{1313} = S_{2323} = \frac{1}{2},$$

All other $S_{ijkl} = 0$.

This can be verified by *Mathematica* as

```
In[30]:= Limit[Sijkl[[3, 3, 3, 3]], t -> 0]
```

Out[30]= 1

```
In[31]:= Limit[Sijkl[[3, 3, 1, 1]], t -> 0]
```

$$\text{Out[31]} = -\frac{\nu}{-1 + \nu}$$

```
In[32]:= Limit[Sijkl[[1, 3, 1, 3]], t -> 0]
```

$$\text{Out[32]} = \frac{1}{2}$$

3.1.2.2 Transversely Isotropic Medium

Transversely isotropic materials are important in engineering as unidirectionally reinforced composites are an example of transversely isotropic materials.

When the elastic modulus, C_{ijkl}, is transversely isotropic, the computation of the Eshelby tensor, S_{ijkl}, is more involved than the isotropic case. Lin and Mura (1973) derived the expression of S_{ijkl} for the transversely isotropic medium analytically except for integrations that had to be carried out numerically. With *Mathematica*, however, it is possible to express S_{ijkl} for the transversely isotropic elastic medium analytically in closed form.

Lin and Mura's (1973) results are cited here as reference. The derivation is elucidated in detail in the original reference and is not duplicated here. Transversely isotropic materials have five independent components. If the x_1–x_2 plane is the transverse plane, the five independent components are

$$C_{1111}, C_{1122}, C_{1313}, C_{3333}, C_{1133}.$$

According to Lin and Mura (1973), the Eshelby tensor for a transversely isotropic medium with a constant ϵ_{ij}^* can be expressed as

$$S_{ijmn} = \frac{1}{8\pi} C_{pqmn}(\bar{G}_{ipjq} + \bar{G}_{jpiq}),$$

where the fourth-rank tensor, $\bar{G}_{ijkl}$, is a function of C_{ijkl} and $\rho \equiv \frac{1}{t}$. The nonzero components of $\bar{G}_{ijkl}$ are given as

$$\bar{G}_{1111} = \bar{G}_{2222} = \frac{\pi}{2}\int_0^1 \Delta(1-x^2)\{[f(1-x^2)+h\rho^2 x^2]$$
$$\times[(3e+d)(1-x^2)+4f\rho^2 x^2] - g^2\rho^2 x^2(1-x^2)\}dx,$$

$$\bar{G}_{3333} = 4\pi\int_0^1 \Delta\rho^2 x^2(d(1-x^2)+f\rho^2 x^2)(e(1-x^2)+f\rho^2 x^2)\,dx,$$

$$\bar{G}_{1122} = \bar{G}_{2211} = \frac{\pi}{2}\int_0^1 \Delta(1-x^2)\{[f(1-x^2)+h\rho^2 x^2]$$
$$\times[(e+3d)(1-x^2)+4f\rho^2 x^2] - 3g^2\rho^2 x^2(1-x^2)\}dx,$$

$$\bar{G}_{1133} = \bar{G}_{2233} = 2\pi\int_0^1 \Delta\rho^2 x^2\{[(d+e)(1-x^2)+2f\rho^2 x^2]$$
$$\times[f(1-x^2)+h\rho^2 x^2] - g^2\rho^2 x^2(1-x^2)\}dx,$$

$$\bar{G}_{3311} = \bar{G}_{3322} = 2\pi\int_0^1 \Delta(1-x^2)[d(1-x^2)+f\rho^2 x^2][e(1-x^2)+f\rho^2 x^2]dx,$$

$$\bar{G}_{1212} = \frac{\pi}{2}\int_0^1 \Delta(1-x^2)^2(g^2\rho^2 x^2 - (d-e)(f(1-x^2)+h\rho^2 x^2))\,dx,$$

$$\bar{G}_{1313} = \bar{G}_{2323} = (-2\pi)\int_0^1 \Delta g\rho^2 x^2(1-x^2)(e(1-x^2)+f\rho^2 x^2)dx,$$

where

$$\Delta^{-1} \equiv [e(1-x^2)+f\rho^2 x^2]\{[d(1-x^2)+f\rho^2 x^2][f(1-x^2)+h\rho^2 x^2] - g^2\rho^2 x^2(1-x^2)\},$$

and

$$d = C_{1111} = C_{2222}, \quad e = (C_{1111}-C_{1122})/2, \quad f = C_{1313} = C_{2323},$$
$$g = C_{1133}+C_{1313}, \quad h = C_{3333}.$$

The aforementioned expressions were the best formulas available before the existence of computer algebra systems, but it is now possible to carry out the integrations explicitly with *Mathematica*.

Inclusions in Infinite Media

Here is a *Mathematica* code to compute S_{ijkl} as a function of C_{1111}, C_{1122}, C_{1313}, C_{3333}, C_{1133}, and ρ. First, the integrands in the integrals are defined as

```
In[35]:= d = c1111;
        e = (c1111 - c1122) / 2; f = c1313;
        g = c1133 + c1313;
        h = c3333;
        Δ = 1 / ((e (1 - x^2) + f ρ^2 x^2)
              ((d (1 - x^2) + f ρ^2 x^2) (f (1 - x^2) + h ρ^2 x^2) -
              g^2 ρ^2 x^2 (1 - x^2)));

        tmp1111 = Pi / 2 Δ (1 - x^2)
              ((f (1 - x^2) + h ρ^2 x^2) ((3 e + d) (1 - x^2) +
              4 f ρ^2 x^2) - g^2 ρ^2 x^2 (1 - x^2));
        tmp3333 = 4 Pi Δ ρ^2 x^2 (d (1 - x^2) + f ρ^2 x^2)
              (e (1 - x^2) + f ρ^2 x^2);
        tmp1122 = Pi / 2 Δ (1 - x^2) ((f (1 - x^2) + h ρ^2 x^2)
              ((e + 3 d) (1 - x^2) + 4 f ρ^2 x^2) -
              3 g^2 ρ^2 x^2 (1 - x^2));
        tmp1133 = 2 Pi Δ ρ^2 x^2 (((d + e) (1 - x^2) + 2 f ρ^2 x^2)
              (f (1 - x^2) + h ρ^2 x^2) - g^2 ρ^2 x^2 (1 - x^2));
        tmp3311 = 2 Pi Δ (1 - x^2) (d (1 - x^2) + f ρ^2 x^2)
              (e (1 - x^2) + f ρ^2 x^2);
        tmp1212 = Pi / 2 Δ (1 - x^2) ^2
              (g^2 ρ^2 x^2 - (d - e) (f (1 - x^2) + h ρ^2 x^2));
        tmp1313 = (-2 Pi) Δ g ρ^2 x^2 (1 - x^2)
              (e (1 - x^2) + f ρ^2 x^2);
        g = {tmp1111, tmp3333, tmp1122, tmp1133,
             tmp3311, tmp1212, tmp1313};
```

Next, the components of S_{ijkl} are computed as

```
tmp1 = Integrate[g, x];
gijkl = (tmp1 /. x → 1) - (tmp1 /. x → 0);
Gijkl = Table[0, {i, 3}, {j, 3}, {k, 3}, {l, 3}];
Gijkl[[1, 1, 1, 1]] = Gijkl[[2, 2, 2, 2]] = gijkl[[1]];
Gijkl[[3, 3, 3, 3]] = gijkl[[2]];
Gijkl[[1, 1, 2, 2]] = Gijkl[[2, 2, 1, 1]] = gijkl[[3]];
Gijkl[[1, 1, 3, 3]] = Gijkl[[2, 2, 3, 3]] = gijkl[[4]];
Gijkl[[3, 3, 1, 1]] = Gijkl[[3, 3, 2, 2]] = gijkl[[5]];
Gijkl[[1, 2, 1, 2]] = Gijkl[[1, 2, 2, 1]] =
     Gijkl[[2, 1, 2, 1]] = Gijkl[[2, 1, 1, 2]] = gijkl[[6]];
```

```
Gijkl[[1, 3, 1, 3]] = Gijkl[[1, 3, 3, 1]] =
    Gijkl[[3, 1, 1, 3]] = Gijkl[[3, 1, 3, 1]] =
        Gijkl[[2, 3, 2, 3]] = Gijkl[[2, 3, 3, 2]] = Gijkl[[3,
            2, 2, 3]] = Gijkl[[3, 2, 3, 2]] = gijkl[[7]];
cijkl = Table[0, {i, 3}, {j, 3}, {k, 3}, {l, 3}];
cijkl[[1, 1, 1, 1]] = cijkl[[2, 2, 2, 2]] = c1111;
cijkl[[3, 3, 3, 3]] = c3333;
cijkl[[1, 1, 2, 2]] = cijkl[[2, 2, 1, 1]] = c1122;
cijkl[[1, 1, 3, 3]] = cijkl[[2, 2, 3, 3]] =
    cijkl[[3, 3, 1, 1]] = cijkl[[3, 3, 2, 2]] = c1133;
cijkl[[1, 2, 1, 2]] = cijkl[[1, 2, 2, 1]] =
    cijkl[[2, 1, 2, 1]] =
    cijkl[[2, 1, 1, 2]] = (c1111 - c1122) / 2;
cijkl[[1, 3, 1, 3]] = cijkl[[1, 3, 3, 1]] =
    cijkl[[3, 1, 1, 3]] = cijkl[[3, 1, 3, 1]] =
        cijkl[[2, 3, 2, 3]] = cijkl[[2, 3, 3, 2]] =
            cijkl[[3, 2, 2, 3]] = cijkl[[3, 2, 3, 2]] = c1313;
Sijkl = Table[Sum[cijkl[[p, q, m, n]]
    (Gijkl[[i, p, j, q]] + Gijkl[[j, p, i, q]]) / (8 Pi),
    {p, 3}, {q, 3}], {i, 3}, {j, 3}, {m, 3}, {n, 3}];
```

The list, `Sijkl[[i,j,k,l]]`, is a nested list and holds all the components of S_{ijkl} with x_3 as the axis of symmetry. As this computation takes some time to execute, the computed result should be saved to a file once the execution is complete so that it is not necessary to repeat the same computation next time the results are needed.

In order to save the computed results to a file, the `SetDirectory` function needs to be used first, which specifies the location of the default directory (folder) where the *Mathematica* output is subsequently saved followed by the `Save` function to specify the file name and the variable(s) to be saved. As can be seen in the following file listing, the size of `eshelby-transverse-isotropic.m` is 256 KB.

In[28]:= **SetDirectory["c:/tmp"]**

Out[28]= c:\tmp

In[29]:= **Save["Eshelby-Transverse-isotropic.m", Sijkl];**

This will set the directory, `c:\tmp`, as the default directory in which files are saved/loaded. Note that the forward slash character (/) must be used as the Windows directory delimiter, instead of the backslash (\) character. The `Save` command saves the contents of the variable `Sijkl` in the directory specified by the `SetDirectory` command under the file name `Eshelby-Transverse-isotropic.m`. The directory command, `dir`, in DOS shows

```
2015-04-23  22:09           <DIR>
2015-04-23  22:09           <DIR>
2015-04-23  22:04           256,405  eshelby-transverse-isotropic.m
            256,405 bytes in 1 file and 2 dirs    258,048 bytes allocated
```

To load the saved result into a new *Mathematica* session, the syntax «filename can be used to load the file filename.

In[1]:= **SetDirectory["c:/tmp"];**
<< eshelby-transverse-isotropic.m ;

It is noted that the components of S_{ijkl} are all in closed form with extremely long expressions. To show the full expressions is neither practical nor of much interest. However, it is possible to see part of the expression using the Short command as

In[63]:= **Short[Sijkl[[1, 2, 1, 2]], 15]**

$$\text{Out[63]//Short=} \frac{1}{8\pi}(c1111 - c1122)$$

$$\left(\frac{1}{2}\pi\left[-\left((2\,(c1133+c1313)^2\,\rho^2 + c1111\,(c1313 - c3333\,\rho^2) + c1122\,(c1313 - c3333\,\rho^2)\right)/\left((c1111 - c1122 - 2\,c1313\,\rho^2)\,(c1111\,(c1313 - c3333\,\rho^2) + \rho^2\,(c1133^2 + 2\,c1133\,c1313 + c1313\,c3333\,\rho^2)\right)\right) - \right.$$

$$\frac{\ll 1\gg}{\sqrt{\ll 1\gg}\ \ll 1\gg} - \frac{\ll 1\gg}{\ll 1\gg} + \left(c1313\,\rho^2\,(\ll 1\gg)\right)$$

$$\text{ArcTan}\left[\frac{\sqrt{2}\ \sqrt{\ll 1\gg} + \ll 1\gg}{\sqrt{\ll 1\gg}}\right]\right/$$

$$\left(\sqrt{2}\ \sqrt{c1133^2 - c1111\ \ll 5\gg}\ \sqrt{\ll 1\gg}\ \sqrt{\ll 1\gg}\right.$$

$$\left.\left.(c1111\,(\ll 1\gg) + \ll 1\gg)^{3/2}\right)\right] + \ll 1\gg\right)$$

The aforementioned results can be shown to be reduced to the Eshelby tensor for isotropic media by setting C_{ijkl} to be isotropic.

3.1.3 Inhomogeneity (Inclusion) Problem

Now that the Eshelby tensor has been obtained when there exists an eigenstrain, ϵ_{ij}^*, inside an ellipsoidal region, it can be shown that this eigenstrain is chosen such that the elastic field induced by an ellipsoidal inhomogeneity can be simulated by choosing an appropriate eigenstrain, ϵ_{ij}^*.

Here, the term "inhomogeneity (inclusion)" is defined as a region, Ω, whose elastic constant, C_{ijkl}^i, is different from the surrounding medium, C_{ijkl}^o. Composite materials and alloys are typical examples of inhomogeneities. Eshelby (1957) showed that an inhomogeneity (inclusion) problem with a distinct elastic constant, C_{ijkl}^i in Ω, from the surrounding medium, C_{ijkl}^o, subject to the far-field constant strain, $<\epsilon_{ij}>$, can be solved by trading this problem with an equivalent eigenstrain problem where an eigenstrain, ϵ_{ij}^*, exists in Ω but with a uniform elastic constant, C_{ijkl}^o throughout the body.

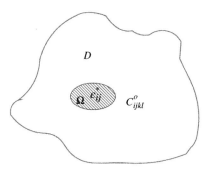

Figure 3.4 Inclusion with ϵ_{ij}^*

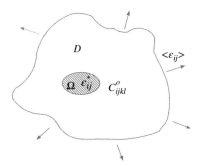

Figure 3.5 Inclusion with ϵ_{ij}^* with $<\epsilon_{ij}>$ at infinity

To elucidate this idea, it is reminded that when an eigenstrain, ϵ_{ij}^*, exists in an ellipsoidal region, Ω, in an infinitely extended medium with an elastic constant, C_{ijkl}^o, throughout, the total strain field, ϵ_{ij}', induced by ϵ_{ij}^* is expressed as (Figure 3.4)

$$\epsilon_{ij}' = S_{ijkl}\epsilon_{kl}^*,$$

where ϵ_{ij}' is the induced strain, which is compatible (the total strain), and S_{ijkl} is the Eshelby tensor.

Now consider the same body as in Figure 3.4 but with an externally applied uniform strain $<\epsilon_{ij}>$ at infinity (Figure 3.5).

Because $<\epsilon_{ij}>$ is uniform and independent of ϵ_{ij}^*, the total strain in Ω, which is compatible,[4] is increased by $<\epsilon_{ij}>$ to be $<\epsilon_{ij}>+\epsilon_{ij}'$. Because ϵ_{ij}^* is the inelastic part out of the total strain, ϵ_{ij}, the elastic part of the total strain, which is proportional to the stress σ_{ij} in Ω, is $<\epsilon_{ij}>+\epsilon_{ij}'-\epsilon_{ij}^*$, and hence, the stress in Ω is expressed as

$$\sigma_{ij} = C_{ijkl}^o(<\epsilon_{kl}>+\epsilon_{kl}'-\epsilon_{kl}^*)$$
$$= C_{ijkl}^o(<\epsilon_{kl}>+S_{ijkl}\epsilon_{kl}^*-\epsilon_{kl}^*). \quad (3.19)$$

---

[4] $\epsilon_{ij} = u_{(i,j)}.$

Inclusions in Infinite Media

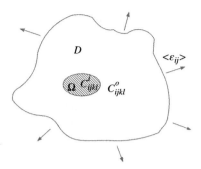

Figure 3.6 Inhomogeneity problem with $<\epsilon_{ij}>$ at infinity

Consider an inhomogeneity (inclusion) problem where the region Ω is occupied by an elastic medium with a different elastic constant, C^i_{ijkl}, from the surrounding medium, C^o_{ijkl}, but without any eigenstrain (Figure 3.6).

If the medium were homogeneous throughout without the inclusion, the stress distribution in the medium would have been uniform and equal to $<\epsilon_{ij}>$. However, because of the presence of the inclusion with C^i_{ijkl}, there must be a disturbance, ϵ'_{ij}, in the strain field. Thus, the total strain, ϵ_{ij}, is expressed as

$$\epsilon_{ij} = <\epsilon_{ij}> + \epsilon'_{ij},$$

and the stress is expressed as

$$\sigma_{ij} = C^i_{ijkl}(<\epsilon_{kl}> + \epsilon'_{kl}). \tag{3.20}$$

The very idea of Eshelby's method is that the eigenstrain, ϵ^*_{ij}, can be chosen such that the stress fields of Equations (3.19) and (3.20) match so that ϵ^*_{ij} can simulate the inhomogeneity (inclusion). This leads to solving the following set of three simultaneous equations:

$$\sigma_{ij} = C^o_{ijkl}(<\epsilon_{kl}> + \epsilon'_{kl} - \epsilon^*_{kl}),$$

$$\sigma_{ij} = C^i_{ijkl}(<\epsilon_{kl}> + \epsilon'_{kl}),$$

$$\epsilon'_{ij} = S_{ijkl}\epsilon^*_{kl}.$$

By solving the aforementioned set of simultaneous equations, the solution for ϵ'_{ij} can be expressed symbolically without using the index notation as

$$\epsilon' = S((C^o - C^i)S - C^o)^{-1}(C^i - C^o)<\epsilon>,$$

where the inverse of a symmetrical, fourth-rank tensor is understood as

$$v_{ijkl}v^{-1}_{klmn} = I_{ijmn}, \quad I_{ijmn} \equiv \frac{1}{2}(\delta_{im}\delta_{jn} + \delta_{in}\delta_{jm}).$$

Therefore, the strain field, ϵ^i, for the inhomogeneity problem with the uniform strain, $<\epsilon>$, at infinity is expressed as

$$\epsilon^i = \epsilon' + <\epsilon>$$

$$= (I + S((C^o - C^i)S - C^o)^{-1}(C^i - C^o)) <\epsilon>,$$
$$= A <\epsilon>, \qquad (3.21)$$

where
$$A \equiv (I + S((C^o - C^i)S - C^o)^{-1}(C^i - C^o)),$$

and is called the strain proportionality factor. The strain, ϵ^i, in Equation (3.21) is uniform when the shape of Ω is ellipsoidal and C^i and C^o are either isotropic or transversely isotropic.

To facilitate the lengthy algebra in computing Equation (3.21) in *Mathematica*, a *Mathematica* package, micromech.m, was developed that contains basic functions to manipulate isotropic and transversely isotropic tensors of second and fourth ranks. This is because carrying out algebra on all the indices of higher rank tensors is redundant and wastes resources. For example, an isotropic, fourth-rank tensor has three independent components (two for elastic moduli tensors) and their product, $A_{ijkl}B_{klmn}$, also has three independent components as

$$A_{ijkl} = a_1 \delta_{ij}\delta_{kl} + a_2 \delta_{ik}\delta_{jl} + a_3 \delta_{il}\delta_{jk},$$
$$B_{ijkl} = b_1 \delta_{ij}\delta_{kl} + b_2 \delta_{ik}\delta_{jl} + b_3 \delta_{il}\delta_{jk}.$$
$$C_{ijmn} = A_{ijkl}B_{klmn}$$
$$= ((a_2 + a_3)b_1 + a_1(3b_1 + b_2 + b_3))\delta_{ij}\delta_{mn} + (a_2 b_2 + a_3 b_3)\delta_{im}\delta_{jn}$$
$$+ (a_3 b_2 + a_2 b_3)\delta_{in}\delta_{jm}.$$

Therefore, if A_{ijkl} and B_{ijkl} are to be represented by (a_1, a_2, a_3) and (b_1, b_2, b_3), respectively, the product, $A_{ijkl}B_{klmn}$, can also be represented as

$$(((a_2 + a_3)b_1 + a_1(3b_1 + b_2 + b_3)), (a_2 b_2 + a_3 b_3), (a_3 b_2 + a_2 b_3)).$$

The source code of micromech.m is listed at the end of this chapter except for the Eshelby tensor for transversely isotropic media as it is too lengthy. The source file is available from the companion web site for downloading.

To utilize the package, micromech.m, save the file, micromech.m, to a working directory and load the package using the << command.

In[1]:= **SetDirectory["c:/tmp"]**

Out[1]= c:\tmp

In[2]:= **<< micromech.m**

The following functions are available in this package:

- TransverseInverse[a] returns the inverse of a transversely isotropic tensor, a_{ijkl}, of the fourth rank. The transverse plane is the x–y plane, and the z axis is the axial direction. The argument is a list of six components that represent independent components of transversely isotropic tensors of the fourth rank, $\{c_{3333}, c_{1111}, c_{3311}, c_{1133}, c_{1122}, c_{1313}\}$ or $\{c_{33}, c_{11}, c_{31}, c_{13}, c_{12}, c_{44}\}$. The return values are also stored in the same format and order.
- TransverseProduct[a, b] returns $a_{ijkl}b_{klmn}$, where both a_{ijkl} and b_{klmn} are transversely isotropic tensors of the fourth rank and the result is also a fourth-rank tensor. The transverse plane is the x–y plane, and the z axis is the axial direction. The argument is a list of six

components that represent independent components of transversely isotropic tensors, $\{c_{3333}, c_{1111}, c_{3311}, c_{1133}, c_{1122}, c_{1313}\}$ or $\{c_{33}, c_{11}, c_{31}, c_{13}, c_{12}, c_{44}\}$. The return values are also stored in the same format and order.

- Transverse24[a, b] returns $a_{ij}b_{ijkl}$, where a_{ij} is a transversely isotropic tensor of the second rank and b_{ijkl} is a transversely isotropic tensor of the fourth rank. The result is a second-rank tensor and is represented by $\{b_{11}, b_{22}, b_{33}, b_{23}, b_{13}, b_{12}\}$.
- Transverse42[a, b] returns $a_{ijkl}b_{kl}$, where a_{ijkl} is a transversely isotropic tensor of the fourth rank and b_{kl} is a second-rank tensor. The result is a second-rank tensor and is represented by $\{b_{11}, b_{22}, b_{33}, b_{23}, b_{13}, b_{12}\}$.
- EngToModulus[E, v] returns c_{ijkl} from E and v where E and v are the Young modulus and Poisson ratio, respectively. The format of c_{ijkl} is $\{c_{3333}, c_{1111}, c_{3311}, c_{1133}, c_{1122}, c_{1313}\}$ or $\{c_{33}, c_{11}, c_{31}, c_{13}, c_{12}, c_{44}\}$.
- IsotropicProduct[c1, c2] computes the product of two isotropic tensors of the fourth rank, c1 and c2. Both c1 and c2 are represented in the format of $\{\lambda, \mu\}$ or $\{c_{1122}, c_{1212}\}$.
- IsotropicInverse[c] computes the inverse of a fourth-rank tensor, c. The tensor c is represented in the format of $\{c_{1122}, c_{1212}\}$.
- Lame[e, nu] converts the Young modulus and Poisson ratio for isotropic materials to λ (c_{1122}) and μ (c_{1212}), respectively.
- IdentityTensor is the identity tensor of the fourth rank.
- EshelbyIsotropic[t, nu] returns the components of the Eshelby tensor for an isotropic medium with the aspect ratio t and the Poisson ratio, v. It returns the result in transversely isotropic format as $\{s_{3333}, s_{1111}, s_{3311}, s_{1133}, s_{1122}, s_{1313}\}$ or $\{s_{33}, s_{11}, s_{31}, s_{13}, s_{12}, s_{44}\}$.
- SphereStrainFactor[$\{\lambda_f, \mu_f\}, \{\lambda_m, \mu_m\}$] returns the proportionality factor of the strain field inside a spherical inclusion ($t = 1$) in the format of $\{A_{1122}, A_{1212}\}$ subject to X_{ij} at infinity.
- SphereStressFactor[$\{\lambda_f, \mu_f\}, \{\lambda_m, \mu_m\}$] returns the proportionality factor of the stress field inside a spherical inclusion ($t = 1$) in the format of $\{B_{1122}, B_{1212}\}$ subject to X_{ij} at infinity.
- CylinderStrainFactor[cf, cm] returns the proportionality factor of strain field inside a cylindrical inclusion with the elastic modulus, cf, embedded in the matrix with the elastic modulus, cm, subject to X_{ij} at infinity. Transversely isotropic tensors can be input as $\{c_{3333}, c_{1111}, c_{3311}, c_{1133}, c_{1122}, c_{1313}\}$ or $\{c_{33}, c_{11}, c_{31}, c_{13}, c_{12}, c_{44}\}$. CylinderStrainFactor × far strain is the strain field inside the inclusion.
- CylinderStressFactor[cf, cm] returns the proportionality factor of the stress field inside a cylindrical inclusion with the elastic modulus, cf, embedded in the matrix with the elastic modulus, cm, subject to X_{ij} at infinity. Transversely isotropic tensors can be input as $\{c_{3333}, c_{1111}, c_{3311}, c_{1133}, c_{1122}, c_{1313}\}$ or $\{c_{33}, c_{11}, c_{31}, c_{13}, c_{12}, c_{44}\}$. CylinderStressFactor × far stress is the stress field inside the inclusion.

The function, SphereStrainFactor, computes A_{ijkl} in Equation (3.21) when Ω is of spherical shape and both the inclusion and the matrix are isotropic. The stress field inside the inclusion with the isotropic elastic modulus, C^i_{ijkl}, embedded in a matrix phase with the isotropic elastic modulus, C^m_{ijkl}, subject to a far-field uniform stress, $<\sigma_{ij}>$, can be obtained by

$$\sigma_{ij} = C^i_{ijkl} A_{klmn} (C^m)^{-1}_{mnpq} <\sigma_{pq}>$$
$$= B_{ijkl} <\sigma_{kl}> . \qquad (3.22)$$

The function SphereStressFactor computes B_{ijkl} in Equation (3.22) when Ω is of spherical shape and both the inclusion and the matrix are isotropic.

In[86]:= `SphereStressFactor[{λf, μf}, {λm, μm}]`

Out[86]= $\left\{ (3(\lambda m + 2\mu m)(\mu f^2(-6\lambda m + 4\mu m) + \lambda f \mu m(9\lambda m + 14\mu m) - \right.$
$\mu f(9\lambda f \lambda m - 6\lambda f \mu m + 14\lambda m \mu m + 4\mu m^2)))/$
$((3\lambda m + 2\mu m)(3\lambda f + 2\mu f + 4\mu m)$
$(6\lambda m \mu f + 9\lambda m \mu m + 16\mu f \mu m + 14\mu m^2)),$
$\left. \frac{15\mu f(\lambda m + 2\mu m)}{2(6\lambda m \mu f + 9\lambda m \mu m + 16\mu f \mu m + 14\mu m^2)} \right\}$

From the aforementioned result, the stress inside the spherical inhomogeneity (Figure 3.7) can be expressed as

$$\sigma_{ij} = 2\beta <\sigma_{ij}> + \alpha \delta_{ij} <\sigma_{kk}>,$$

where

$$\alpha = \frac{15\mu_f(\lambda_m + 2\mu_m)}{2(\lambda_m(6\mu_f + 9\mu_m) + 2\mu_m(8\mu_f + 7\mu_m))},$$

and

$$\beta = \frac{3(\lambda_m + 2\mu_m)}{(3\lambda_m + 2\mu_m)(3\lambda f + 2\mu_f + 4\mu_m)(\lambda_m(6\mu_f + 9\mu_m) + 2\mu_m(8\mu_f + 7\mu_m))}$$
$$\times \frac{\lambda_f(-9\lambda_m \mu_f + 9\lambda_m \mu_m + 6\mu_f \mu_m + 14\mu_m^2) - 2\mu_f(\lambda_m(3\mu_f + 7\mu_m) + 2\mu_m(\mu_m - \mu_f))}{(3\lambda_m + 2\mu_m)(3\lambda f + 2\mu_f + 4\mu_m)(\lambda_m(6\mu_f + 9\mu_m) + 2\mu_m(8\mu_f + 7\mu_m))}.$$

For example, when the matrix is made of epoxy (the Young modulus = 4.3 GPa, the Poisson ratio = 0.35) and the inhomogeneity is E-glass (the Young modulus = 73 GPa, the Poisson ratio = 0.3), the stress proportionality factor can be computed as

In[8]:= `cm = Lame[4.3, 0.35]; cf = Lame[73, 0.3];`

In[9]:= `SphereStressFactor [cf, cm]`

Out[9]= $\{-0.212153, 1.01609\}$

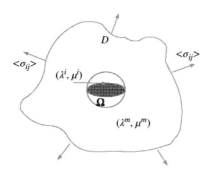

Figure 3.7 Stress inside a spherical inclusion

Inclusions in Infinite Media

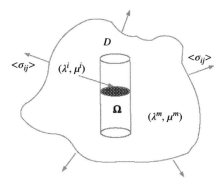

Figure 3.8 Stress inside a cylindrical inclusion ($t \to \inf$)

so that the stress inside the sphere can be expressed as

$$\sigma_{ij} = 2 \times 1.01609 <\sigma_{ij}> -0.212153\, \delta_{ij} <\sigma_{kk}>.$$

Similarly, the stress field in a cylindrical inclusion ($t \to \infty$) can be obtained by using the `CylinderStressFactor` function (Figure 3.8).

By assuming that both the matrix and inclusion are transversely isotropic and their Lamé constants are (λ_m, μ_m) and (λ_i, μ_i), respectively, the stress field inside the cylindrical inclusion is expressed as

$$\sigma_{ij} = B_{ijkl} <\sigma_{kl}>.$$

The components of B_{ijkl} can be obtained by the following code:

```
In[87]:= CylinderStressFactor[{cf33, cf11, cf31, cf31, cf12, cf44},
        {cm33, cm11, cm31, cm31, cm12, cm44}]

Out[87]= {((-2 cf31^2 + cf33 (cf11 + cf12 + cm11 - cm12)) (cm11 + cm12) +
        2 cf31 (-cm11 + cm12) cm31) /
        ((cf11 + cf12 + cm11 - cm12) (-2 cm31^2 + (cm11 + cm12) cm33)),
        ((cm11 (3 cf11 (cm11 - cm12) (cm11 + cm12) +
        cf11^2 (5 cm11 + cm12) -
        cf12 (5 cf12 cm11 + cm11^2 + cf12 cm12 - cm12^2))) /
        (3 cf11 cm11 - 3 cf12 cm11 + cm11^2 - cf11 cm12 +
        cf12 cm12 - cm12^2) - ((cm11 - cm12) cm31
        (cf31 (cm11 + cm12) - (cf11 + cf12) cm31)) /
        (-2 cm31^2 + (cm11 + cm12) cm33)) /
        ((cf11 + cf12 + cm11 - cm12) (cm11 + cm12)),
        (-cm31 (cf33 (cf11 + cf12 + cm11 - cm12) +
        2 cf31 (-cf31 + cm31)) + 2 cf31 cm11 cm33) /
        ((cf11 + cf12 + cm11 - cm12) (-2 cm31^2 + (cm11 + cm12) cm33)),
        ((cm11 - cm12) (cf31 (cm11 + cm12) - (cf11 + cf12) cm31)) /
```

$$\Big(\big((cf11 + cf12 + cm11 - cm12\big)\big(-2\,cm31^2 + (cm11 + cm12)\,cm33\big)\Big),$$
$$\Big(\big(cm11\,\big(cm11\,(cf11^2 - cf12\,(cf12 - 3\,cm11) - cf11\,cm11\big) + $$
$$3\,\big(-cf11^2 + cf12^2\big)\,cm12 + (cf11 - 3\,cf12)\,cm12^2\big)\Big) / $$
$$\big(3\,cf11\,cm11 - 3\,cf12\,cm11 + cm11^2 - cf11\,cm12 + $$
$$cf12\,cm12 - cm12^2\big) - ((cm11 - cm12)\,cm31$$
$$(cf31\,(cm11 + cm12) - (cf11 + cf12)\,cm31)) / $$
$$\big(-2\,cm31^2 + (cm11 + cm12)\,cm33\big)\Big) / $$
$$\big((cf11 + cf12 + cm11 - cm12)\,(cm11 + cm12)\big),$$
$$\frac{cf44}{cf44 + cm44}\Big\}$$

The function CylinderStressFactor takes the material properties of both the cylindrical inclusion and the matrix as input. The material properties are entered in the order of $\{c_{3333}, c_{1111}, c_{3311}, c_{1133}, c_{1122}, c_{1313}\}$ in index notation or $\{c_{33}, c_{11}, c_{31}, c_{13}, c_{12}, c_{44}\}$ in Voigt notation. The aforementioned output has six components that correspond to $B_{3333}, B_{1111}, B_{3311}, B_{1133}, B_{1122}, B_{1313}$.

If both the cylindrical inclusion and the matrix are isotropic, the stress proportionality factor can be computed as

In[89]:= **CylinderStressFactor[{2 μf + λf, 2 μf + λf, λf, λf, λf, μf},**
{2 μm + λm, 2 μm + λm, λm, λm, λm, μm}] // Simplify

Out[89]= $\Big\{\dfrac{2\,\mu f\,(\lambda m + \mu m)\,(\mu f + \mu m) + \lambda f\,(3\,\lambda m\,\mu f + \mu m\,(3\,\mu f + \mu m))}{\mu m\,(\lambda f + \mu f + \mu m)\,(3\,\lambda m + 2\,\mu m)},$

$\Big(\lambda f\,(\lambda m + \mu m)\,(3\,\lambda m\,(3\,\mu f + \mu m) + 4\,\mu m\,(5\,\mu f + \mu m)) + 2\,\mu f\,\big(5\,\lambda m^2$
$(\mu f + \mu m) + 4\,\lambda m\,\mu m\,(4\,\mu f + 3\,\mu m) + 2\,\mu m^2\,(5\,\mu f + 3\,\mu m)\big)\Big) /$
$(2\,(\lambda f + \mu f + \mu m)\,(3\,\lambda m + 2\,\mu m)\,(\lambda m\,(\mu f + \mu m) + \mu m\,(3\,\mu f + \mu m))),$

$\dfrac{-2\,\lambda m\,\mu f\,(\mu f + \mu m) + \lambda f\,\big(-3\,\lambda m\,\mu f + 3\,\lambda m\,\mu m + 4\,\mu m^2\big)}{2\,\mu m\,(\lambda f + \mu f + \mu m)\,(3\,\lambda m + 2\,\mu m)},$

$\dfrac{-\lambda m\,\mu f + \lambda f\,\mu m}{(\lambda f + \mu f + \mu m)\,(3\,\lambda m + 2\,\mu m)},$

$\Big(-2\,\mu f\,\big(4\,\lambda m\,\mu m^2 + 2\,\mu m^2\,(-\mu f + \mu m) + \lambda m^2\,(\mu f + \mu m)\big) +$
$\lambda f\,\big(-3\,\lambda m^2\,(\mu f - \mu m) + 4\,\mu m^2\,(\mu f + \mu m) + \lambda m\,\mu m\,(-3\,\mu f + 7\,\mu m)\big)\Big) /$
$(2\,(\lambda f + \mu f + \mu m)\,(3\,\lambda m + 2\,\mu m)\,(\lambda m\,(\mu f + \mu m) + \mu m\,(3\,\mu f + \mu m))),$

$\dfrac{\mu f}{\mu f + \mu m}\Big\}$

Note that $C_{1111} = 2\mu + \lambda$, $C_{1122} = \lambda$, and $C_{1212} = \mu$.

For aspect ratios other than $t = 1$ and $t \to \infty$, the Eshelby tensor for isotropic media can be evaluated by the EshelbyIsotropic[t, nu] function where t is the aspect ratio and v is the Poisson ratio. The return values of EshelbyIsotropic[t, nu] are $\{s_{3333}, s_{1111}, S_{3311}, S_{1133}, S_{1122}, S_{1313}\}$. As can be seen, the output is rather long, and it is not possible to show all the output in one page. The function Short can display part of the lengthy formula as

In[62]:= `Short[EshelbyIsotropic[t, nu], 7]`

Out[62]//Short=
$$\left\{-\frac{1}{2(1-nu)(-1+t^2)^3} + \ll 21 \gg + \right.$$

$$i\left(\frac{3t^3\left((-1+t^2)^2\right)^{1/4}\text{Arg}[t+\sqrt{-1+t}\sqrt{1+t}]\text{Cos}[\frac{1}{2}\text{Arg}[-1+t^2]]}{2(1-nu)(-1+t^2)^3} - \right.$$

$$\left(t\text{Cos}[\frac{3}{2}\text{Arg}[1-t^2]]\right.$$

$$\text{Log}\left[\sqrt{\sqrt{(1-t^2)^2}\text{Cos}[\frac{1}{2}\text{Arg}[1-t^2]]^2 + \left(t+(\ll 1\gg^2)^{1/4}\text{Sin}[\frac{1}{2}\text{Arg}[1-\ll 1\gg]\right)^2}\right]\right) / \left(2(1-nu)\left((1-t^2)^2\right)^{3/4}\right) +$$

$$\frac{nu\, t \ll 1\gg \text{Log}\left[\sqrt{\sqrt{(\ll 1\gg \ll 1\gg^2)}\ll 1\gg^2+\ll 1\gg}\right]}{(1-nu)(\ll 1\gg^2)^{3/4}} + \ll 8\gg + \frac{\ll 1\gg}{\ll 1\gg} -$$

$$\frac{\ll 1\gg}{\ll 1\gg} + \frac{nu\,((\ll 1\gg)^2)^{1/4}\,t\,\text{Cos}[\frac{1}{2}\text{Arg}[-1+\frac{1}{t^2}]]\text{Sin}[\frac{3}{2}\text{Arg}[1-t^2]]}{(1-nu)\,((1-\ll 1\gg)^2)^{3/4}} +$$

$$\left(3t^3\left((-1+t^2)^2\right)^{1/4}\text{Log}\left[\sqrt{\left[((-1+t)^2)^{1/4}((\ll 1\gg)^2)^{1/4}\text{Cos}[\frac{1}{2}\text{Arg}[1+t]]\right.}\right.\right.$$

$$\left.\left.\text{Sin}[\frac{1}{2}\text{Arg}[-1+t]] + \ll 1\gg^{\ll 1\gg}\ll 3\gg\right)^2 + (\ll 1\gg)^2\right]$$

$$\left.\text{Sin}[\frac{1}{2}\text{Arg}[-1+t^2]]\right) / \left(2(1-nu)(-1+t^2)^3\right), \ll 4\gg, \ll 1\gg\right\}$$

The strain proportionality factor, A_{ijkl}, in Equation (3.21) is even longer and is not practical to be included in the package.

The following code implements Equation (3.21) using the functions available in the package:

```
In[82]:= ei = EshelbyIsotropic[t, nu] /. {nu → λm / (λm + μm) / 2};
        cm = {2 μm + λm, 2 μm + λm, λm, λm, λm, μm};
        ci = {2 μi + λi, 2 μi + λi, λi, λi, λi, μi};
        afactor = IdentityTensor + TransverseProduct[
            TransverseProduct[ei, TransverseInverse[
                TransverseProduct[cm - ci, ei] - cm]], ci - cm];
```

For example, the A_{3333} component for $t = 2$ is computed as

```
In[86]:= afactor[[1]] /. t -> 2 // Simplify
```

$$\text{Out[86]} = \left(3\,(\lambda m + 2\,\mu m)\,\left(\lambda m\,\mu m\,\left(18\,\mu i + 6\,\mu m + 7\,\sqrt{3}\,\mu i\,\text{Log}[2+\sqrt{3}\,] + \sqrt{3}\,\mu m\,\text{Log}[2+\sqrt{3}\,] - 6\,\lambda i\,(-6+\sqrt{3}\,\text{Log}[2+\sqrt{3}\,])\right)\right.\right.$$
$$+ 2\,\mu m^2\,\left(18\,\mu i - \sqrt{3}\,\mu i\,\text{Log}[2+\sqrt{3}\,] + \sqrt{3}\,\mu m\,\text{Log}[2+\sqrt{3}\,] - 4\,\lambda i\,(-6+\sqrt{3}\,\text{Log}[2+\sqrt{3}\,])\right)$$
$$+ 3\,\lambda m^2\,\left(\mu m\,(3 - \sqrt{3}\,\text{Log}[2+\sqrt{3}\,]) + 3\,\mu i\,(-2 + \sqrt{3}\,\text{Log}[2+\sqrt{3}\,])\right)\Big)\Big) \Big/$$
$$\left(3\,\lambda i\,\left(27\,\lambda m^2\,(\mu m\,(3 - \sqrt{3}\,\text{Log}[2+\sqrt{3}\,]) + \mu i\,(-2 + \sqrt{3}\,\text{Log}[2+\sqrt{3}\,]))\right.\right.$$
$$+ 2\,\mu m^2\,\left(\mu i\,(-66 + 45\,\sqrt{3}\,\text{Log}[2+\sqrt{3}\,] - 14\,\text{Log}[2+\sqrt{3}\,]^2) + \mu m\,(108 - 49\,\sqrt{3}\,\text{Log}[2+\sqrt{3}\,] + 14\,\text{Log}[2+\sqrt{3}\,]^2)\right) + \lambda m$$
$$\mu m\,\left(\mu i\,(-186 + 117\,\sqrt{3}\,\text{Log}[2+\sqrt{3}\,] - 28\,\text{Log}[2+\sqrt{3}\,]^2) + \mu m\,(282 - 121\,\sqrt{3}\,\text{Log}[2+\sqrt{3}\,] + 28\,\text{Log}[2+\sqrt{3}\,]^2)\right)\Big)$$
$$+ 2\,\Big(3\,\lambda m^2\,\left(9\,\mu i^2\,(-2 + \sqrt{3}\,\text{Log}[2+\sqrt{3}\,])\right.$$
$$+ 2\,\mu m^2\,(-6 + 9\,\sqrt{3}\,\text{Log}[2+\sqrt{3}\,] - 7\,\text{Log}[2+\sqrt{3}\,]^2)$$
$$+ \mu i\,\mu m\,(45 - 25\,\sqrt{3}\,\text{Log}[2+\sqrt{3}\,] + 14\,\text{Log}[2+\sqrt{3}\,]^2)\Big)$$
$$+ 2\,\mu m^2\,\left(\mu i^2\,(-66 + 45\,\sqrt{3}\,\text{Log}[2+\sqrt{3}\,] - 14\,\text{Log}[2+\sqrt{3}\,]^2)\right.$$
$$+ 2\,\mu m^2\,(-6 + 9\,\sqrt{3}\,\text{Log}[2+\sqrt{3}\,] - 7\,\text{Log}[2+\sqrt{3}\,]^2)$$
$$+ \mu i\,\mu m\,(132 - 63\,\sqrt{3}\,\text{Log}[2+\sqrt{3}\,] + 28\,\text{Log}[2+\sqrt{3}\,]^2)\Big)$$
$$+ \lambda m\,\mu m\,\left(\mu i^2\,(-186 + 117\,\sqrt{3}\,\text{Log}[2+\sqrt{3}\,] - 28\,\text{Log}[2+\sqrt{3}\,]^2)\right.$$
$$+ 10\,\mu m^2\,(-6 + 9\,\sqrt{3}\,\text{Log}[2+\sqrt{3}\,] - 7\,\text{Log}[2+\sqrt{3}\,]^2) + \mu i$$
$$\mu m\,(390 - 195\,\sqrt{3}\,\text{Log}[2+\sqrt{3}\,] + 98\,\text{Log}[2+\sqrt{3}\,]^2)\Big)\Big)\Big)$$

Eshelby also showed how to obtain the elastic field outside the inclusion (Eshelby 1959). However, it is not easy to implement the procedure and numerical integration is inevitable.

3.2 Multilayered Inclusions

3.2.1 Background

Eshelby demonstrated a sophisticated approach to obtain the elastic field inside an ellipsoidal inclusion in an infinite medium. The most significant contribution of Eshelby is that the stress and strain fields inside the inclusion were shown to be uniform when the material properties are either isotropic or transversely isotropic. Some of the limitations of the Eshelby approach include difficulty in obtaining the stress and strain field outside the inclusion although Eshelby

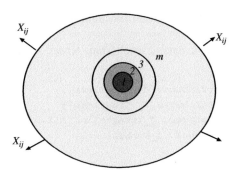

Figure 3.9 Multilayered material

(1959) described a method for such a problem that was not in closed form inevitably involving numerical integrations.

The Eshelby method cannot be used when an inclusion is coated by another layer or layers as shown in Figure 3.9, and different approaches must be developed for this type of problems. Christensen and Lo (1979) showed a procedure to obtain the stress fields for three-phase composites where a spherical inclusion is surrounded by another layer embedded in an infinite medium. They derived a set of nonlinear equations that need to be solved numerically. It is in principle possible to extend their approach to a problem of an inclusion surrounded by more than one layer, but due to the huge amount of algebra, no work has been reported. Herve and Zaoui (1993) extended the approach of Christensen and Lo (1979) for n-layered materials.

Oshima and Nomura (1985) derived the elastic field when a spherical inclusion is surrounded by an arbitrary number of layers in an infinite matrix subject to a constant strain at the far field. This section is based on their methodology where Navier's equation for the displacement is solved exactly, which requires a considerable amount of algebra. This problem is one of the prime examples of how *Mathematica* can be used to carry out time-consuming and error-prone tensor algebra on computer. As such, derivation of the formulas along with its implementation in *Mathematica* is elaborated in detail. The results for three-phase materials are in closed form, which have never been reported earlier, and the results for four-phase materials are rather intriguing because of the length of the closed-form solution.

3.2.2 *Implementation of Index Manipulation in* Mathematica

Although *Mathematica* does not come with rules of performing algebra for index manipulation, it is straightforward to implement them so that time-consuming index manipulation can be facilitated automatically. This subsection introduces some of the useful functions and rules that can be used throughout the analysis for layered inclusion problems including the adaption of summation convention and differentiation rules with the coordinate systems.

3.2.2.1 Symmetrical Tensors

Many tensorial quantities in mechanics are second-rank tensors that are symmetrical, including the stress and strain tensors. The symmetrical property of tensors can be implemented in *Mathematica* using the `SetAttributes` function, which automatically sorts the indices in order.

The Kronecker delta, δ_{ij}, and a general second-rank tensor, X_{ij}, can be defined as functions instead of lists as

```
In[1]:= SetAttributes[δ, Orderless];
        SetAttributes[X, Orderless];
        δ[i_Integer, j_Integer] := If[i == j, 1, 0];
        δ[i_Symbol, i_Symbol] := 3;

In[5]:= X[3, 2]

Out[5]= X[2, 3]

In[6]:= δ[2, 1]

Out[6]= 0

In[7]:= δ[k, k]

Out[7]= 3
```

The `SetAttributes` function automatically converts indexed quantities such as X_{ji} into X_{ij} as ordered indices. The rule `δ[i_Integer, j_Integer]` means that matching occurs only when both i and j are of integer type. The rule `δ[i_Symbol, j_Symbol]` means that matching occurs only when both i and j are symbols. Therefore, `δ[k, k]` returns 3 while `δ[3, 3]` returns 1.

3.2.2.2 Summation Convention Involving δ_{ij}

To implement the summation convention that involves δ_{ij}, it is necessary to add new rules to the built-in function, `Times`, which is protected by default and needs to be overridden by using the `Unprotect` function. After adding new rules, it is necessary to use the `Protect` function so that no further rules can be added.

```
Unprotect[Times];
Times[x[j_Symbol], δ[i_, j_Symbol]] := x[i];
Times[X[i_Symbol, j_], δ[i_Symbol, k_]] := X[k, j];
Times[δ[i_Symbol, j_], δ[i_Symbol, k_]] := δ[j, k];
Protect[Times];
```

This allows simplification such as $\delta_{ij} x_j = x_i$ and $\delta_{ij}\delta_{jk} = \delta_{ik}$

```
In[37]:= δ[i, k] x[k]

Out[37]= x[i]

In[38]:= δ[i, k] x[k] y[i]

Out[38]= x[i] y[i]

In[39]:= δ[i, j] δ[j, k] δ[k, 1]

Out[39]= δ[i, 1]
```

Inclusions in Infinite Media

To implement the rule $x_i x_i = r^2$, it is necessary to modify the built-in function of Power using the Unprotect function, add a new rule, and then use the Protect function so that no further rules can be added.

```
In[40]:= Unprotect[Power];
         Power[x[i_Symbol], 2] := r^2;
         Protect[Power];

In[43]:= x[k] x[k]

Out[43]= r²

In[44]:= δ[i, j] x[i] x[j]

Out[44]= r²
```

To implement the rules for differentiation of tensorial quantities that involve functions of $r(=\sqrt{x_i x_i})$, it is necessary to modify the built-in function, D (differentiation), using the Unprotect function, add new rules, and then use the Protect function so that no further rules can be added. The following rules are needed to be added. The last rule is needed to interchange a list with a differentiation.

$$(ar^n + b)_{,i} = a_{,i} r^n + an(n-1)\frac{x_i}{r} + b_{,i},$$

$$(ax_i + b)_{,j} = a_{,j} x_i + a\delta_{ij} + b_{,j},$$

$$(ax_i^n + b)_{,j} = a_{,j} x_i^n + an x_i^{n-1}\delta_{ij} + b_{,j},$$

$$(af(r)^n + b)_{,i} = a_{,i} f(r)^n + an f(r)^{n-1} f'(r)\frac{x_i}{r} + b_{,i},$$

$$(ab)_{,i} = a_{,i} b + a b_{,i},$$

$$\{a_1, a_2, a_3, \cdots\}_{,i} = \{a_{1,i}, a_{2,i}, a_{3,i}, \cdots\}.$$

```
In[1]:= Unprotect[D];
   D[a_. r^n_. + b_., x[i_]] := D[a, x[i]] r^n + n r^(n-1) x[i] / r a + D[b, x[i]];
   D[a_. x[i_] + b_., x[j_]] := D[a, x[j]] x[i] + a δ[i, j] + D[b, x[j]];
   D[a_. x[i_Integer]^n_. + b_., x[j_]] :=
     D[a, x[j]] x[i]^n + a n x[i]^(n-1) δ[i, j] + D[b, x[j]];
   D[a_. f_[r]^n_. + b_., x[i_]] :=
     D[a, x[i]] f[r]^n + a n f[r]^(n-1) f'[r] x[i] / r + D[b, x[i]];
   D[Times[a_, b_], x[i_]] := D[a, x[i]] b + a D[b, x[i]];
   D[a_List, b_] := Map[D[#, b] &, a]
   Protect[D];
```

The aforementioned rules enable differentiation of tensorial quantities that are functions of r such as

$$\left(ar^4 + \frac{1}{r}\right)_{,i} = -\frac{x_i}{r^3} + 4ar^2 x_i,$$

$$\left(ar^4 + \frac{1}{r}\right)_{,ij} = 8ax_i x_j + \frac{3x_i x_j}{r^5} - \frac{\delta_{ij}}{r^3} + 4ar^2 \delta_{ij},$$

$$\left(\frac{x_i x_j g(r)^3}{r}\right)_{,j} = \frac{x_i(3rg(r)^2 g'(r) + 3g(r)^3) + g(r)^3 x_i}{r} - \frac{g(r)^3 x_i}{r}.$$

In[48]:= `D[a r^4 + 1/r, x[i]]`

Out[48]= $-\dfrac{x[i]}{r^3} + 4\,a\,r^2\,x[i]$

In[49]:= `D[%, x[j]]`

Out[49]= $8\,a\,x[i]\,x[j] + \dfrac{3\,x[i]\,x[j]}{r^5} - \dfrac{\delta[i,j]}{r^3} + 4\,a\,r^2\,\delta[i,j]$

In[51]:= `D[x[i] x[j] g[r]^3 / r, x[j]]`

Out[51]= $-\dfrac{g[r]^3 x[i]}{r} + \dfrac{1}{r}\left(g[r]^3 x[i] + x[i]\left(3\,g[r]^3 + 3\,r\,g[r]^2\,g'[r]\right)\right)$

The aforementioned new rules should be saved and loaded using the « command as necessary for later use.

3.2.3 General Formulation

The goal of this section is to analytically derive the elastic field for multilayered inclusion problems in which multiple concentric inclusions of spherical shape are embedded in an infinitely extended matrix by solving the field equations directly with *Mathematica*. First, a single inclusion problem is solved. Even though this problem is a special case of the work done by Eshelby for the aspect ratio of 1, it is noted that Eshelby's approach cannot be used to handle multilayered inclusion problems.

Consider an isotropic elastic body that consists of a spherical inclusion at the center possibly coated by multiple layers around the inclusion subject to a constant far-field strain field at infinity as shown in Figure 3.9. Navier's equation at each phase for the displacement without body force is expressed as

$$\mu u_{i,jj} + (\mu + \lambda)\theta_{,i} = 0, \qquad (3.23)$$

where λ and μ are the Lamé constants, u_i is the displacement, and θ is the trace of the strain defined as

$$\theta = \epsilon_{jj} = u_{j,j}.$$

The boundary condition is such that the strain, ϵ_{ij}, at the far field, X_{ij}, is constant, i.e.,

$$\epsilon_{ij} \to X_{ij} \quad \text{as} \quad r \to \infty. \qquad (3.24)$$

Finding the general solution to Equation (3.23) is difficult in general. However, a common technique used in solid mechanics is to decompose a field quantity into the deviatoric part and the hydrostatic part. The resulting two equations for the deviatoric part and for the hydrostatic part are simpler, and no coupling exists between the two equations.

First, the far-field strain, X_{ij}, is split into the deviatoric part and the hydrostatic part as

$$X_{ij} = \hat{X}_{ij} + \frac{1}{3} X \delta_{ij},$$

where

$$X = X_{ii},$$

which accounts for the hydrostatic part of the far-field strain, and $\hat{X}_{ij}$ is the deviatoric (shear) part of the far-field strain. Note that

$$\hat{X}_{ii} = 0.$$

As there is no coupling between the hydrostatic part and the shear part of the deformation, the solution to Equation (3.23) can be derived separately for X and $\hat{X}_{ij}$.

3.2.3.1 Solution to Equation (3.23) for X

First, the solution to Equation (3.23) subject to X as $r \to \infty$ as the boundary condition is sought. When the body is subject to the hydrostatic part of the far-field strain, X, the displacement in Equation (3.23) must be proportional to X. Hence, the general form of the displacement field must be written as

$$u_i = h(r) x_i X, \tag{3.25}$$

where $h(r)$ is a function of $r(=\sqrt{x_i x_i})$ alone yet to be determined. The term x_i in Equation (3.25) is necessary because of the requirement that both sides of Equation (3.25) need to be of the same rank of tensors (first-rank tensors). Differentiation of Equation (3.25) with respect to x_i followed by subsequent differentiations yields

$$\theta = u_{j,j} = (3h(r) + r h'(r)) X,$$

$$\theta_{,i} = u_{j,ji} = \left(h''(r) + \frac{4 h'(r)}{r} \right) x_i X.$$

This can be entered into *Mathematica* as

In[25]:= `ui = h[r] x[i] X`

Out[25]= $X\, h[r]\, x[i]$

In[26]:= `θ = D[ui, x[i]]`

Out[26]= $3\, X\, h[r] + r\, X\, h'[r]$

In[27]:= `θi = D[θ, x[i]] // Simplify`

Out[27]= $\dfrac{X\, x[i]\, (4\, h'[r] + r\, h''[r])}{r}$

Substituting the aforementioned into Equation (3.23) yields

$$(2\mu + \lambda) \left(\frac{r h''(r) + 4 h'(r)}{r} \right) x_i X = 0.$$

A *Mathematica* code follows:

```
In[28]:= μ D[D[ui, x[j]], x[j]] + (μ + λ) D[θ, x[i]] // Simplify
```

$$\text{Out[28]=} \quad \frac{X (\lambda + 2\mu) x[i] (4 h'[r] + r h''[r])}{r}$$

Therefore, $h(r)$, must satisfy

$$rh''(r) + 4h'(r) = 0.$$

This differential equation for $h(r)$ can be solved as

```
In[29]:= sol1 = DSolve[4 h'[r] + r h''[r] == 0, h[r], r][[1]]
```

$$\text{Out[29]=} \quad \left\{ h[r] \to -\frac{C[1]}{3 r^3} + C[2] \right\}$$

$$h(r) = -\frac{c_1}{3r^3} + c_2,$$

where c_1 and c_2 are integral constants.

Therefore, the general solutions of the displacement, u_i, the strain, ϵ_{ij}, the stress, σ_{ij}, and the traction force, t_i, corresponding to X are expressed using two unknown coefficients as

$$u_i = \left(-\frac{c_1}{3r^3} + c_2 \right) x_i X, \qquad (3.26)$$

$$\epsilon_{ij} = c_1 \left(\frac{x_i x_j}{r^5} - \frac{\delta_{ij}}{3r^3} \right) X + c_2 \delta_{ij} X, \qquad (3.27)$$

$$\sigma_{ij} = -\frac{2c_1 \mu (r^2 \delta_{ij} - 3x_i x_j)}{3r^5} X + c_2 (3\lambda + 2\mu) \delta_{ij} X, \qquad (3.28)$$

$$t_i = \frac{4c_1 \mu x_i}{3r^4} X + \frac{c_2 x_i (3\lambda + 2\mu)}{r} X. \qquad (3.29)$$

The *Mathematica* code to derive the aforementioned follows:

```
In[30]:= Ui = ui /. sol1
```

$$\text{Out[30]=} \quad X \left(-\frac{C[1]}{3 r^3} + C[2] \right) x[i]$$

```
In[31]:= Uj = Ui /. i → j
```

$$\text{Out[31]=} \quad X \left(-\frac{C[1]}{3 r^3} + C[2] \right) x[j]$$

```
In[39]:= Eijtmp = (D[Ui, x[j]] + D[Uj, x[i]]) / 2;

In[40]:= Eij = Collect[Eijtmp, {C[1], C[2], Simplify}]
```

$$\text{Out[40]=} \quad X C[2] \delta[i, j] + \frac{1}{2} C[1] \left(\frac{2 X x[i] x[j]}{r^5} - \frac{2 X \delta[i, j]}{3 r^3} \right)$$

```
In[41]:= Sijtmp = 2 μ Eij + λ δ[i, j] (Eij /. i → j);

In[42]:= Sij = Collect[Sijtmp, {C[1], C[2]}, Simplify]
```

```
Out[42]= X (3 λ + 2 μ) C[2] δ[i, j] - (2 X μ C[1] (-3 x[i] x[j] + r² δ[i, j])) / (3 r⁵)

In[43]:= Titmp = Sij x[j] / r;

In[44]:= Ti = Collect[Titmp, {C[1], C[2]}, Simplify]

Out[44]= (4 X μ C[1] x[i]) / (3 r⁴) + (X (3 λ + 2 μ) C[2] x[i]) / r
```

The Collect function collects terms involving the same powers of objects and thus helps formatting the output.

For multiphase materials, each phase contains two unknown constants that need to be determined from the traction and displacement continuity condition at the interface and the far-field boundary condition. The determination of these constants is fully discussed in the next subsection.

3.2.3.2 Solution for $\hat{X}_{ij}$

In this subsection, the displacement solution, u_i, to Equation (3.23) subject to $\hat{X}_{ij}$ as $r \to \infty$ is considered.

Because of the linearity nature of Navier's equation, the displacement field, u_i, must be proportional to $\hat{X}_{ij}$ throughout the body. As the displacement, u_i, is a first-rank tensor and $\hat{X}_{ij}$ is a second-rank tensor, the proportionality factor, U_{ijk}, between u_i and $\hat{X}_{jk}$ must be a third-rank tensor according to the quotient rule as

$$u_i = U_{ijk}\hat{X}_{jk}.$$

As U_{ijk} is a third-rank tensor, which must be a combination of x_i and δ_{ij}, all the possible combinations that U_{ijk} can take are restricted to the following terms:

$$U_{ijk} = \frac{f_1(r)}{r^2}x_ix_jx_k + f_2(r)\delta_{ij}x_k + f_3(r)\delta_{ik}x_j + f_4(r)\delta_{jk}x_i,$$

where $f_1(r) \sim f_4(r)$ are functions of r only. The first function, $f_1(r)$, is divided by r^2 to have the same dimension with the rest. It follows that

$$u_i = U_{ijk}\hat{X}_{jk}$$
$$= \frac{f_1(r)}{r^2}x_ix_jx_k\hat{X}_{jk} + f_2(r)x_k\hat{X}_{ik} + f_3(r)x_j\hat{X}_{ij} + f_4(r)x_i\hat{X}_{jj}.$$

As $\hat{X}_{jj} = 0$, the last term is dropped. Also from $\hat{X}_{ij} = \hat{X}_{ji}$, the second and third terms can be combined into one. Therefore, the proper form for u_i is expressed as

$$u_i = U_{ijk}\hat{X}_{jk}$$
$$= \left(\frac{f_1(r)}{r^2}x_ix_j + f_2(r)\delta_{ij}\right)x_k\hat{X}_{jk}$$
$$\equiv \left(\frac{f(r)}{r^2}x_ix_j + g(r)\delta_{ij}\right)x_k\hat{X}_{jk}, \qquad (3.30)$$

where $f_1(r)$ and $f_2(r)$ were replaced by $f(r)$ and $g(r)$, respectively. By substituting Equation (3.30) into Equation (3.23), the differential equations for $f(r)$ and $g(r)$ are derived. This requires extensive index manipulation by hand but a mere exercise for *Mathematica*.

The following rule is useful when multiplications of indexed quantities are involved. The purpose of applying this rule is so that a term such as $x_i x_j \hat{X}_{ij}$ can be isolated with nonoverlapped indices, p and q, to avoid conflict with commonly used indices, such as i and j.

```
In[133]:= myrule = {x[i_] x[j_] X[i_, j_] -> x[p] x[q] X[p, q],
           X[k_, k_] -> 0, x[i_] X[i_, j_] -> x[m] X[j, m]};
         x[a] x[b] X[a, b] /. myrule
Out[134]= x[p] x[q] X[p, q]

In[135]:= x[i] X[j, i] /. myrule
Out[135]= x[m] X[j, m]
```

Equation (3.30) is entered as

```
In[136]:= ui =
   f[r] x[i] x[j] x[k] /r^2 X[j, k] + g[r] δ[i, j] x[k] X[j, k];
```

The quantities, $\theta = u_{j,j}$ and $\theta_{,i}$, can be evaluated as

```
In[152]:= θ = (D[ui, x[i]] // Expand) /. myrule
```

$$\text{Out[152]}= \frac{3 f[r] x[p] x[q] X[p, q]}{r^2} +$$
$$\frac{x[p] x[q] X[p, q] f'[r]}{r} + \frac{x[p] x[q] X[p, q] g'[r]}{r}$$

```
In[153]:= θi = Expand[D[θ, x[i]]] /. myrule
```

$$\text{Out[153]}= \frac{6 f[r] x[m] X[i, m]}{r^2} - \frac{6 f[r] x[i] x[p] x[q] X[p, q]}{r^4} +$$
$$\frac{2 x[m] X[i, m] f'[r]}{r} + \frac{2 x[i] x[p] x[q] X[p, q] f'[r]}{r^3} +$$
$$\frac{2 x[m] X[i, m] g'[r]}{r} - \frac{x[i] x[p] x[q] X[p, q] g'[r]}{r^3} +$$
$$\frac{x[i] x[p] x[q] X[p, q] f''[r]}{r^2} + \frac{x[i] x[p] x[q] X[p, q] g''[r]}{r^2}$$

The quantity, $u_{i,jj}$, can be evaluated as

```
In[155]:= Δui = (D[D[ui, x[n]], x[n]] // Expand) /. myrule
```

$$\text{Out[155]}= \frac{4 f[r] x[m] X[i, m]}{r^2} - \frac{10 f[r] x[i] x[p] x[q] X[p, q]}{r^4} +$$
$$\frac{4 x[i] x[p] x[q] X[p, q] f'[r]}{r^3} + \frac{4 x[m] X[i, m] g'[r]}{r} +$$
$$\frac{x[i] x[p] x[q] X[p, q] f''[r]}{r^2} + x[m] X[i, m] g''[r]$$

Equation (3.30) is now expressed in terms of $f(r)$ and $g(r)$ as

In[156]:= **navier = μ Δui + (μ + λ) θi**

Out[156]= $\mu \left(\dfrac{4\,\text{f[r]\,x[m]\,X[i, m]}}{r^2} - \dfrac{10\,\text{f[r]\,x[i]\,x[p]\,x[q]\,X[p, q]}}{r^4} + \right.$
$\dfrac{4\,\text{x[i]\,x[p]\,x[q]\,X[p, q]\,f'[r]}}{r^3} + \dfrac{4\,\text{x[m]\,X[i, m]\,g'[r]}}{r} +$
$\dfrac{\text{x[i]\,x[p]\,x[q]\,X[p, q]\,f''[r]}}{r^2} + \text{x[m]\,X[i, m]\,g''[r]} \Big) +$

$(\lambda + \mu) \left(\dfrac{6\,\text{f[r]\,x[m]\,X[i, m]}}{r^2} - \dfrac{6\,\text{f[r]\,x[i]\,x[p]\,x[q]\,X[p, q]}}{r^4} + \right.$
$\dfrac{2\,\text{x[m]\,X[i, m]\,f'[r]}}{r} + \dfrac{2\,\text{x[i]\,x[p]\,x[q]\,X[p, q]\,f'[r]}}{r^3} +$
$\dfrac{2\,\text{x[m]\,X[i, m]\,g'[r]}}{r} - \dfrac{\text{x[i]\,x[p]\,x[q]\,X[p, q]\,g'[r]}}{r^3} +$
$\dfrac{\text{x[i]\,x[p]\,x[q]\,X[p, q]\,f''[r]}}{r^2} +$
$\dfrac{\text{x[i]\,x[p]\,x[q]\,X[p, q]\,g''[r]}}{r^2} \Big)$

The aforementioned equation consists of two independent terms, one proportional to $x_m X_{im}$ and the other proportional to $x_i x_p x_q X_{pq}$. They can be separated using the Coefficient command as

In[158]:= **eq1 = Coefficient[navier, x[m] X[i, m]]**

Out[158]= $(\lambda + \mu) \left(\dfrac{6\,\text{f[r]}}{r^2} + \dfrac{2\,\text{f'[r]}}{r} + \dfrac{2\,\text{g'[r]}}{r} \right) + \mu \left(\dfrac{4\,\text{f[r]}}{r^2} + \dfrac{4\,\text{g'[r]}}{r} + \text{g''[r]} \right)$

In[159]:= **eq2 = Coefficient[navier, x[i] x[p] x[q] X[p, q]]**

Out[159]= $\mu \left(-\dfrac{10\,\text{f[r]}}{r^4} + \dfrac{4\,\text{f'[r]}}{r^3} + \dfrac{\text{f''[r]}}{r^2} \right) +$
$(\lambda + \mu) \left(-\dfrac{6\,\text{f[r]}}{r^4} + \dfrac{2\,\text{f'[r]}}{r^3} - \dfrac{\text{g'[r]}}{r^3} + \dfrac{\text{f''[r]}}{r^2} + \dfrac{\text{g''[r]}}{r^2} \right)$

Therefore, the differential equations that $f(r)$ and $g(r)$ must satisfy are derived as

$$(\lambda + \mu)\left(\frac{2f'(r)}{r} + \frac{6f(r)}{r^2} + \frac{2g'(r)}{r}\right) + \mu\left(\frac{4f(r)}{r^2} + g''(r) + \frac{4g'(r)}{r}\right) = 0, \qquad (3.31)$$

$$(\lambda + \mu)\left(\frac{f''(r)}{r^2} + \frac{2f'(r)}{r^3} - \frac{6f(r)}{r^4} + \frac{g''(r)}{r^2} - \frac{g'(r)}{r^3}\right)$$
$$+ \mu\left(\frac{f''(r)}{r^2} + \frac{4f'(r)}{r^3} - \frac{10f(r)}{r^4}\right) = 0. \qquad (3.32)$$

A set of two simultaneous differential equations for $f(r)$ and $g(r)$, Equations (3.31) and (3.32), can be solved using *Mathematica*'s differential equation solver, DSolve, as

In[160]:= **soll = DSolve[{eq1 == 0, eq2 == 0}, {f[r], g[r]}, r][[1]]**

Out[160]= $\left\{f[r] \to \frac{C[1]}{r^5} + \frac{C[2]}{r^3} + r^2\, C[3],\right.$

$\left. g[r] \to -\frac{2\,C[1]}{5\,r^5} + \frac{2\,\mu\,C[2]}{3\,r^3\,(\lambda+\mu)} - \frac{r^2\,(5\,\lambda+7\,\mu)\,C[3]}{2\,\lambda+7\,\mu} + C[4]\right\}$

$$f(r) = \frac{c_1}{r^5} + \frac{c_2}{r^3} + c_3 r^2, \tag{3.33}$$

$$g(r) = -\frac{2c_1}{5r^5} + \frac{2c_2\mu}{3r^3(\lambda+\mu)} - \frac{c_3 r^2(5\lambda+7\mu)}{2\lambda+7\mu} + c_4, \tag{3.34}$$

where c_1–c_4 are integral constants.

The aforementioned solution for $f(r)$ and $g(r)$ implies that there are four independent solutions for the displacement, strain, stress, and traction field. The four terms must be combined so that the boundary condition and the continuity condition are satisfied. Equations (3.33) and (3.34) are substituted into Equation (3.30) to derive four independent solutions for the displacement as

In[161]:= **Ui = ui /. soll /. myrule**

Out[161]= $\left(-\frac{2\,C[1]}{5\,r^5} + \frac{2\,\mu\,C[2]}{3\,r^3\,(\lambda+\mu)} - \frac{r^2\,(5\,\lambda+7\,\mu)\,C[3]}{2\,\lambda+7\,\mu} + C[4]\right) x[m]\, X[i, m]\, +$

$\dfrac{\left(\frac{C[1]}{r^5} + \frac{C[2]}{r^3} + r^2\,C[3]\right) x[i]\, x[p]\, x[q]\, X[p, q]}{r^2}$

In[162]:= **Uj = Ui /. i → j**

Out[162]= $\left(-\frac{2\,C[1]}{5\,r^5} + \frac{2\,\mu\,C[2]}{3\,r^3\,(\lambda+\mu)} - \frac{r^2\,(5\,\lambda+7\,\mu)\,C[3]}{2\,\lambda+7\,\mu} + C[4]\right) x[m]\, X[j, m]\, +$

$\dfrac{\left(\frac{C[1]}{r^5} + \frac{C[2]}{r^3} + r^2\,C[3]\right) x[j]\, x[p]\, x[q]\, X[p, q]}{r^2}$

Therefore, the strain components, ϵ_{ij}, are expressed as

In[163]:= **Eij = Expand[(D[Ui, x[j]] + D[Uj, x[i]])/2] /. myrule**

Out[163]= $-\dfrac{2\,C[1]\,X[i, j]}{5\,r^5} + \dfrac{2\,\mu\,C[2]\,X[i, j]}{3\,r^3\,(\lambda+\mu)} -$

$\dfrac{5\,r^2\,\lambda\,C[3]\,X[i, j]}{2\,\lambda+7\,\mu} - \dfrac{7\,r^2\,\mu\,C[3]\,X[i, j]}{2\,\lambda+7\,\mu} + C[4]\,X[i, j]\, +$

$\dfrac{2\,C[1]\,x[j]\,x[m]\,X[i, m]}{r^7} + \dfrac{C[2]\,x[j]\,x[m]\,X[i, m]}{r^5} -$

$\dfrac{\mu\,C[2]\,x[j]\,x[m]\,X[i, m]}{r^5\,(\lambda+\mu)} + C[3]\,x[j]\,x[m]\,X[i, m]\, -$

$\dfrac{5\,\lambda\,C[3]\,x[j]\,x[m]\,X[i, m]}{2\,\lambda+7\,\mu} - \dfrac{7\,\mu\,C[3]\,x[j]\,x[m]\,X[i, m]}{2\,\lambda+7\,\mu} +$

$\dfrac{2\,C[1]\,x[i]\,x[m]\,X[j, m]}{r^7} + \dfrac{C[2]\,x[i]\,x[m]\,X[j, m]}{r^5} -$

Inclusions in Infinite Media

$$\frac{\mu C[2] x[i] x[m] X[j, m]}{r^5 (\lambda+\mu)} + C[3] x[i] x[m] X[j, m] -$$

$$\frac{5 \lambda C[3] x[i] x[m] X[j, m]}{2\lambda+7\mu} - \frac{7\mu C[3] x[i] x[m] X[j, m]}{2\lambda+7\mu} -$$

$$\frac{7 C[1] x[i] x[j] x[p] x[q] X[p, q]}{r^9} -$$

$$\frac{5 C[2] x[i] x[j] x[p] x[q] X[p, q]}{r^7} +$$

$$\frac{C[1] x[p] x[q] X[p, q] \delta[i, j]}{r^7} +$$

$$\frac{C[2] x[p] x[q] X[p, q] \delta[i, j]}{r^5} + C[3] x[p] x[q] X[p, q] \delta[i, j]$$

The stress components, σ_{ij}, are expressed as

In[184]:= **Eii = Expand[Eij /. i → j] /. myrule // Simplify**

Out[184]= $(\mu (-4 \lambda C[2] - 14 \mu C[2] + 21 r^5 \lambda C[3] + 21 r^5 \mu C[3])$
$\quad x[p] x[q] X[p, q]) / (r^5 (\lambda+\mu) (2\lambda+7\mu))$

In[185]:= **Sij = Expand[2 μ Eij + λ δ[i, j] Eii] /. myrule // Simplify**

Out[185]= $-((\mu (2 r^4 (\lambda\mu (54 C[1] - 20 r^2 C[2] + 180 r^7 C[3] - 135 r^5 C[4]) +$
$\quad 7\mu^2 (6 C[1] - 10 r^2 C[2] + 15 r^7 C[3] - 15 r^5 C[4]) +$
$\quad 3\lambda^2 (4 C[1] + 25 r^7 C[3] - 10 r^5 C[4]))$
$\quad X[i, j] + 15 (2 x[j] (r^2 (-14\mu^2 C[1] +$
$\quad \lambda\mu (-18 C[1] - 7 r^2 C[2] + 3 r^7 C[3]) +$
$\quad \lambda^2 (-4 C[1] - 2 r^2 C[2] + 3 r^7 C[3])) x[m]$
$\quad X[i, m] + (2\lambda^2 + 9\lambda\mu + 7\mu^2) (7 C[1] + 5 r^2 C[2])$
$\quad x[i] x[p] x[q] X[p, q]) + r^2$
$\quad (-2 (14\mu^2 C[1] + \lambda^2 (4 C[1] + 2 r^2 C[2] - 3 r^7 C[3]) +$
$\quad \lambda\mu (18 C[1] + 7 r^2 C[2] - 3 r^7 C[3])) x[i]$
$\quad x[m] X[j, m] - (14\mu^2 (C[1] + r^2 C[2] + r^7 C[3]) +$
$\quad \lambda^2 (4 C[1] + 25 r^7 C[3]) +$
$\quad \lambda\mu (18 C[1] + 4 r^2 C[2] + 39 r^7 C[3]))$
$\quad x[p] x[q] X[p, q] \delta[i, j]))))/$
$\quad (15 r^9 (\lambda+\mu) (2\lambda+7\mu)))$

The traction components, t_i, are expressed as

In[186]:= **Ti = (Sij x[j] / r // Expand) /. myrule // Simplify**

Out[186]= $(\mu (2 r^2 (6\lambda^2 (8 C[1] + 5 r^2 (C[2] - 4 r^5 C[3] + r^3 C[4])) +$
$\quad 7\mu^2 (24 C[1] + 5 r^2 (2 C[2] - 3 r^5 C[3] + 3 r^3 C[4])) +$
$\quad \lambda\mu (216 C[1] + 5 r^2 (25 C[2] - 45 r^5 C[3] + 27 r^3 C[4])))$
$\quad x[m] X[i, m] - 15 (\lambda+\mu)$
$\quad (\lambda (16 C[1] + 16 r^2 C[2] - 19 r^7 C[3]) +$
$\quad 14\mu (4 C[1] + 4 r^2 C[2] - r^7 C[3])) x[i]$
$\quad x[p] x[q] X[p, q])) / (15 r^8 (\lambda+\mu) (2\lambda+7\mu))$

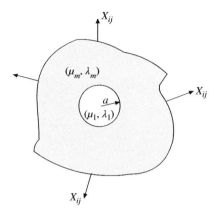

Figure 3.10 Two-phase material

The expressions derived are valid for each phase, and the unknown coefficients must be determined from the continuity condition and the far-field boundary condition.

3.2.4 Exact Solution for Two-Phase Materials

By using the elastic field solution derived, it is possible to obtain the elastic field where a single inclusion is coated by multiple layers with different elastic moduli in an infinite matrix subject to a constant strain at infinity.

When the inclusion has no coated layer, this problem is reduced to the Eshelby solution for a spherical inclusion with the aspect ratio being unity. However, it is difficult to directly apply the Eshelby approach for the elastic field outside the inclusion unlike the present one. It is also noted that the Eshelby method cannot be applied to multiphase coated layers.

First, consider a two-phase material where a spherical inclusion with a radius a is surrounded by an infinitely extended material (Figure 3.10). The Lamé constants for the inclusion phase are denoted as (λ_1, μ_1), and the Lamé constants for the matrix phase are denoted as (λ_m, μ_m).

3.2.4.1 Exact Solution for X

The unknown coefficients in Equations (3.26) and (3.29) for each phase can be determined so that the displacement and traction are continuous at $r = a$.

The general forms for the displacement and the traction force are restated here as

$$u_i = \left(-\frac{c_1}{3r^3} + c_2\right) x_i X,$$

$$t_i = \frac{4c_1 \mu x_i}{3r^4} X + \frac{c_2 x_i (3\lambda + 2\mu)}{r} X.$$

The unknowns, c_1 and c_2, must be solved for each phase. Let the displacement and the traction force for each phase be denoted as

$$u_i^{in} = \left(-\frac{c_1^{in}}{3r^3} + c_2^{in}\right) x_i X,$$

$$t_i^{in} = \frac{4c_1^{in}\mu_1 x_i}{3r^4}X + \frac{c_2^{in}x_i(3\lambda_1 + 2\mu_1)}{r}X,$$

$$u_i^{out} = \left(-\frac{c_1^{out}}{3r^3} + c_2^{out}\right)x_i X,$$

$$t_i^{out} = \frac{4c_1^{out}\mu_m x_i}{3r^4}X + \frac{c_2^{out}x_i(3\lambda_m + 2\mu_m)}{r}X.$$

From the condition that the displacement, u_i^{in}, must remain finite at $r = 0$, c_1^{in} must be 0. From the condition that as $r \to \infty$, $u_i^{out} \to x_i X$, c_2^{out} must be 1. Therefore, the unknown coefficients are limited to c_2^{in} and c_1^{out}, and they can be determined from the following continuity conditions of the displacement and traction at $r = a$ as

$$u_i^{in} = u_i^{out} \quad r = a,$$

$$t_i^{in} = t_i^{out} \quad r = a.$$

The displacement for each phase can be entered as

In[40]:= **DispIn = Ui /. C → c1**

Out[40]= $X \left(-\dfrac{c1[1]}{3\,r^3} + c1[2] \right) x[i]$

In[41]:= **DispOut = Ui /. C → cm**

Out[41]= $X \left(-\dfrac{cm[1]}{3\,r^3} + cm[2] \right) x[i]$

The traction for each phase can be entered as

In[42]:= **TractIn = Ti /. {C → c1, μ → μ1, λ → λ1}**

Out[42]= $\dfrac{4\,X\,\mu1\,c1[1]\,x[i]}{3\,r^4} + \dfrac{X\,(3\,\lambda1 + 2\,\mu1)\,c1[2]\,x[i]}{r}$

In[43]:= **TractOut = Ti /. {C → cm, μ → μm, λ → λm}**

Out[43]= $\dfrac{4\,X\,\mu m\,cm[1]\,x[i]}{3\,r^4} + \dfrac{X\,(3\,\lambda m + 2\,\mu m)\,cm[2]\,x[i]}{r}$

The continuity conditions of the displacement and traction are entered as

In[179]:= **eq1 = (DispIn - DispOut) /. r → a**

Out[179]= $X \left(-\dfrac{c1[1]}{3\,a^3} + c1[2] \right) x[i] - X \left(-\dfrac{cm[1]}{3\,a^3} + cm[2] \right) x[i]$

In[180]:= **eq2 = (TractIn - TractOut) /. r → a**

Out[180]= $\dfrac{4\,X\,\mu1\,c1[1]\,x[i]}{3\,a^4} + \dfrac{X\,(3\,\lambda1 + 2\,\mu1)\,c1[2]\,x[i]}{a} - \dfrac{4\,X\,\mu m\,cm[1]\,x[i]}{3\,a^4} - \dfrac{X\,(3\,\lambda m + 2\,\mu m)\,cm[2]\,x[i]}{a}$

The unknown coefficients are now solved as

In[181]:= `c1[1] = 0; cm[2] = 1;`

In[182]:= `sol2 = Solve[{eq1 == 0, eq2 == 0}, {c1[2], cm[1]}][[1]]`

Out[182]= $\left\{ c1[2] \to \dfrac{3\,(\lambda m + 2\,\mu m)}{3\,\lambda 1 + 2\,\mu 1 + 4\,\mu m},\right.$

$\left. cm[1] \to -\dfrac{3\,a^3\,(-3\,\lambda 1 + 3\,\lambda m - 2\,\mu 1 + 2\,\mu m)}{3\,\lambda 1 + 2\,\mu 1 + 4\,\mu m} \right\}$

$$c_1^{in} = 0,\ c_2^{in} = \dfrac{3(\lambda_m + 2\mu_m)}{3\lambda_1 + 2\mu_1 + 4\mu_m},$$

$$c_1^{out} = 1,\ c_2^{out} = -\dfrac{3a^3(-3\lambda_1 + 3\lambda_m - 2\mu_1 + 2\mu_m)}{3\lambda_1 + 2\mu_1 + 4\mu_m}.$$

Therefore, the displacement, strain, stress, and traction force for each phase can be computed as

1. Inside ($0 < r < a$)

In[49]:= `uiin = DispIn /. sol2`

Out[49]= $\dfrac{3\,X\,(\lambda m + 2\,\mu m)\,x[i]}{3\,\lambda 1 + 2\,\mu 1 + 4\,\mu m}$

In[50]:= `εijin = Eij /. C → c1 /. sol2`

Out[50]= $\dfrac{3\,X\,(\lambda m + 2\,\mu m)\,\delta[i,j]}{3\,\lambda 1 + 2\,\mu 1 + 4\,\mu m}$

In[51]:= `σijin = Sij /. {C → c1, λ → λ1, μ → μ1} /. sol2`

Out[51]= $\dfrac{3\,X\,(3\,\lambda 1 + 2\,\mu 1)\,(\lambda m + 2\,\mu m)\,\delta[i,j]}{3\,\lambda 1 + 2\,\mu 1 + 4\,\mu m}$

In[52]:= `tiin = TractIn /. sol2`

Out[52]= $\dfrac{3\,X\,(3\,\lambda 1 + 2\,\mu 1)\,(\lambda m + 2\,\mu m)\,x[i]}{r\,(3\,\lambda 1 + 2\,\mu 1 + 4\,\mu m)}$

2. Outside ($a < r < \infty$)

In[190]:= `uiout = DispOut /. sol2`

Out[190]= $X\left(1 + \dfrac{a^3\,(-3\,\lambda 1 + 3\,\lambda m - 2\,\mu 1 + 2\,\mu m)}{r^3\,(3\,\lambda 1 + 2\,\mu 1 + 4\,\mu m)}\right) x[i]$

In[191]:= `εijout = Eij /. {C → cm} /. sol2`

Out[191]= $X\,\delta[i,j] - \dfrac{3\,a^3\,(-3\,\lambda 1 + 3\,\lambda m - 2\,\mu 1 + 2\,\mu m)\left(\dfrac{2\,X\,x[i]\,x[j]}{r^5} - \dfrac{2\,X\,\delta[i,j]}{3\,r^3}\right)}{2\,(3\,\lambda 1 + 2\,\mu 1 + 4\,\mu m)}$

In[195]:= **σijout = Sij /. {C → cm, λ → λm, μ → μm} /. sol2**

Out[195]= $X (3 \lambda m + 2 \mu m) \delta[i, j] + \left(2 a^3 X \mu m (-3 \lambda 1 + 3 \lambda m - 2 \mu 1 + 2 \mu m)\right.$
$\left.(-3 x[i] x[j] + r^2 \delta[i, j])\right) / (r^5 (3 \lambda 1 + 2 \mu 1 + 4 \mu m))$

In[196]:= **tiout = TractOut /. sol2**

Out[196]= $\dfrac{X (3 \lambda m + 2 \mu m) x[i]}{r} - \dfrac{4 a^3 X \mu m (-3 \lambda 1 + 3 \lambda m - 2 \mu 1 + 2 \mu m) x[i]}{r^4 (3 \lambda 1 + 2 \mu 1 + 4 \mu m)}$

3.2.4.2 Exact Solution for X'_{ij}

When the body is subject to the deviatoric part of the strain, X'_{ij}, the elastic field can be obtained by the same procedure described earlier. However, unlike in the previous subsection where the body is subject to X, which is a scalar (a zeroth-rank tensor), this analysis is more involved as the source of the elastic field is X'_{ij}, which is a second-rank tensor. The amount of algebra involved is significantly larger and without *Mathematica*, it is not possible to carry on all the algebra.

First, it is recalled that the displacement, u_i, and the traction, t_i, are expressed by combinations of four independent functions as

In[215]:= **Ui = ui /. sol1 /. myrule**

Out[215]= $\left(-\dfrac{2 C[1]}{5 r^5} + \dfrac{2 \mu C[2]}{3 r^3 (\lambda + \mu)} - \dfrac{r^2 (5 \lambda + 7 \mu) C[3]}{2 \lambda + 7 \mu} + C[4]\right) x[m] X[i, m] +$
$\dfrac{\left(\dfrac{C[1]}{r^5} + \dfrac{C[2]}{r^3} + r^2 C[3]\right) x[i] x[p] x[q] X[p, q]}{r^2}$

In[216]:= **Ti = (Sij x[j] / r // Expand) /. myrule // Simplify**

Out[216]= $\left(\mu \left(2 r^2 \left(6 \lambda^2 \left(8 C[1] + 5 r^2 \left(C[2] - 4 r^5 C[3] + r^3 C[4]\right)\right) + \right.\right.\right.$
$7 \mu^2 \left(24 C[1] + 5 r^2 \left(2 C[2] - 3 r^5 C[3] + 3 r^3 C[4]\right)\right) +$
$\lambda \mu \left(216 C[1] + 5 r^2 \left(25 C[2] - 45 r^5 C[3] + 27 r^3 C[4]\right)\right)\right)$
$x[m] X[i, m] - 15 (\lambda + \mu)$
$\left(\lambda \left(16 C[1] + 16 r^2 C[2] - 19 r^7 C[3]\right) + \right.$
$\left.14 \mu \left(4 C[1] + 4 r^2 C[2] - r^7 C[3]\right)\right) x[i]$
$\left.\left. x[p] x[q] X[p, q]\right)\right) / \left(15 r^8 (\lambda + \mu) (2 \lambda + 7 \mu)\right)$

Therefore, the displacement and traction inside the inclusion can be assumed to be a linear combination of the four independent functions as

In[217]:= **DispIn = Collect[**
(Ui /. {C → c1, λ → λ1, μ → μ1}), Table[c1[i], {i, 1, 4}]]

Out[217]= $\left(\dfrac{2 \mu 1 \, c1[2]}{3 r^3 (\lambda 1 + \mu 1)} - \dfrac{r^2 (5 \lambda 1 + 7 \mu 1) c1[3]}{2 \lambda 1 + 7 \mu 1} + c1[4]\right) x[m] X[i, m] +$
$\dfrac{\left(\dfrac{c1[2]}{r^3} + r^2 c1[3]\right) x[i] x[p] x[q] X[p, q]}{r^2}$

```
In[220]:= DispOut = Collect[
    (Ui /. {C → c2, λ → λm, μ → μm}), Table[c2[i], {i, 1, 4}]]
```

Out[220]= $c2[4] \, x[m] \, X[i, m] + c2[3]$
$\left(-\dfrac{r^2 \, (5 \lambda m + 7 \mu m) \, x[m] \, X[i, m]}{2 \lambda m + 7 \mu m} + x[i] \, x[p] \, x[q] \, X[p, q] \right) +$
$c2[1] \left(-\dfrac{2 \, x[m] \, X[i, m]}{5 \, r^5} + \dfrac{x[i] \, x[p] \, x[q] \, X[p, q]}{r^7} \right) +$
$c2[2] \left(\dfrac{2 \mu m \, x[m] \, X[i, m]}{3 \, r^3 \, (\lambda m + \mu m)} + \dfrac{x[i] \, x[p] \, x[q] \, X[p, q]}{r^5} \right)$

```
In[221]:= TractIn = Ti /. {C → c1, λ → λ1, μ → μ1}
```

Out[221]= $\Big(\mu 1 \, \big(2 \, r^2 \, \big(30 \, r^2 \, \lambda 1^2 \, (c1[2] - 4 \, r^5 \, c1[3] + r^3 \, c1[4]) +$
$35 \, r^2 \, \mu 1^2 \, (2 \, c1[2] - 3 \, r^5 \, c1[3] + 3 \, r^3 \, c1[4]) +$
$5 \, r^2 \, \lambda 1 \, \mu 1 \, (25 \, c1[2] - 45 \, r^5 \, c1[3] + 27 \, r^3 \, c1[4]) \big) \, x[m]$
$X[i, m] - 15 \, (\lambda 1 + \mu 1) \, \big(\lambda 1 \, (16 \, r^2 \, c1[2] - 19 \, r^7 \, c1[3]) +$
$14 \, \mu 1 \, (4 \, r^2 \, c1[2] - r^7 \, c1[3]) \big) \, x[i] \, x[p]$
$x[q] \, X[p, q] \big) \big) / \big(15 \, r^8 \, (\lambda 1 + \mu 1) \, (2 \, \lambda 1 + 7 \, \mu 1) \big)$

```
In[222]:= TractOut = Ti /. {C → c2, λ → λm, μ → μm}
```

Out[222]= $\Big(\mu m \, \big(2 \, r^2 \, \big(6 \, \lambda m^2$
$\big(8 \, c2[1] + 5 \, r^2 \, (c2[2] - 4 \, r^5 \, c2[3] + r^3 \, c2[4]) \big) + 7 \, \mu m^2$
$\big(24 \, c2[1] + 5 \, r^2 \, (2 \, c2[2] - 3 \, r^5 \, c2[3] + 3 \, r^3 \, c2[4]) \big) +$
$\lambda m \, \mu m \, \big(216 \, c2[1] + 5 \, r^2 \, (25 \, c2[2] - 45 \, r^5 \, c2[3] +$
$27 \, r^3 \, c2[4]) \big) \big) \, x[m] \, X[i, m] -$
$15 \, (\lambda m + \mu m) \, \big(\lambda m \, (16 \, c2[1] + 16 \, r^2 \, c2[2] - 19 \, r^7 \, c2[3]) +$
$14 \, \mu m \, (4 \, c2[1] + 4 \, r^2 \, c2[2] - r^7 \, c2[3]) \big) \big)$
$x[i] \, x[p] \, x[q] \, X[p, q] \big) \big) /$
$\big(15 \, r^8 \, (\lambda m + \mu m) \, (2 \, \lambda m + 7 \, \mu m) \big)$

As the displacement, u_i^{in}, has to remain finite at $r = 0$, it follows that $c_1^{in} = 0$ and $c_2^{in} = 0$. Also, as $r \to \infty$, $u_i^{out} \to x_i \hat{X}_{ij}$, it follows that $c_4^{out} = 1$ and $c_3^{out} = 0$. Therefore, the rest of the unknown coefficients can be determined from the following two continuity conditions of the displacement and the traction at $r = a$:

$$u_i^{in} = u_i^{out} \quad r = a,$$
$$t_i^{in} = t_i^{out} \quad r = a.$$

The following *Mathematica* code sets up a set of equations and solves for the unknown coefficients.

```
In[48]:= c1[1] = 0; c1[2] = 0; c2[4] = 1; c2[3] = 0;
```

```
In[48]:= tmp1 = ((DispIn - DispOut) /. r → a) // Expand
```

Out[48]= $-x[m] \, X[i, m] - \dfrac{5 \, a^2 \, \lambda 1 \, c1[3] \, x[m] \, X[i, m]}{2 \, \lambda 1 + 7 \, \mu 1} -$
$\dfrac{7 \, a^2 \, \mu 1 \, c1[3] \, x[m] \, X[i, m]}{2 \, \lambda 1 + 7 \, \mu 1} + c1[4] \, x[m] \, X[i, m] + \dfrac{2 \, c2[1] \, x[m] \, X[i, m]}{5 \, a^5} -$
$\dfrac{2 \, \mu m \, c2[2] \, x[m] \, X[i, m]}{3 \, a^3 \, (\lambda m + \mu m)} + c1[3] \, x[i] \, x[p] \, x[q] \, X[p, q] -$
$\dfrac{c2[1] \, x[i] \, x[p] \, x[q] \, X[p, q]}{a^7} - \dfrac{c2[2] \, x[i] \, x[p] \, x[q] \, X[p, q]}{a^5}$

In[50]:= `eq1 = Coefficient[tmp1, x[m] X[i, m]]`

Out[50]= $-1 - \dfrac{5 a^2 \lambda 1\, c1[3]}{2\lambda 1 + 7\mu 1} - \dfrac{7 a^2 \mu 1\, c1[3]}{2\lambda 1 + 7\mu 1} + c1[4] + \dfrac{2\, c2[1]}{5 a^5} - \dfrac{2\mu m\, c2[2]}{3 a^3 (\lambda m + \mu m)}$

In[51]:= `eq2 = Coefficient[tmp1, x[i] x[p] x[q] X[p, q]]`

Out[51]= $c1[3] - \dfrac{c2[1]}{a^7} - \dfrac{c2[2]}{a^5}$

In[52]:= `tmp2 = ((TractIn - TractOut) /. r -> a)`

Out[52]= $\big(\mu 1\, (2 a^2\, (30 a^2 \lambda 1^2\, (-4 a^5 c1[3] + a^3 c1[4]) + 35 a^2 \mu 1^2\, (-3 a^5 c1[3] + 3 a^3 c1[4]) + 5 a^2 \lambda 1\, \mu 1\, (-45 a^5 c1[3] + 27 a^3 c1[4]))\, x[m] X[i, m] - 15\, (\lambda 1 + \mu 1)\, (-19 a^7 \lambda 1\, c1[3] - 14 a^7 \mu 1\, c1[3])\, x[i] x[p] x[q] X[p, q])\big) / \big(15 a^8\, (\lambda 1 + \mu 1)\, (2 \lambda 1 + 7 \mu 1)\big) -$
$\big(\mu m\, \{2 a^2\, (6\lambda m^2\, (8\, c2[1] + 5 a^2\, (a^3 + c2[2])) + 7\mu m^2\, (24\, c2[1] + 5 a^2\, (3 a^3 + 2\, c2[2])) + \lambda m\, \mu m\, (216\, c2[1] + 5 a^2\, (27 a^3 + 25\, c2[2])))\, x[m] X[i, m] - 15\, (\lambda m + \mu m)\, (14\mu m\, (4\, c2[1] + 4 a^2\, c2[2]) + \lambda m\, (16\, c2[1] + 16 a^2\, c2[2]))\, x[i] x[p] x[q] X[p, q]\}\big) / \big(15 a^8\, (\lambda m + \mu m)\, (2 \lambda m + 7 \mu m)\big)$

In[53]:= `eq3 = Coefficient[tmp2, x[m] X[i, m]]`

Out[53]= $\big(2\mu 1\, (30 a^2 \lambda 1^2\, (-4 a^5 c1[3] + a^3 c1[4]) + 35 a^2 \mu 1^2\, (-3 a^5 c1[3] + 3 a^3 c1[4]) + 5 a^2 \lambda 1\, \mu 1\, (-45 a^5 c1[3] + 27 a^3 c1[4]))\big) / \big(15 a^6\, (\lambda 1 + \mu 1)\, (2 \lambda 1 + 7 \mu 1)\big) - \big(2\mu m\, (6\lambda m^2\, (8\, c2[1] + 5 a^2\, (a^3 + c2[2])) + 7\mu m^2\, (24\, c2[1] + 5 a^2\, (3 a^3 + 2\, c2[2])) + \lambda m\, \mu m\, (216\, c2[1] + 5 a^2\, (27 a^3 + 25\, c2[2])))\big) / \big(15 a^6\, (\lambda m + \mu m)\, (2 \lambda m + 7 \mu m)\big)$

In[54]:= `eq4 = Coefficient[tmp2, x[i] x[p] x[q] X[p, q]]`

Out[54]= $-\dfrac{\mu 1\, (-19 a^7 \lambda 1\, c1[3] - 14 a^7 \mu 1\, c1[3])}{a^8\, (2\lambda 1 + 7\mu 1)} + \dfrac{\mu m\, (14\mu m\, (4\, c2[1] + 4 a^2\, c2[2]) + \lambda m\, (16\, c2[1] + 16 a^2\, c2[2]))}{a^8\, (2\lambda m + 7\mu m)}$

In[55]:= `sol2 = Solve[{eq1 == 0, eq2 == 0, eq3 == 0, eq4 == 0}, {c1[3], c1[4], c2[1], c2[2]}][[1]]`

Out[55]= $\Big\{c1[3] \to 0,\ c1[4] \to \dfrac{15\, (\lambda m\, \mu m + 2\mu m^2)}{6\lambda m\, \mu 1 + 9\lambda m\, \mu m + 16\mu 1\, \mu m + 14\mu m^2},$
$c2[1] \to \dfrac{15 a^5\, (\mu 1 - \mu m)\, (\lambda m + \mu m)}{6\lambda m\, \mu 1 + 9\lambda m\, \mu m + 16\mu 1\, \mu m + 14\mu m^2},$
$c2[2] \to -\dfrac{15 a^3\, (\mu 1 - \mu m)\, (\lambda m + \mu m)}{6\lambda m\, \mu 1 + 9\lambda m\, \mu m + 16\mu 1\, \mu m + 14\mu m^2}\Big\}$

Therefore, the displacement, strain, stress, and traction fields inside the inclusion ($0 < r < a$) are expressed in closed form as

In[56]:= `uiin = DispIn /. sol2`

Out[56]= $\dfrac{15\, (\lambda m\, \mu m + 2\mu m^2)\, x[m] X[i, m]}{6\lambda m\, \mu 1 + 9\lambda m\, \mu m + 16\mu 1\, \mu m + 14\mu m^2}$

In[57]:= `eijin = Eij /. {C -> c1, \lambda -> \lambda 1, \mu -> \mu 1} /. sol2`

Out[57]= $\dfrac{15\, (\lambda m\, \mu m + 2\mu m^2)\, X[i, j]}{6\lambda m\, \mu 1 + 9\lambda m\, \mu m + 16\mu 1\, \mu m + 14\mu m^2}$

In[58]:= **σijin = Sij /. {C → c1, λ → λ1, μ → μ1} /. sol2 // Simplify**

Out[58]= $\dfrac{30\,\mu 1\,\mu m\,(\lambda m + 2\,\mu m)\,X[i,\,j]}{2\,\mu m\,(8\,\mu 1 + 7\,\mu m) + \lambda m\,(6\,\mu 1 + 9\,\mu m)}$

In[59]:= **tiin = TractIn /. sol2 // Simplify**

Out[59]= $\dfrac{30\,\mu 1\,\mu m\,(\lambda m + 2\,\mu m)\,x[m]\,X[i,\,m]}{r\,(2\,\mu m\,(8\,\mu 1 + 7\,\mu m) + \lambda m\,(6\,\mu 1 + 9\,\mu m))}$

The displacement, strain, stress, and traction fields outside the inclusion ($a < r < \infty$) are expressed in closed form as

In[237]:= **uiout = Collect[DispOut /. sol2,
   {x[m] X[i, m], x[i] x[p] x[q] X[p, q]}, Simplify]**

Out[237]= $\left(1 - \dfrac{10\,a^3\,(\mu 1 - \mu m)\,\mu m}{r^3\,(2\,\mu m\,(8\,\mu 1 + 7\,\mu m) + \lambda m\,(6\,\mu 1 + 9\,\mu m))} - \dfrac{6\,a^5\,(\mu 1 - \mu m)\,(\lambda m + \mu m)}{r^5\,(2\,\mu m\,(8\,\mu 1 + 7\,\mu m) + \lambda m\,(6\,\mu 1 + 9\,\mu m))}\right) x[m]\,X[i,\,m] + \dfrac{15\,(a^5 - a^3\,r^2)\,(\mu 1 - \mu m)\,(\lambda m + \mu m)\,x[i]\,x[p]\,x[q]\,X[p,\,q]}{r^7\,(2\,\mu m\,(8\,\mu 1 + 7\,\mu m) + \lambda m\,(6\,\mu 1 + 9\,\mu m))}$

In[238]:= **εijout = Eij /. {C → c2, λ → λm, μ → μm} /. sol2 // Simplify**

Out[238]= $\bigl(r^4\,(-10\,a^3\,r^2\,(\mu 1 - \mu m)\,\mu m - 6\,a^5\,(\mu 1 - \mu m)\,(\lambda m + \mu m) + r^5\,(2\,\mu m\,(8\,\mu 1 + 7\,\mu m) + \lambda m\,(6\,\mu 1 + 9\,\mu m)))\,X[i,\,j] + 15\,a^3\,(\mu 1 - \mu m)\,(x[j]\,(r^2\,(-r^2\,\lambda m + 2\,a^2\,(\lambda m + \mu m))\,x[m]\,X[i,\,m] - (7\,a^2 - 5\,r^2)\,(\lambda m + \mu m)\,x[i]\,x[p]\,x[q]\,X[p,\,q]) + r^2\,((-r^2\,\lambda m + 2\,a^2\,(\lambda m + \mu m))\,x[i]\,x[m]\,X[j,\,m] + (a^2 - r^2)\,(\lambda m + \mu m)\,x[p]\,x[q]\,X[p,\,q]\,\delta[i,\,j]))\bigr)/\bigl(r^9\,(2\,\mu m\,(8\,\mu 1 + 7\,\mu m) + \lambda m\,(6\,\mu 1 + 9\,\mu m))\bigr)$

In[239]:= **σijout = Sij /. {C → c2, λ → λm, μ → μm} /. sol2 // Simplify**

Out[239]= $\bigl(2\,\mu m\,(r^4\,(-10\,a^3\,r^2\,(\mu 1 - \mu m)\,\mu m - 6\,a^5\,(\mu 1 - \mu m)\,(\lambda m + \mu m) + r^5\,(2\,\mu m\,(8\,\mu 1 + 7\,\mu m) + \lambda m\,(6\,\mu 1 + 9\,\mu m)))\,X[i,\,j] + 15\,a^3\,(\mu 1 - \mu m)\,(x[j]\,(r^2\,(-r^2\,\lambda m + 2\,a^2\,(\lambda m + \mu m))\,x[m]\,X[i,\,m] - (7\,a^2 - 5\,r^2)\,(\lambda m + \mu m)\,x[i]\,x[p]\,x[q]\,X[p,\,q]) + r^2\,((-r^2\,\lambda m + 2\,a^2\,(\lambda m + \mu m))\,x[i]\,x[m]\,X[j,\,m] + (-r^2\,\mu m + a^2\,(\lambda m + \mu m))\,x[p]\,x[q]\,X[p,\,q]\,\delta[i,\,j])))\bigr)/\bigl(r^9\,(2\,\mu m\,(8\,\mu 1 + 7\,\mu m) + \lambda m\,(6\,\mu 1 + 9\,\mu m))\bigr)$

In[241]:= **tiout = TractOut /. sol2 // Simplify**

Out[241]= $\bigl(2\,\mu m\,(r^2\,(24\,a^5\,(\mu 1 - \mu m)\,(\lambda m + \mu m) - 5\,a^3\,r^2\,(\mu 1 - \mu m)\,(3\,\lambda m + 2\,\mu m) + r^5\,(2\,\mu m\,(8\,\mu 1 + 7\,\mu m) + \lambda m\,(6\,\mu 1 + 9\,\mu m)))\,x[m]\,X[i,\,m] - 60\,a^3\,(a^2 - r^2)\,(\mu 1 - \mu m)\,(\lambda m + \mu m)\,x[i]\,x[p]\,x[q]\,X[p,\,q])\bigr)/\bigl(r^8\,(2\,\mu m\,(8\,\mu 1 + 7\,\mu m) + \lambda m\,(6\,\mu 1 + 9\,\mu m))\bigr)$

The aforementioned results naturally coincide with the result from Eshelby's approach.

3.2.5 Exact Solution for Three-Phase Materials

A medium that contains a spherical inclusion coated by a layer (three-phase material) is of practical importance as it finds many applications in industrial materials (Figure 3.11). Obviously, the analysis for such a material is much more involved than the two-phase material discussed in the previous subsection. Unlike the single inclusion problem employed by the Eshelby method, the stress inside the inclusion is no longer uniform.

Christensen and Lo (1979) derived the equations to be solved for the stress field in a three-phase model using the spherical coordinate system. They did not show the explicit expression of the stress field though. By using the approach shown, it is possible to obtain the stress field exactly.

Consider a medium that has a spherical inclusion at the center with a radius of a_1 coated by another concentric inclusion (layer) with a radius of a_2 embedded in an infinitely extended matrix. The material properties for the inclusion, coating, and matrix are denoted as (λ_1, μ_1), (λ_2, μ_2), and (λ_m, μ_m), respectively.

3.2.5.1 Exact Solution for X

The elastic field of the three-phase material subject to X at the far field can be determined so that the displacement and the traction are continuous at $r = a_1$ and $r = a_2$.

The expressions for the displacement and the traction force have been derived as

$$u_i = \left(-\frac{c_1}{3r^3} + c_2\right) x_i X,$$

$$t_i = \frac{4c_1 \mu x_i}{3r^4} X + \frac{c_2 x_i (3\lambda + 2\mu)}{r} X.$$

Let the displacement and the traction force for each phase be denoted as follows:

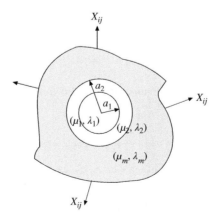

Figure 3.11 Three-phase material

First inclusion (core):

$$u_i^1 = \left(-\frac{c_1^1}{3r^3} + c_2^1\right) x_i X,$$

$$t_i^1 = \frac{4c_1^1 \mu^1 x_i}{3r^4} X + \frac{c_2^1 x_i (3\lambda^1 + 2\mu^1)}{r} X.$$

Second inclusion (layer):

$$u_i^2 = \left(-\frac{c_1^2}{3r^3} + c_2^2\right) x_i X,$$

$$t_i^2 = \frac{4c_1^2 \mu^2 x_i}{3r^4} X + \frac{c_2^2 x_i (3\lambda^2 + 2\mu^2)}{r} X.$$

Matrix:

$$u_i^m = \left(-\frac{c_1^m}{3r^3} + c_2^m\right) x_i X,$$

$$t_i^m = \frac{4c_1^m \mu^m x_i}{3r^4} X + \frac{c_2^m x_i (3\lambda^m + 2\mu^m)}{r} X,$$

where the superscript "1" refers to the innermost inclusion, "2" refers to the coating material, and "m" refer to the matrix phase. From the condition that the displacement must remain finite, c_1^1 must be 0. From the condition that as $r \to \infty$, $u_i^{out} \to x_i X$, c_2^m must be 1. Therefore, the unknown coefficients can be determined from the following continuity conditions at $r = a_1$ and $r = a_2$:

$$u_i^1 = u_i^2 \quad r = a_1,$$
$$u_i^2 = u_i^m \quad r = a_2,$$
$$t_i^1 = t_i^2 \quad r = a_1,$$
$$t_i^2 = t_i^m \quad r = a_2.$$

A *Mathematica* code to solve for the unknowns is shown as follows:

```
In[44]:= Displ = Ui /. C → c1
```

$$\text{Out[44]= } X\left(-\frac{c1[1]}{3\,r^3} + c1[2]\right) x[i]$$

```
In[45]:= Disp2 = Ui /. C → c2
```

$$\text{Out[45]= } X\left(-\frac{c2[1]}{3\,r^3} + c2[2]\right) x[i]$$

```
In[46]:= Dispm = Ui /. C → cm
```

$$\text{Out[46]= } X\left(-\frac{cm[1]}{3\,r^3} + cm[2]\right) x[i]$$

In[47]:= **Tract1 = Ti /. {μ → μ1, λ → λ1, C → c1}**

Out[47]= $\dfrac{4 X \mu 1\, c1[1]\, x[i]}{3\, r^4} + \dfrac{X\,(3\,\lambda 1 + 2\,\mu 1)\, c1[2]\, x[i]}{r}$

In[48]:= **Tract2 = Ti /. {μ → μ2, λ → λ2, C → c2}**

Out[48]= $\dfrac{4 X \mu 2\, c2[1]\, x[i]}{3\, r^4} + \dfrac{X\,(3\,\lambda 2 + 2\,\mu 2)\, c2[2]\, x[i]}{r}$

In[49]:= **Tractm = Ti /. {μ → μm, λ → λm, C → cm}**

Out[49]= $\dfrac{4 X \mu m\, cm[1]\, x[i]}{3\, r^4} + \dfrac{X\,(3\,\lambda m + 2\,\mu m)\, cm[2]\, x[i]}{r}$

The unknown coefficients, C_2^1, C_1^2, C_2^2, and C_1^3, are solved as

In[50]:= **c1[1] = 0; cm[2] = 1;**

In[51]:= **eq1 = (Disp1 - Disp2) /. r → a1**

Out[51]= $X\, c1[2]\, x[i] - X\left(-\dfrac{c2[1]}{3\, a1^3} + c2[2]\right) x[i]$

In[52]:= **eq2 = (Disp2 - Dispm) /. r → a2**

Out[52]= $X\left(-\dfrac{c2[1]}{3\, a2^3} + c2[2]\right) x[i] - X\left(1 - \dfrac{cm[1]}{3\, a2^3}\right) x[i]$

In[53]:= **eq3 = (Tract1 - Tract2) /. r → a1**

Out[53]= $\dfrac{X\,(3\,\lambda 1 + 2\,\mu 1)\, c1[2]\, x[i]}{a1} - \dfrac{4 X \mu 2\, c2[1]\, x[i]}{3\, a1^4} - \dfrac{X\,(3\,\lambda 2 + 2\,\mu 2)\, c2[2]\, x[i]}{a1}$

In[54]:= **eq4 = (Tract2 - Tractm) /. r → a2**

Out[54]= $-\dfrac{X\,(3\,\lambda m + 2\,\mu m)\, x[i]}{a2} + \dfrac{4 X \mu 2\, c2[1]\, x[i]}{3\, a2^4} + \dfrac{X\,(3\,\lambda 2 + 2\,\mu 2)\, c2[2]\, x[i]}{a2} - \dfrac{4 X \mu m\, cm[1]\, x[i]}{3\, a2^4}$

In[55]:= **sol = Solve[{eq1 == 0, eq2 == 0, eq3 == 0, eq4 == 0}, {c1[2], c2[1], c2[2], cm[1]}][[1]]**

Out[55]= {c1[2] →
$-\bigl(9\, a2^3\,(\lambda 2 + 2\,\mu 2)\,(\lambda m + 2\,\mu m)\bigr) / \bigl(-9\, a2^3\, \lambda 1\, \lambda 2 - 6\, a2^3\, \lambda 2\, \mu 1 - 12\, a1^3\, \lambda 1\, \mu 2 - 6\, a2^3\, \lambda 1\, \mu 2 + 12\, a1^3\, \lambda 2\, \mu 2 - 12\, a2^3\, \lambda 2\, \mu 2 - 8\, a1^3\, \mu 1\, \mu 2 - 4\, a2^3\, \mu 1\, \mu 2 + 8\, a1^3\, \mu 2^2 - 8\, a2^3\, \mu 2^2 + 12\, a1^3\, \lambda 1\, \mu m - 12\, a2^3\, \lambda 1\, \mu m - 12\, a1^3\, \lambda 2\, \mu m + 8\, a1^3\, \mu 1\, \mu m - 8\, a2^3\, \mu 1\, \mu m - 8\, a1^3\, \mu 2\, \mu m - 16\, a2^3\, \mu 2\, \mu m\bigr)$,

c2[1] → $-\bigl(9\, a1^3\, a2^3\,(-3\,\lambda 1 + 3\,\lambda 2 - 2\,\mu 1 + 2\,\mu 2)\,(\lambda m + 2\,\mu m)\bigr) / \bigl(9\, a2^3\, \lambda 1\, \lambda 2 + 6\, a2^3\, \lambda 2\, \mu 1 + 12\, a1^3\, \lambda 1\, \mu 2 + 6\, a2^3\, \lambda 1\, \mu 2 - 12\, a1^3\, \lambda 2\, \mu 2 + 12\, a2^3\, \lambda 2\, \mu 2 + 8\, a1^3\, \mu 1\, \mu 2 + 4\, a2^3\, \mu 1\, \mu 2 - 8\, a1^3\, \mu 2^2 + 8\, a2^3\, \mu 2^2 - 12\, a1^3\, \lambda 1\, \mu m + 12\, a2^3\, \lambda 1\, \mu m + 12\, a1^3\, \lambda 2\, \mu m - 8\, a1^3\, \mu 1\, \mu m + 8\, a2^3\, \mu 1\, \mu m + 8\, a1^3\, \mu 2\, \mu m + 16\, a2^3\, \mu 2\, \mu m\bigr)$,

c2[2] → $\bigl(3\, a2^3\,(3\,\lambda 1 + 2\,\mu 1 + 4\,\mu 2)\,(\lambda m + 2\,\mu m)\bigr) / \bigl(9\, a2^3\, \lambda 1\, \lambda 2 + 6\, a2^3\, \lambda 2\, \mu 1 + 12\, a1^3\, \lambda 1\, \mu 2 + 6\, a2^3\, \lambda 1\, \mu 2 - 12\, a1^3\, \lambda 2\, \mu 2 + 12\, a2^3\, \lambda 2\, \mu 2 + 8\, a1^3\, \mu 1\, \mu 2 + 4\, a2^3\, \mu 1\, \mu 2 - 8\, a1^3\, \mu 2^2 + 8\, a2^3\, \mu 2^2 - 12\, a1^3\, \lambda 1\, \mu m + 12\, a2^3\, \lambda 1\, \mu m + 12\, a1^3\, \lambda 2\, \mu m - 8\, a1^3\, \mu 1\, \mu m + 8\, a2^3\, \mu 1\, \mu m + 8\, a1^3\, \mu 2\, \mu m + 16\, a2^3\, \mu 2\, \mu m\bigr)$, cm[1] →
$3\, a2^3 + \bigl(9\, a2^3\,(-3\, a1^3\, \lambda 1 + 3\, a2^3\, \lambda 1 + 3\, a1^3\, \lambda 2 - 2\, a1^3\, \mu 1 + 2\, a2^3\, \mu 1 + 2\, a1^3\, \mu 2 + 4\, a2^3\, \mu 2)\,(\lambda m + 2\,\mu m)\bigr) / \bigl(-9\, a2^3\, \lambda 1\, \lambda 2 - 6\, a2^3\, \lambda 2\, \mu 1 - 12\, a1^3\, \lambda 1\, \mu 2 - 6\, a2^3\, \lambda 1\, \mu 2 + 12\, a1^3\, \lambda 2\, \mu 2 - 12\, a2^3\, \lambda 2\, \mu 2 - 8\, a1^3\, \mu 1\, \mu 2 - 4\, a2^3\, \mu 1\, \mu 2 + 8\, a1^3\, \mu 2^2 - 8\, a2^3\, \mu 2^2 + 12\, a1^3\, \lambda 1\, \mu m - 12\, a2^3\, \lambda 1\, \mu m - 12\, a1^3\, \lambda 2\, \mu m + 8\, a1^3\, \mu 1\, \mu m - 8\, a2^3\, \mu 1\, \mu m - 8\, a1^3\, \mu 2\, \mu m - 16\, a2^3\, \mu 2\, \mu m\bigr)$}

The displacement for each phase can be obtained as

In[56]:= **ui1 = Displ /. sol // Simplify**

Out[56]= $\left(9\, a2^3\, X\, (\lambda 2 + 2\, \mu 2)\, (\lambda m + 2\, \mu m)\, x[i]\right) /$
$\left(4\, a1^3\, (3\, \lambda 1 - 3\, \lambda 2 + 2\, \mu 1 - 2\, \mu 2)\, (\mu 2 - \mu m) + a2^3\, (3\, \lambda 1 + 2\, \mu 1 + 4\, \mu 2)\, (3\, \lambda 2 + 2\, \mu 2 + 4\, \mu m)\right)$

In[57]:= **ui2 = Disp2 /. sol // Simplify**

Out[57]= $\left(3\, a2^3\, X\, \left(a1^3\, (-3\, \lambda 1 + 3\, \lambda 2 - 2\, \mu 1 + 2\, \mu 2) + r^3\, (3\, \lambda 1 + 2\, \mu 1 + 4\, \mu 2)\right)\, (\lambda m + 2\, \mu m)\, x[i]\right) /$
$\left(r^3\, \left(4\, a1^3\, (3\, \lambda 1 - 3\, \lambda 2 + 2\, \mu 1 - 2\, \mu 2)\, (\mu 2 - \mu m) + a2^3\, (3\, \lambda 1 + 2\, \mu 1 + 4\, \mu 2)\, (3\, \lambda 2 + 2\, \mu 2 + 4\, \mu m)\right)\right)$

In[58]:= **ui3 = Dispm /. sol // Simplify**

Out[58]= $X\, \Big(1 - \big(a2^3\, \big(a2^3\, (3\, \lambda 1 + 2\, \mu 1 + 4\, \mu 2)\, (3\, \lambda 2 - 3\, \lambda m + 2\, \mu 2 - 2\, \mu m) +$
$a1^3\, (3\, \lambda 1 - 3\, \lambda 2 + 2\, \mu 1 - 2\, \mu 2)\, (3\, \lambda m + 4\, \mu 2 + 2\, \mu m)\big)\big) /$
$\left(r^3\, \left(4\, a1^3\, (3\, \lambda 1 - 3\, \lambda 2 + 2\, \mu 1 - 2\, \mu 2)\, (\mu 2 - \mu m) + a2^3\, (3\, \lambda 1 + 2\, \mu 1 + 4\, \mu 2)\, (3\, \lambda 2 + 2\, \mu 2 + 4\, \mu m)\right)\right)\Big)\, x[i]$

The strain for each phase can be obtained as

In[59]:= **eij1 = Eij /. C → c1 /. sol**

Out[59]= $-\left(9\, a2^3\, X\, (\lambda 2 + 2\, \mu 2)\, (\lambda m + 2\, \mu m)\, \delta[i, j]\right) /$
$\big(-9\, a2^3\, \lambda 1\, \lambda 2 - 6\, a2^3\, \lambda 2\, \mu 1 - 12\, a1^3\, \lambda 1\, \mu 2 - 6\, a2^3\, \lambda 1\, \mu 2 + 12\, a1^3\, \lambda 2\, \mu 2 -$
$12\, a2^3\, \lambda 2\, \mu 2 - 8\, a1^3\, \mu 1\, \mu 2 - 4\, a2^3\, \mu 1\, \mu 2 + 8\, a1^3\, \mu 2^2 - 8\, a2^3\, \mu 2^2 + 12\, a1^3\, \lambda 1\, \mu m -$
$12\, a2^3\, \lambda 1\, \mu m - 12\, a1^3\, \lambda 2\, \mu m + 8\, a1^3\, \mu 1\, \mu m - 8\, a2^3\, \mu 1\, \mu m - 8\, a1^3\, \mu 2\, \mu m - 16\, a2^3\, \mu 2\, \mu m\big)$

In[60]:= **eij2 = Eij /. C → c2 /. sol**

Out[60]= $\left(3\, a2^3\, X\, (3\, \lambda 1 + 2\, \mu 1 + 4\, \mu 2)\, (\lambda m + 2\, \mu m)\, \delta[i, j]\right) /$
$\big(9\, a2^3\, \lambda 1\, \lambda 2 + 6\, a2^3\, \lambda 2\, \mu 1 + 12\, a1^3\, \lambda 1\, \mu 2 + 6\, a2^3\, \lambda 1\, \mu 2 - 12\, a1^3\, \lambda 2\, \mu 2 +$
$12\, a2^3\, \lambda 2\, \mu 2 + 8\, a1^3\, \mu 1\, \mu 2 + 4\, a2^3\, \mu 1\, \mu 2 - 8\, a1^3\, \mu 2^2 + 8\, a2^3\, \mu 2^2 - 12\, a1^3\, \lambda 1\, \mu m +$
$12\, a2^3\, \lambda 1\, \mu m + 12\, a1^3\, \lambda 2\, \mu m - 8\, a1^3\, \mu 1\, \mu m + 8\, a2^3\, \mu 1\, \mu m + 8\, a1^3\, \mu 2\, \mu m + 16\, a2^3\, \mu 2\, \mu m\big) -$
$\left(9\, a1^3\, a2^3\, (-3\, \lambda 1 + 3\, \lambda 2 - 2\, \mu 1 + 2\, \mu 2)\, (\lambda m + 2\, \mu m)\, \left(\frac{2\, X\, x[i]\, x[j]}{r^5} - \frac{2\, X\, \delta[i, j]}{3\, r^3}\right)\right) /$
$\big(2\, \big(9\, a2^3\, \lambda 1\, \lambda 2 + 6\, a2^3\, \lambda 2\, \mu 1 + 12\, a1^3\, \lambda 1\, \mu 2 + 6\, a2^3\, \lambda 1\, \mu 2 - 12\, a1^3\, \lambda 2\, \mu 2 +$
$12\, a2^3\, \lambda 2\, \mu 2 + 8\, a1^3\, \mu 1\, \mu 2 + 4\, a2^3\, \mu 1\, \mu 2 - 8\, a1^3\, \mu 2^2 + 8\, a2^3\, \mu 2^2 - 12\, a1^3\, \lambda 1\, \mu m +$
$12\, a2^3\, \lambda 1\, \mu m + 12\, a1^3\, \lambda 2\, \mu m - 8\, a1^3\, \mu 1\, \mu m + 8\, a2^3\, \mu 1\, \mu m + 8\, a1^3\, \mu 2\, \mu m + 16\, a2^3\, \mu 2\, \mu m\big)\big)$

In[61]:= **eijm = Eij /. C → cm /. sol**

Out[61]= $X\, \delta[i, j] +$
$\frac{1}{2}\, \big(3\, a2^3 + \big(9\, a2^3\, \big(-3\, a1^3\, \lambda 1 + 3\, a2^3\, \lambda 1 + 3\, a1^3\, \lambda 2 - 2\, a1^3\, \mu 1 + 2\, a2^3\, \mu 1 + 2\, a1^3\, \mu 2 + 4\, a2^3\, \mu 2\big)\, (\lambda m + 2\, \mu m)\big) /$
$\big(-9\, a2^3\, \lambda 1\, \lambda 2 - 6\, a2^3\, \lambda 2\, \mu 1 - 12\, a1^3\, \lambda 1\, \mu 2 - 6\, a2^3\, \lambda 1\, \mu 2 + 12\, a1^3\, \lambda 2\, \mu 2 - 12\, a2^3\, \lambda 2\, \mu 2 -$
$8\, a1^3\, \mu 1\, \mu 2 - 4\, a2^3\, \mu 1\, \mu 2 + 8\, a1^3\, \mu 2^2 - 8\, a2^3\, \mu 2^2 + 12\, a1^3\, \lambda 1\, \mu m - 12\, a2^3\, \lambda 1\, \mu m - 12\, a1^3\, \lambda 2\, \mu m +$
$8\, a1^3\, \mu 1\, \mu m - 8\, a2^3\, \mu 1\, \mu m - 8\, a1^3\, \mu 2\, \mu m - 16\, a2^3\, \mu 2\, \mu m\big)\big)\, \left(\frac{2\, X\, x[i]\, x[j]}{r^5} - \frac{2\, X\, \delta[i, j]}{3\, r^3}\right)$

The stress for each phase can be obtained as

In[62]:= **σij1 = Sij /. {C → c1, μ → μ1, λ → λ1} /. sol // Simplify**

Out[62]= $\left(9\, a2^3\, X\, (3\, \lambda 1 + 2\, \mu 1)\, (\lambda 2 + 2\, \mu 2)\, (\lambda m + 2\, \mu m)\, \delta[i, j]\right) /$
$\left(4\, a1^3\, (3\, \lambda 1 - 3\, \lambda 2 + 2\, \mu 1 - 2\, \mu 2)\, (\mu 2 - \mu m) + a2^3\, (3\, \lambda 1 + 2\, \mu 1 + 4\, \mu 2)\, (3\, \lambda 2 + 2\, \mu 2 + 4\, \mu m)\right)$

In[63]:= **σij2 = Sij /. {C → c2, μ → μ2, λ → λ2} /. sol // Simplify**

Out[63]= $\left(3\, a2^3\, X\, (\lambda m + 2\, \mu m)\, \big(6\, a1^3\, (3\, \lambda 1 - 3\, \lambda 2 + 2\, \mu 1 - 2\, \mu 2)\, \mu 2\, x[i]\, x[j] +$
$r^2\, \big(2\, a1^3\, \mu 2\, (-3\, \lambda 1 + 3\, \lambda 2 - 2\, \mu 1 + 2\, \mu 2) + r^3\, (3\, \lambda 2 + 2\, \mu 2)\, (3\, \lambda 1 + 2\, \mu 1 + 4\, \mu 2)\big)\, \delta[i, j]\big)\big) /$
$\left(r^5\, \left(4\, a1^3\, (3\, \lambda 1 - 3\, \lambda 2 + 2\, \mu 1 - 2\, \mu 2)\, (\mu 2 - \mu m) + a2^3\, (3\, \lambda 1 + 2\, \mu 1 + 4\, \mu 2)\, (3\, \lambda 2 + 2\, \mu 2 + 4\, \mu m)\right)\right)$

In[64]:= **σijm = Sij /. {C → cm, μ → μm, λ → λm} /. sol // Simplify**

Out[64]= X (3 λm + 2 μm) δ[i, j] +
  (2 a2³ X μm (-a2³ (3 λ1 + 2 μ1 + 4 μ2) (3 λ2 - 3 λm + 2 μ2 - 2 μm) - a1³ (3 λ1 - 3 λ2 + 2 μ1 - 2 μ2)
    (3 λm + 4 μ2 + 2 μm)) (-3 x[i] x[j] + r² δ[i, j]))/
  (r⁵ (4 a1³ (3 λ1 - 3 λ2 + 2 μ1 - 2 μ2) (μ2 - μm) + a2³ (3 λ1 + 2 μ1 + 4 μ2) (3 λ2 + 2 μ2 + 4 μm)))

Finally, the traction field for each phase can be obtained as

In[65]:= ti1 = Tract1 /. sol

Out[65]= -(9 a2³ X (3 λ1 + 2 μ1) (λ2 + 2 μ2) (λm + 2 μm) x[i])/
  (r (-9 a2³ λ1 λ2 - 6 a2³ λ2 μ1 - 12 a1³ λ1 μ2 - 6 a2³ λ1 μ2 + 12 a1³ λ2 μ2 -
    12 a2³ λ2 μ2 - 8 a1³ μ1 μ2 - 4 a2³ μ1 μ2 + 8 a1³ μ2² - 8 a2³ μ2² + 12 a1³ λ1 μm -
    12 a2³ λ1 μm - 12 a1³ λ2 μm + 8 a1³ μ1 μm - 8 a2³ μ1 μm - 8 a1³ μ2 μm - 16 a2³ μ2 μm))

In[66]:= ti2 = Tract2 /. sol

Out[66]= -(12 a1³ a2³ X μ2 (-3 λ1 + 3 λ2 - 2 μ1 + 2 μ2) (λm + 2 μm) x[i])/
  (r⁴ (9 a2³ λ1 λ2 + 6 a2³ λ2 μ1 + 12 a1³ λ1 μ2 + 6 a2³ λ1 μ2 - 12 a1³ λ2 μ2 +
    12 a2³ λ2 μ2 + 8 a1³ μ1 μ2 + 4 a2³ μ1 μ2 - 8 a1³ μ2² + 8 a2³ μ2² - 12 a1³ λ1 μm +
    12 a2³ λ1 μm + 12 a1³ λ2 μm - 8 a1³ μ1 μm + 8 a2³ μ1 μm + 8 a1³ μ2 μm + 16 a2³ μ2 μm)) +
  (3 a2³ X (3 λ2 + 2 μ2) (3 λ1 + 2 μ1 + 4 μ2) (λm + 2 μm) x[i])/
  (r (9 a2³ λ1 λ2 + 6 a2³ λ2 μ1 + 12 a1³ λ1 μ2 + 6 a2³ λ1 μ2 - 12 a1³ λ2 μ2 +
    12 a2³ λ2 μ2 + 8 a1³ μ1 μ2 + 4 a2³ μ1 μ2 - 8 a1³ μ2² + 8 a2³ μ2² - 12 a1³ λ1 μm +
    12 a2³ λ1 μm + 12 a1³ λ2 μm - 8 a1³ μ1 μm + 8 a2³ μ1 μm + 8 a1³ μ2 μm + 16 a2³ μ2 μm))

In[67]:= ti3 = Tractm /. sol

Out[67]= $\frac{X (3 λm + 2 μm) x[i]}{r}$ + $\frac{1}{3 r^4}$ 4 X μm
  (3 a2³ + (9 a2³ (-3 a1³ λ1 + 3 a2³ λ1 + 3 a1³ λ2 - 2 a1³ μ1 + 2 a2³ μ1 + 2 a1³ μ2 + 4 a2³ μ2) (λm + 2 μm))/
  (-9 a2³ λ1 λ2 - 6 a2³ λ2 μ1 - 12 a1³ λ1 μ2 - 6 a2³ λ1 μ2 + 12 a1³ λ2 μ2 - 12 a2³ λ2 μ2 -
    8 a1³ μ1 μ2 - 4 a2³ μ1 μ2 + 8 a1³ μ2² - 8 a2³ μ2² + 12 a1³ λ1 μm - 12 a2³ λ1 μm -
    12 a1³ λ2 μm + 8 a1³ μ1 μm - 8 a2³ μ1 μm - 8 a1³ μ2 μm - 16 a2³ μ2 μm)) x[i]

It should be noted that the stress field inside the innermost inclusion is constant.

3.2.5.2 Exact Solution for X'_{ij}

The elastic fields when the three-phase material is subject to X'_{ij} at the far-field can be obtained with a similar method described. The resulting expressions are obviously lengthy as more symbolic parameters are involved. Nevertheless, the obtained expressions are analytic, and they are new results not reported previously.

By using the same notation as in the previous subsection, the general expressions for the displacement and the traction have been derived as

In[44]:= Ui

Out[44]= $\left\{-\frac{2 x[m] X[i, m]}{5 r^5} + \frac{x[i] x[p] x[q] X[p, q]}{r^7}, \frac{2 μ x[m] X[i, m]}{3 r^3 (λ + μ)} + \frac{x[i] x[p] x[q] X[p, q]}{r^5},\right.$
$\left.-\frac{5 r^2 λ x[m] X[i, m]}{2 λ + 7 μ} - \frac{7 r^2 μ x[m] X[i, m]}{2 λ + 7 μ} + x[i] x[p] x[q] X[p, q], x[m] X[i, m]\right\}$

In[45]:= Ti

Out[45]= $\left\{\frac{16 μ x[m] X[i, m]}{5 r^6} - \frac{8 μ x[i] x[p] x[q] X[p, q]}{r^8},\right.$
$\left.\frac{2 λ μ x[m] X[i, m]}{r^4 (λ + μ)} + \frac{4 μ² x[m] X[i, m]}{3 r^4 (λ + μ)} - \frac{8 λ μ x[i] x[p] x[q] X[p, q]}{r^6 (λ + μ)} - \frac{8 μ² x[i] x[p] x[q] X[p, q]}{r^6 (λ + μ)}\right.$,

$$-\frac{16 r \lambda \mu x[m] X[i, m]}{2\lambda+7\mu} - \frac{14 r \mu^2 x[m] X[i, m]}{2\lambda+7\mu} +$$

$$\frac{19 \lambda \mu x[i] x[p] x[q] X[p, q]}{r(2\lambda+7\mu)} + \frac{14 \mu^2 x[i] x[p] x[q] X[p, q]}{r(2\lambda+7\mu)}, \frac{2\mu x[m] X[i, m]}{r}\}$$

By using these expressions, the displacement field for each of the three phases is assumed with unknown coefficients as

In[47]:= `Disp1 = (Ui.Table[ c1[i], {i, 4}]) /. {λ → λ1, μ → μ1}`

Out[47]= $c1[4] \, x[m] \, X[i, m] +$

$c1[3] \left(-\dfrac{5 r^2 \lambda 1 \, x[m] \, X[i, m]}{2 \lambda 1 + 7 \mu 1} - \dfrac{7 r^2 \mu 1 \, x[m] \, X[i, m]}{2 \lambda 1 + 7 \mu 1} + x[i] \, x[p] \, x[q] \, X[p, q] \right) +$

$c1[1] \left(-\dfrac{2 \, x[m] \, X[i, m]}{5 r^5} + \dfrac{x[i] \, x[p] \, x[q] \, X[p, q]}{r^7} \right) +$

$c1[2] \left(\dfrac{2 \mu 1 \, x[m] \, X[i, m]}{3 r^3 (\lambda 1 + \mu 1)} + \dfrac{x[i] \, x[p] \, x[q] \, X[p, q]}{r^5} \right)$

In[48]:= `Disp2 = (Ui.Table[ c2[i], {i, 4}]) /. {λ → λ2, μ → μ2}`

Out[48]= $c2[4] \, x[m] \, X[i, m] +$

$c2[3] \left(-\dfrac{5 r^2 \lambda 2 \, x[m] \, X[i, m]}{2 \lambda 2 + 7 \mu 2} - \dfrac{7 r^2 \mu 2 \, x[m] \, X[i, m]}{2 \lambda 2 + 7 \mu 2} + x[i] \, x[p] \, x[q] \, X[p, q] \right) +$

$c2[1] \left(-\dfrac{2 \, x[m] \, X[i, m]}{5 r^5} + \dfrac{x[i] \, x[p] \, x[q] \, X[p, q]}{r^7} \right) +$

$c2[2] \left(\dfrac{2 \mu 2 \, x[m] \, X[i, m]}{3 r^3 (\lambda 2 + \mu 2)} + \dfrac{x[i] \, x[p] \, x[q] \, X[p, q]}{r^5} \right)$

In[49]:= `Dispm = (Ui.Table[ cm[i], {i, 4}]) /. {λ → λm, μ → μm}`

Out[49]= $cm[4] \, x[m] \, X[i, m] +$

$cm[3] \left(-\dfrac{5 r^2 \lambda m \, x[m] \, X[i, m]}{2 \lambda m + 7 \mu m} - \dfrac{7 r^2 \mu m \, x[m] \, X[i, m]}{2 \lambda m + 7 \mu m} + x[i] \, x[p] \, x[q] \, X[p, q] \right) +$

$cm[1] \left(-\dfrac{2 \, x[m] \, X[i, m]}{5 r^5} + \dfrac{x[i] \, x[p] \, x[q] \, X[p, q]}{r^7} \right) +$

$cm[2] \left(\dfrac{2 \mu m \, x[m] \, X[i, m]}{3 r^3 (\lambda m + \mu m)} + \dfrac{x[i] \, x[p] \, x[q] \, X[p, q]}{r^5} \right)$

where `c1[i]`, `c2[i]`, and `cm[i]` are unknown coefficients for the inclusion, coating material, and matrix, respectively. The material properties, ($\mu 1$, $\lambda 1$), ($\mu 2$, $\lambda 2$), and (μm, λm), are the Lamé constants for the inclusion, coating material, and matrix, respectively.

Similarly, the traction field for each of the three phases is assumed with unknown coefficients as

In[50]:= `Tract1 = (Ti.Table[c1[i], {i, 4}]) /. {λ → λ1, μ → μ1}`

Out[50]= $\dfrac{2 \mu 1 \, c1[4] \, x[m] \, X[i, m]}{r} + c1[1] \left(\dfrac{16 \mu 1 \, x[m] \, X[i, m]}{5 r^6} - \dfrac{8 \mu 1 \, x[i] \, x[p] \, x[q] \, X[p, q]}{r^8} \right) +$

$c1[2] \left(\dfrac{2 \lambda 1 \mu 1 \, x[m] \, X[i, m]}{r^4 (\lambda 1 + \mu 1)} + \dfrac{4 \mu 1^2 \, x[m] \, X[i, m]}{3 r^4 (\lambda 1 + \mu 1)} - \dfrac{8 \lambda 1 \mu 1 \, x[i] \, x[p] \, x[q] \, X[p, q]}{r^6 (\lambda 1 + \mu 1)} - \dfrac{8 \mu 1^2 \, x[i] \, x[p] \, x[q] \, X[p, q]}{r^6 (\lambda 1 + \mu 1)} \right) +$

$$c1[3]\left(-\frac{16\,r\,\lambda1\,\mu1\,x[m]\,X[i,m]}{2\,\lambda1+7\,\mu1}-\frac{14\,r\,\mu1^2\,x[m]\,X[i,m]}{2\,\lambda1+7\,\mu1}+\right.$$
$$\left.\frac{19\,\lambda1\,\mu1\,x[i]\,x[p]\,x[q]\,X[p,q]}{r\,(2\,\lambda1+7\,\mu1)}+\frac{14\,\mu1^2\,x[i]\,x[p]\,x[q]\,X[p,q]}{r\,(2\,\lambda1+7\,\mu1)}\right)$$

In[51]:= **Tract2 = (Ti.Table[c2[i], {i, 4}]) /. {λ → λ2, μ → μ2}**

Out[51]= $\dfrac{2\,\mu2\,c2[4]\,x[m]\,X[i,m]}{r} + c2[1]\left(\dfrac{16\,\mu2\,x[m]\,X[i,m]}{5\,r^6} - \dfrac{8\,\mu2\,x[i]\,x[p]\,x[q]\,X[p,q]}{r^8}\right) +$

$$c2[2]\left(\frac{2\,\lambda2\,\mu2\,x[m]\,X[i,m]}{r^4\,(\lambda2+\mu2)}+\frac{4\,\mu2^2\,x[m]\,X[i,m]}{3\,r^4\,(\lambda2+\mu2)}-\right.$$
$$\left.\frac{8\,\lambda2\,\mu2\,x[i]\,x[p]\,x[q]\,X[p,q]}{r^6\,(\lambda2+\mu2)}-\frac{8\,\mu2^2\,x[i]\,x[p]\,x[q]\,X[p,q]}{r^6\,(\lambda2+\mu2)}\right)+$$

$$c2[3]\left(-\frac{16\,r\,\lambda2\,\mu2\,x[m]\,X[i,m]}{2\,\lambda2+7\,\mu2}-\frac{14\,r\,\mu2^2\,x[m]\,X[i,m]}{2\,\lambda2+7\,\mu2}+\right.$$
$$\left.\frac{19\,\lambda2\,\mu2\,x[i]\,x[p]\,x[q]\,X[p,q]}{r\,(2\,\lambda2+7\,\mu2)}+\frac{14\,\mu2^2\,x[i]\,x[p]\,x[q]\,X[p,q]}{r\,(2\,\lambda2+7\,\mu2)}\right)$$

In[62]:= **Tractm = (Ti.Table[cm[i], {i, 4}]) /. {λ → λm, μ → μm}**

Out[52]= $\dfrac{2\,\mu m\,cm[4]\,x[m]\,X[i,m]}{r} + cm[1]\left(\dfrac{16\,\mu m\,x[m]\,X[i,m]}{5\,r^6} - \dfrac{8\,\mu m\,x[i]\,x[p]\,x[q]\,X[p,q]}{r^8}\right) +$

$$cm[2]\left(\frac{2\,\lambda m\,\mu m\,x[m]\,X[i,m]}{r^4\,(\lambda m+\mu m)}+\frac{4\,\mu m^2\,x[m]\,X[i,m]}{3\,r^4\,(\lambda m+\mu m)}-\right.$$
$$\left.\frac{8\,\lambda m\,\mu m\,x[i]\,x[p]\,x[q]\,X[p,q]}{r^6\,(\lambda m+\mu m)}-\frac{8\,\mu m^2\,x[i]\,x[p]\,x[q]\,X[p,q]}{r^6\,(\lambda m+\mu m)}\right)+$$

$$cm[3]\left(-\frac{16\,r\,\lambda m\,\mu m\,x[m]\,X[i,m]}{2\,\lambda m+7\,\mu m}-\frac{14\,r\,\mu m^2\,x[m]\,X[i,m]}{2\,\lambda m+7\,\mu m}+\right.$$
$$\left.\frac{19\,\lambda m\,\mu m\,x[i]\,x[p]\,x[q]\,X[p,q]}{r\,(2\,\lambda m+7\,\mu m)}+\frac{14\,\mu m^2\,x[i]\,x[p]\,x[q]\,X[p,q]}{r\,(2\,\lambda m+7\,\mu m)}\right)$$

From the condition that the displacement must remain finite inside the inclusion, it follows that $c1[1] = c1[2] = 0$. Also, from the condition that the traction does not diverge as $r \to \infty$, it follows that $cm[4] = 1$ and $cm[3] = 0$. The continuity condition that $u_i^1 = u_i^2$ at $r = a_1$ can be entered as

In[52]:= **c1[1] = 0; c1[2] = 0; cm[4] = 1; cm[3] = 0;**

In[53]:= **tmp1 = ((Disp1 - Disp2) /. r → a1)**

Out[53]= $c1[4]\,x[m]\,X[i,m] - c2[4]\,x[m]\,X[i,m] +$

$$c1[3]\left(-\frac{5\,a1^2\,\lambda1\,x[m]\,X[i,m]}{2\,\lambda1+7\,\mu1}-\frac{7\,a1^2\,\mu1\,x[m]\,X[i,m]}{2\,\lambda1+7\,\mu1}+x[i]\,x[p]\,x[q]\,X[p,q]\right)-$$

$$c2[3]\left(-\frac{5\,a1^2\,\lambda2\,x[m]\,X[i,m]}{2\,\lambda2+7\,\mu2}-\frac{7\,a1^2\,\mu2\,x[m]\,X[i,m]}{2\,\lambda2+7\,\mu2}+x[i]\,x[p]\,x[q]\,X[p,q]\right)-$$

$$c2[1]\left(-\frac{2\,x[m]\,X[i,m]}{5\,a1^5}+\frac{x[i]\,x[p]\,x[q]\,X[p,q]}{a1^7}\right)-$$

$$c2[2]\left(\frac{2\,\mu2\,x[m]\,X[i,m]}{3\,a1^3\,(\lambda2+\mu2)}+\frac{x[i]\,x[p]\,x[q]\,X[p,q]}{a1^5}\right)$$

Note that the aforementioned expressions consist of two independent terms, $x_m X_{im}$ and $x_i x_p x_q X_{pq}$, and hence, there are two independent equations that can be extracted as

In[55]:= `eq1 = Coefficient[tmp1, x[m] X[i, m]] = 0`

Out[55]= $\left(-\dfrac{5\,a1^2\,\lambda 1}{2\,\lambda 1 + 7\,\mu 1} - \dfrac{7\,a1^2\,\mu 1}{2\,\lambda 1 + 7\,\mu 1}\right) c1[3] + c1[4] + \dfrac{2\,c2[1]}{5\,a1^5} - \dfrac{2\,\mu 2\,c2[2]}{3\,a1^3\,(\lambda 2 + \mu 2)} - \left(-\dfrac{5\,a1^2\,\lambda 2}{2\,\lambda 2 + 7\,\mu 2} - \dfrac{7\,a1^2\,\mu 2}{2\,\lambda 2 + 7\,\mu 2}\right) c2[3] - c2[4] = 0$

In[56]:= `eq2 = Coefficient[tmp1, x[i] x[p] x[q] X[p, q]] = 0`

Out[56]= $c1[3] - \dfrac{c2[1]}{a1^7} - \dfrac{c2[2]}{a1^5} - c2[3] = 0$

The continuity condition that $u_i^z = u_i^m$ at $r = a_2$ can be entered and two independent equations can be obtained. The output is suppressed, however, as it is lengthy.

In[56]:= `tmp2 = ((Disp2 - Dispm) /. r -> a2);`

In[57]:= `eq3 = Coefficient[tmp2, x[m] X[i, m]] = 0`

Out[57]= $-1 - \dfrac{2\,c2[1]}{5\,a2^5} + \dfrac{2\,\mu 2\,c2[2]}{3\,a2^3\,(\lambda 2 + \mu 2)} + \left(-\dfrac{5\,a2^2\,\lambda 2}{2\,\lambda 2 + 7\,\mu 2} - \dfrac{7\,a2^2\,\mu 2}{2\,\lambda 2 + 7\,\mu 2}\right) c2[3] + c2[4] + \dfrac{2\,cm[1]}{5\,a2^5} - \dfrac{2\,\mu m\,cm[2]}{3\,a2^3\,(\lambda m + \mu m)} = 0$

In[58]:= `eq4 = Coefficient[tmp2, x[i] x[p] x[q] X[p, q]] = 0`

Out[58]= $\dfrac{c2[1]}{a2^7} + \dfrac{c2[2]}{a2^5} + c2[3] - \dfrac{cm[1]}{a2^7} - \dfrac{cm[2]}{a2^5} = 0$

The traction continuity conditions at $r = a_1$ and $r = a_2$ can be entered and independent equations can be extracted as

In[59]:= `tmp3 = ((Tract1 - Tract2) /. r -> a1);`

In[60]:= `eq5 = Coefficient[tmp3, x[m] X[i, m]] = 0`

Out[60]= $\left(-\dfrac{16\,a1\,\lambda 1\,\mu 1}{2\,\lambda 1 + 7\,\mu 1} - \dfrac{14\,a1\,\mu 1^2}{2\,\lambda 1 + 7\,\mu 1}\right) c1[3] + \dfrac{2\,\mu 1\,c1[4]}{a1} - \dfrac{16\,\mu 2\,c2[1]}{5\,a1^6} - \left(\dfrac{2\,\lambda 2\,\mu 2}{a1^4\,(\lambda 2 + \mu 2)} + \dfrac{4\,\mu 2^2}{3\,a1^4\,(\lambda 2 + \mu 2)}\right) c2[2] - \left(-\dfrac{16\,a1\,\lambda 2\,\mu 2}{2\,\lambda 2 + 7\,\mu 2} - \dfrac{14\,a1\,\mu 2^2}{2\,\lambda 2 + 7\,\mu 2}\right) c2[3] - \dfrac{2\,\mu 2\,c2[4]}{a1} = 0$

In[61]:= `eq6 = Coefficient[tmp3, x[i] x[p] x[q] X[p, q]] = 0`

Out[61]= $\left(\dfrac{19\,\lambda 1\,\mu 1}{a1\,(2\,\lambda 1 + 7\,\mu 1)} + \dfrac{14\,\mu 1^2}{a1\,(2\,\lambda 1 + 7\,\mu 1)}\right) c1[3] + \dfrac{8\,\mu 2\,c2[1]}{a1^8} - \left(-\dfrac{8\,\lambda 2\,\mu 2}{a1^6\,(\lambda 2 + \mu 2)} - \dfrac{8\,\mu 2^2}{a1^6\,(\lambda 2 + \mu 2)}\right) c2[2] - \left(\dfrac{19\,\lambda 2\,\mu 2}{a1\,(2\,\lambda 2 + 7\,\mu 2)} + \dfrac{14\,\mu 2^2}{a1\,(2\,\lambda 2 + 7\,\mu 2)}\right) c2[3] = 0$

Inclusions in Infinite Media

In[82]:= `tmp4 = ((Tract2 - Tractm) /. r → a2);`

In[83]:= `eq7 = Coefficient[tmp4, x[m] X[i, m]] == 0`

Out[83]= $-\dfrac{2\,\mu m}{a2} + \dfrac{16\,\mu 2\,c2[1]}{5\,a2^6} + \left(\dfrac{2\,\lambda 2\,\mu 2}{a2^4\,(\lambda 2+\mu 2)} + \dfrac{4\,\mu 2^2}{3\,a2^4\,(\lambda 2+\mu 2)}\right) c2[2] + \left(-\dfrac{16\,a2\,\lambda 2\,\mu 2}{2\,\lambda 2+7\,\mu 2} - \dfrac{14\,a2\,\mu 2^2}{2\,\lambda 2+7\,\mu 2}\right) c2[3] + \dfrac{2\,\mu 2\,c2[4]}{a2} - \dfrac{16\,\mu m\,cm[1]}{5\,a2^6} - \left(\dfrac{2\,\lambda m\,\mu m}{a2^4\,(\lambda m+\mu m)} + \dfrac{4\,\mu m^2}{3\,a2^4\,(\lambda m+\mu m)}\right) cm[2] = 0$

In[84]:= `eq8 = Coefficient[tmp4, x[i] x[p] x[q] X[p, q]] = 0`

Out[84]= $-\dfrac{8\,\mu 2\,c2[1]}{a2^8} + \left(-\dfrac{8\,\lambda 2\,\mu 2}{a2^6\,(\lambda 2+\mu 2)} - \dfrac{8\,\mu 2^2}{a2^6\,(\lambda 2+\mu 2)}\right) c2[2] + \left(\dfrac{19\,\lambda 2\,\mu 2}{a2\,(2\,\lambda 2+7\,\mu 2)} + \dfrac{14\,\mu 2^2}{a2\,(2\,\lambda 2+7\,\mu 2)}\right) c2[3] + \dfrac{8\,\mu m\,cm[1]}{a2^8} - \left(-\dfrac{8\,\lambda m\,\mu m}{a2^6\,(\lambda m+\mu m)} - \dfrac{8\,\mu m^2}{a2^6\,(\lambda m+\mu m)}\right) cm[2] = 0$

A set of eight simultaneous equations can be solved as

In[65]:= `sol = Solve[{eq1, eq2, eq3, eq4, eq5, eq6, eq7, eq8},`
        `{c1[3], c1[4], c2[1], c2[2], c2[3], c2[4], cm[1], cm[2]}][[1]];`

The output is too lengthy to be printed. Finally, the displacement, strain, stress, and traction in each phase can be obtained as

In[66]:= `ui1 = Disp1 /. sol;`
        `ui2 = Disp2 /. sol;`
        `uim = Dispm /. sol;`

In[69]:= `eij1 = Sum[Eij[[ii]] c1[ii], {ii, 1, 4}] /. {λ → λ1, μ → μ1} /. sol;`
        `eij2 = Sum[Eij[[ii]] c2[ii], {ii, 1, 4}] /. {λ → λ2, μ → μ2} /. sol;`
        `eijm = Sum[Eij[[ii]] cm[ii], {ii, 1, 4}] /. {λ → λm, μ → μm} /. sol;`

In[72]:= `σij1 = Sum[Sij[[ii]] c1[ii], {ii, 1, 4}] /. {λ → λ1, μ → μ1} /. sol;`
        `σij2 = Sum[Sij[[ii]] c2[ii], {ii, 1, 4}] /. {λ → λ2, μ → μ2} /. sol;`
        `σijm = Sum[Sij[[ii]] cm[ii], {ii, 1, 4}] /. {λ → λm, μ → μm} /. sol;`

In[75]:= `ti1 = Tract1 /. sol;`
        `ti2 = Tract2 /. sol;`
        `ti3 = Tractm /. sol;`

For instance, the strain field inside the innermost inclusion is expressed as

In[86]:= `eij1 // Simplify`

Out[86]= $\big(225\,a2^3\,\mu 2\,(\lambda 2+2\,\mu 2)\,\mu m\,(\lambda m+2\,\mu m)\,\big(\big(280\,a1^3\,a2^2\,r^2\,(5\,\lambda 1+7\,\mu 1)\,(\mu 1-\mu 2)\,(\lambda 2+\mu 2)\,(\mu 2-\mu m) - $
$56\,a1^5\,(21\,a2^2\,(\lambda 1+\mu 1) + 5\,r^2\,(5\,\lambda 1+7\,\mu 1))\,(\mu 1-\mu 2)\,(\lambda 2+\mu 2)\,(\mu 2-\mu m) + $
$40\,a1^7\,(\lambda 1\,(37\,\lambda 2\,(\mu 1-\mu 2) + 7\,(8\,\mu 1-5\,\mu 2)\,\mu 2) + 7\,\mu 1\,(\lambda 2\,(5\,\mu 1-8\,\mu 2) + 7\,(\mu 1-\mu 2)\,\mu 2))$
$(\mu 2-\mu m) + a2^7\,(14\,\mu 1\,(\mu 1+4\,\mu 2) + \lambda 1\,(19\,\mu 1+16\,\mu 2))$
$(14\,\mu 2\,(\mu 2+4\,\mu m) + \lambda 2\,(19\,\mu 2+16\,\mu m))\big)\,X[i, j] + $
$280\,a1^3\,\big(a1^2-a2^2\big)\,(\mu 1-\mu 2)\,(\lambda 2+\mu 2)\,(\mu 2-\mu m)\,(-3\,\lambda 1\,x[j]\,x[m]\,X[i, m] - $
$3\,\lambda 1\,x[i]\,x[m]\,X[j, m] + (2\,\lambda 1+7\,\mu 1)\,x[p]\,x[q]\,X[p, q]\,\delta[i, j])\big)\big) / $
$\big(-1008\,a1^5\,a2^5\,(\mu 1-\mu 2)\,(\lambda 2+\mu 2)^2\,(14\,\mu 1\,(\mu 1+4\,\mu 2) + \lambda 1\,(19\,\mu 1+16\,\mu 2))$
$(\mu 2-\mu m)\,(2\,\mu m\,(8\,\mu 2+7\,\mu m) + \lambda m\,(6\,\mu 2+9\,\mu m)) + $
$200\,a1^7\,a2^3\,\big(\lambda 1\,(\lambda 2\,\mu 2\,(152\,\mu 1^2 + 23\,\mu 1\,\mu 2 - 112\,\mu 2^2) + \lambda 2^2\,(57\,\mu 1^2 - 3\,\mu 1\,\mu 2 - 54\,\mu 2^2) + $
$7\,\mu 2^2\,(19\,\mu 1^2 + 7\,\mu 1\,\mu 2 - 8\,\mu 2^2)) + 7\,\mu 1\,(2\,\lambda 2\,\mu 2\,(8\,\mu 1^2 + 11\,\mu 1\,\mu 2 - 28\,\mu 2^2) + $
$3\,\lambda 2^2\,(2\,\mu 1^2 + 4\,\mu 1\,\mu 2 - 9\,\mu 2^2) + 14\,\mu 2^2\,(\mu 1^2 + \mu 1\,\mu 2 - 2\,\mu 2^2))\big)\,(\mu 2-\mu m)$
$(2\,\mu m\,(8\,\mu 2+7\,\mu m) + \lambda m\,(6\,\mu 2+9\,\mu m)) + a2^{10}\,(2\,\mu 2\,(8\,\mu 1+7\,\mu 2) + \lambda 2\,(6\,\mu 1+9\,\mu 2))$
$(14\,\mu 1\,(\mu 1+4\,\mu 2) + \lambda 1\,(19\,\mu 1+16\,\mu 2))$
$(2\,\mu m\,(8\,\mu 2+7\,\mu m) + \lambda m\,(6\,\mu 2+9\,\mu m))$
$(14\,\mu 2\,(\mu 2+4\,\mu m) + \lambda 2\,(19\,\mu 2+16\,\mu m)) + 16\,a1^{10}\,(\mu 1-\mu 2)$
$(\lambda 1\,(38\,\lambda 2\,(\mu 1-\mu 2) + 7\,(19\,\mu 1-4\,\mu 2)\,\mu 2) + 7\,\mu 1\,(\lambda 2\,(4\,\mu 1-19\,\mu 2) + 14\,(\mu 1-\mu 2)\,\mu 2))\,(\mu 2-\mu m)$
$(3\,\lambda 2\,(9\,\lambda m\,(\mu 2-\mu m) + 2\,(12\,\mu 2-7\,\mu m)\,\mu m) + 2\,\mu 2\,(3\,\lambda m\,(7\,\mu 2-12\,\mu m) + 56\,(\mu 2-\mu m)\,\mu m)) + $

$$50\,a1^3\,a2^7\,(\mu1-\mu2)\,(14\,\mu1\,(\mu1+4\,\mu2)+\lambda1\,(19\,\mu1+16\,\mu2))$$
$$\bigl(2\,\lambda2\,\mu2\,\bigl(3\,\lambda m\,\bigl\{28\,\mu2^2+13\,\mu2\,\mu m-48\,\mu m^2\bigr\}+14\,\mu m\,\bigl(16\,\mu2^2+3\,\mu2\,\mu m-16\,\mu m^2\bigr)\bigr)+$$
$$28\,\mu2^2\,\bigl(2\,\mu m\,\bigl(4\,\mu2^2+3\,\mu2\,\mu m-7\,\mu m^2\bigr)+3\,\lambda m\,\bigl(\mu2^2+\mu2\,\mu m-3\,\mu m^2\bigr)\bigr)+$$
$$3\,\lambda2^2\,\bigl(9\,\lambda m\,\bigl(3\,\mu2^2+\mu2\,\mu m-4\,\mu m^2\bigr)-2\,\mu m\,\bigl(-36\,\mu2^2+\mu2\,\mu m+28\,\mu m^2\bigr)\bigr)\bigr)\bigr)$$

The strain field in the matrix is lengthy and part of the expressed is listed as

In[89]:= ϵijm

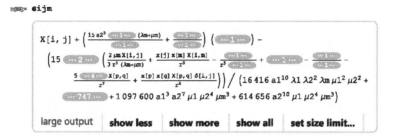

3.2.6 Exact Solution for Four-Phase Materials

The analysis for a four-phase material where a spherical inclusion is surrounded by two extra coating layers embedded in an infinite matrix as shown in Figure 3.12 is straightforward following the procedure described earlier. As the solution procedure is spelled out in the previous subsection, this problem is an exercise of the employed method. As expected, the analytical solution of the elastic field for such a material is expected to be extremely large, which is verified by an output from *Mathematica*. This inevitably poses a fundamental question of how useful an analytical solution is over a numerical solution if the length of the analytical solution is too large, which is only traceable by computer.

Nevertheless, analytical solutions are desirable when used in parametric study that requires all the relevant parameters to be present. In this subsection, instead of printing the output from each process, only the logic and the corresponding *Mathematica* code are shown. The complete program and its output can be downloaded from the companion web page.

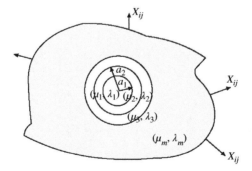

Figure 3.12 Four-phase material

3.2.6.1 Exact Solution for X

The general form of the displacement and traction when a body is subject to X at the far field can be expressed as

In[31]:= `Ui = ui /. sol1`

Out[31]= $X \left(-\dfrac{C[1]}{3\,r^3} + C[2] \right) x[i]$

In[38]:= `Ti = Collect[Titmp, {C[1], C[2]}, Simplify]`

Out[38]= $\dfrac{4\,X\,\mu\,C[1]\,x[i]}{3\,r^4} + \dfrac{X\,(3\,\lambda + 2\,\mu)\,C[2]\,x[i]}{r}$

Using the aforementioned expression, the displacement and the traction at each phase can be expressed as

In[44]:=
```
Disp1 = Ui /. C → c1;
Disp2 = Ui /. C → c2;
Disp3 = Ui /. C → c3;
Dispm = Ui /. C → cm;
Tract1 = Ti /. {μ → μ1, λ → λ1, C → c1};
Tract2 = Ti /. {μ → μ2, λ → λ2, C → c2};
Tract3 = Ti /. {μ → μ3, λ → λ3, C → c3};
Tractm = Ti /. {μ → μm, λ → λm, C → cm};
```

where the numbers "1–3" refer to the inclusion and the two layers and "m" refers to the matrix. The unknown coefficients are denoted as `c1[1]–c1[4]`, `c2[1]–c2[4]`, `c3[1]–c3[4]`, and `cm[1]–cm[4]`. From the requirement that u_i must remain finite at $r = 0$, it follows that `c1[1] = 0`. From the requirement that $\epsilon_{ij} \to \hat{X}_{ij}$ as $r \to \infty$, it follows that `cm[2] = 1`. Therefore, the continuity condition for the displacement and the traction at $r = a_1$, $r = a_2$, and $r = a_3$ are entered as

In[54]:= `c1[1] = 0; cm[2] = 1;`

In[55]:=
```
eq1 = Coefficient[(Disp1 - Disp2) /. r → a1, x[i] X];
eq2 = Coefficient[(Disp2 - Disp3) /. r → a2, x[i] X];
eq3 = Coefficient[(Disp3 - Dispm) /. r → a3, x[i] X];
eq4 = Coefficient[(Tract1 - Tract2) /. r → a1, x[i] X];
eq5 = Coefficient[(Tract2 - Tract3) /. r → a2, x[i] X];
eq6 = Coefficient[(Tract3 - Tractm) /. r → a3, x[i] X];
```

In[61]:=
```
sol = Solve[{eq1 == 0, eq2 == 0, eq3 == 0, eq4 == 0, eq5 == 0, eq6 == 0},
    {c1[2], c2[1], c2[2], c3[1], c3[2], cm[1]}][[1]];
```

The displacement, strain, stress, and traction for each phase are now computed from the following code:

```
In[62]:= ui1 = Disp1 /. sol // Simplify;
         ui2 = Disp2 /. sol // Simplify;
         ui3 = Disp3 /. sol // Simplify;
         uim = Dispm /. sol // Simplify;

In[66]:= eij1 = Eij /. C → c1 /. sol;
         eij2 = Eij /. C → c2 /. sol;
         eij3 = Eij /. C → c3 /. sol;
         eijm = Eij /. C → cm /. sol;

In[70]:= σij1 = Sij /. {C → c1, μ → μ1, λ → λ1} /. sol // Simplify;
         σij2 = Sij /. {C → c2, μ → μ2, λ → λ2} /. sol // Simplify;
         σij3 = Sij /. {C → c3, μ → μ3, λ → λ3} /. sol // Simplify;
         σijm = Sij /. {C → cm, μ → μm, λ → λm} /. sol // Simplify;

In[74]:= ti1 = Tract1 /. sol;
         ti2 = Tract2 /. sol;
         ti3 = Tract3 /. sol;
         tim = Tractm4 /. sol;
```

The strain inside the inclusion, ϵ_{ij}^I, can be expressed as

```
In[88]:= eij1 // Simplify

Out[88]= (27 a2³ a3³ x (λ2 + 2 μ2) (λ3 + 2 μ3) (λm + 2 μm) δ[i, j]) /
         (4 a1³ (3 λ1 - 3 λ2 + 2 μ1 - 2 μ2) (a2³ (3 λ3 + 4 μ2 + 2 μ3) (μ3 - μm) + a3³ (μ2 - μ3) (3 λ3 + 2 μ3 + 4 μm)) +
          a2³ (3 λ1 + 2 μ1 + 4 μ2)
          (4 a2³ (3 λ2 - 3 λ3 + 2 μ2 - 2 μ3) (μ3 - μm) + a3³ (3 λ2 + 2 μ2 + 4 μ3) (3 λ3 + 2 μ3 + 4 μm)))
```

The strains in the second and third layers are expressed as

```
In[89]:= eij2 // Simplify

Out[89]= (9 a2³ a3³ x (λ3 + 2 μ3) (λm + 2 μm) (3 a1³ (3 λ1 - 3 λ2 + 2 μ1 - 2 μ2) x[i] x[j] +
         r² (a1³ (-3 λ1 + 3 λ2 - 2 μ1 + 2 μ2) + r³ (3 λ1 + 2 μ1 + 4 μ2)) δ[i, j])) /
         (r⁵ (4 a2⁶ (3 λ1 + 2 μ1 + 4 μ2) (3 λ2 - 3 λ3 + 2 μ2 - 2 μ3) (μ3 - μm) + 4 a1³ a3³ (3 λ1 - 3 λ2 + 2 μ1 - 2 μ2)
         (μ2 - μ3) (3 λ3 + 2 μ3 + 4 μm) + a2³ (4 a1³ (3 λ1 - 3 λ2 + 2 μ1 - 2 μ2) (3 λ3 + 4 μ2 + 2 μ3) (μ3 - μm) +
         a3³ (3 λ1 + 2 μ1 + 4 μ2) (3 λ2 + 2 μ2 + 4 μ3) (3 λ3 + 2 μ3 + 4 μm))))


In[90]:= eij3 // Simplify

Out[90]= - (3 a2³ a3³ x (λm + 2 μm)
         (-1/a2³ (4 a1³ (3 λ1 - 3 λ2 + 2 μ1 - 2 μ2) (μ2 - μ3) + a2³ (3 λ1 + 2 μ1 + 4 μ2) (3 λ2 + 2 μ2 + 4 μ3))
         δ[i, j] - 1/r⁵ (-a2³ (3 λ1 + 2 μ1 + 4 μ2) (3 λ2 - 3 λ3 + 2 μ2 - 2 μ3) -
         a1³ (3 λ1 - 3 λ2 + 2 μ1 - 2 μ2) (3 λ3 + 4 μ2 + 2 μ3)) (-3 x[i] x[j] + r² δ[i, j]))) /
         (4 a1³ (3 λ1 - 3 λ2 + 2 μ1 - 2 μ2) (a2³ (3 λ3 + 4 μ2 + 2 μ3) (μ3 - μm) + a3³ (μ2 - μ3) (3 λ3 + 2 μ3 + 4 μm)) +
          a2³ (3 λ1 + 2 μ1 + 4 μ2)
          (4 a2³ (3 λ2 - 3 λ3 + 2 μ2 - 2 μ3) (μ3 - μm) + a3³ (3 λ2 + 2 μ2 + 4 μ3) (3 λ3 + 2 μ3 + 4 μm)))
```

The strain in the matrix is expressed as

In[91]:= **eijm // Simplify**

Out[91]= $X \left(\delta[i,j] - \dfrac{1}{r^5} \right.$
$a3^3 \left(1 - \left(3\, a1^3\, a2^3\, (3\,\lambda1 - 3\,\lambda2 + 2\,\mu1 - 2\,\mu2)\, \left(\dfrac{1}{a2^3}(4\,a3^3\,(\mu2-\mu3) - a2^3\,(3\,\lambda3 + 4\,\mu2 + 2\,\mu3)\right) - \right.\right.$
$((3\,\lambda1 + 2\,\mu1 + 4\,\mu2)\,(a2^3\,(3\,\lambda2 - 3\,\lambda3 + 2\,\mu2 - 2\,\mu3) - a3^3\,(3\,\lambda2 + 2\,\mu2 + 4\,\mu3)))/$
$\left(a1^3\,(3\,\lambda1 - 3\,\lambda2 + 2\,\mu1 - 2\,\mu2)\right)\,(\lambda m + 2\,\mu m)\Big)\Big/\Big(4\,a1^3\,(3\,\lambda1 - 3\,\lambda2 + 2\,\mu1 - 2\,\mu2)$
$\left(a2^3\,(3\,\lambda3 + 4\,\mu2 + 2\,\mu3)\,(\mu3 - \mu m) + a3^3\,(\mu2 - \mu3)\,(3\,\lambda3 + 2\,\mu3 + 4\,\mu m)\right) +$
$a2^3\,(3\,\lambda1 + 2\,\mu1 + 4\,\mu2)\,(4\,a2^3\,(3\,\lambda2 - 3\,\lambda3 + 2\,\mu2 - 2\,\mu3)\,(\mu3 - \mu m) +$
$a3^3\,(3\,\lambda2 + 2\,\mu2 + 4\,\mu3)\,(3\,\lambda3 + 2\,\mu3 + 4\,\mu m))\Big)\Big)\,(-3\,x[i]\,x[j] + r^2\,\delta[i,j])\Big)$

3.2.6.2 Exact Solution for X'_{ij}

The *Mathematica* code is shown next but without output in this subsection, as the typical output is too long. The displacement (Disp1-Dispm) and the traction (Tract1-Tractm) in each of the four phases are set up first. Here, the numbers 1–3 refer to the material properties of the inclusion and the surrounding two coating materials, and "m" refers to the matrix. They are assumed to contain unknown coefficients (c1[i]–cm[i]) for each phase.

```
In[48]:= Disp1 = (Ui.Table[ c1[i], {i, 4}]) /. {λ → λ1, μ → μ1};
         Disp2 = (Ui.Table[ c2[i], {i, 4}]) /. {λ → λ2, μ → μ2};
         Disp3 = (Ui.Table[ c3[i], {i, 4}]) /. {λ → λ3, μ → μ3};
         Dispm = (Ui.Table[ cm[i], {i, 4}]) /. {λ → λm, μ → μm};
         Tract1 = (Ti.Table[c1[i], {i, 4}]) /. {λ → λ1, μ → μ1};
         Tract2 = (Ti.Table[c2[i], {i, 4}]) /. {λ → λ2, μ → μ2};
         Tract3 = (Ti.Table[c3[i], {i, 4}]) /. {λ → λ3, μ → μ3};
         Tractm = (Ti.Table[cm[i], {i, 4}]) /. {λ → λm, μ → μm};
```

The following code sets up the equations for the unknowns (c1[i]–c4[i]) by applying the continuity conditions across each phase for the displacement and the traction. As each continuity condition contains two independent terms, there are 12 equations for 12 unknowns excluding c1[1], c1[2], c4[3], and c4[4]. The solution for all the unknown coefficients is obtained by the Solve function. The output from the Solve command is omitted as it is too lengthy.

```
In[54]:= c1[1] = 0; c1[2] = 0; cm[4] = 1; cm[3] = 0;
In[55]:= tmp1 = ((Disp1 - Disp2) /. r → a1) ;
         eq1 = Coefficient[tmp1, x[m] X[i, m]] = 0;
         eq2 = Coefficient[tmp1, x[i] x[p] x[q] X[p, q]] == 0;
In[58]:= tmp2 = ((Disp2 - Disp3) /. r → a2);
         eq3 = Coefficient[tmp2, x[m] X[i, m]] = 0;
         eq4 = Coefficient[tmp2, x[i] x[p] x[q] X[p, q]] == 0;
In[61]:= tmp3 = ((Disp3 - Dispm) /. r → a3);
         eq5 = Coefficient[tmp3, x[m] X[i, m]] = 0;
         eq6 = Coefficient[tmp3, x[i] x[p] x[q] X[p, q]] == 0;
```

```
In[64]:= tmp4 = ((Tract1 - Tract2) /. r → a1);
        eq7 = Coefficient[tmp4, x[m] X[i, m]] == 0;
        eq8 = Coefficient[tmp4, x[i] x[p] x[q] X[p, q]] == 0;
In[67]:= tmp5 = ((Tract2 - Tract3) /. r → a2);
        eq9 = Coefficient[tmp5, x[m] X[i, m]] == 0;
        eq10 = Coefficient[tmp5, x[i] x[p] x[q] X[p, q]] == 0;
In[70]:= tmp6 = ((Tract3 - Tractm) /. r → a3);
        eq11 = Coefficient[tmp6, x[m] X[i, m]] == 0;
        eq12 = Coefficient[tmp6, x[i] x[p] x[q] X[p, q]] == 0;
In[73]:= sol = Solve[{eq1, eq2, eq3, eq4, eq5, eq6, eq7, eq8, eq9, eq10, eq11, eq12}, {c1[3], c1[4],
        c2[1], c2[2], c2[3], c2[4], c3[1], c3[2], c3[3], c3[4], cm[1], cm[2]}][[1]];
```

The closed-form solutions for the displacement, strain, stress, and traction for each phase are stored in ui1-ui4, ϵij1-ϵij4, σij1-σij4, and ti1-ti4, respectively.

```
In[74]:= ui1 = Disp1 /. sol;
        ui2 = Disp2 /. sol;
        ui3 = Disp3 /. sol;
        uim = Dispm /. sol;
In[78]:= εij1 = Sum[Eij[[ii]] c1[ii], {ii, 1, 4}] /. {λ → λ1, μ → μ1} /. sol;
        εij2 = Sum[Eij[[ii]] c2[ii], {ii, 1, 4}] /. {λ → λ2, μ → μ3} /. sol;
        εij3 = Sum[Eij[[ii]] c3[ii], {ii, 1, 4}] /. {λ → λ3, μ → μ3} /. sol;
        εijm = Sum[Eij[[ii]] cm[ii], {ii, 1, 4}] /. {λ → λm, μ → μm} /. sol;
In[112]:= σij1 = Sum[Sij[[ii]] c1[ii], {ii, 1, 4}] /. {λ → λ1, μ → μ1} /. sol;
        σij2 = Sum[Sij[[ii]] c2[ii], {ii, 1, 4}] /. {λ → λ2, μ → μ3} /. sol;
        σij3 = Sum[Sij[[ii]] c3[ii], {ii, 1, 4}] /. {λ → λ3, μ → μ3} /. sol;
        σijm = Sum[Sij[[ii]] cm[ii], {ii, 1, 4}] /. {λ → λm, μ → μm} /. sol;
In[86]:= ti1 = Tract1 /. sol;
        ti2 = Tract2 /. sol;
        ti3 = Tract3 /. sol;
        tim = Tractm /. sol;
```

To save the result for later use, use the SetDirectory command to specify the location to be saved and Save function for the filename and variables to be saved.

```
In[98]:= SetDirectory["c:/tmp"];
        Save["4phase.m", ui1, ui2, ui3, ui4, εij1,
         εij2, εij3, εij4, σij1, σij2, σij3, σij4, ti1, ti2, ti3, ti4];
```

It should be noted that the size of the file, 4phase.m, is 10 MB.

```
Volume in drive C is unlabeled      Serial number is de1b:91ef
Directory of   c:\tmp\4phase.m

2012/03/23  14:15      10,455,160  4phase.m
       10,455,160 bytes in 1 file and 0 dirs  10,457,088 bytes allocated
  331,125,198,848 bytes free
```

3.2.7 Exact Solution for 2-D Multiphase Materials

The procedure described is also applicable to the stress analysis where a cylindrical-shaped inclusion with the aspect ratio of infinity is embedded in an infinite matrix phase. In this case, the state of plane strain is assumed in which $\epsilon_{13} = \epsilon_{23} = \epsilon_{33} = 0$ with x_3 as the axis of symmetry.

All the derivation processes are the same as in the previous subsections except that the computation is limited to x_1–x_2.

3.2.7.1 Solution for X

The displacement and the traction force are expressed by

$$u_i = \left(-\frac{c_1}{2r^2} + c_2\right) x_j X,$$

$$t_i = \frac{\mu}{r^3} c_1 x_i X + \frac{2(\lambda + \mu)}{r} c_2 x_i.$$

Unknown coefficients, c_1–c_2, are determined for each phase to satisfy the continuity condition of the displacement and traction.

3.2.7.2 Solution for $\hat{X}_{ij}$

The displacement and the traction force are expressed by

$$u_i = c_1 \left(\frac{x_i x_j x_k X_{jk}}{r^6} - \frac{x_k X_{ik}}{2r^4}\right) + c_2 \left(\frac{x_i x_j x_k X_{jk}}{r^4} + \frac{\mu x_k X_{ik}}{r^2(\lambda + \mu)}\right)$$

$$+ c_3 \left(x_i x_j x_k X_{jk} - \frac{r^2(2\lambda + 3\mu) x_k X_{ik}}{\lambda + 3\mu}\right) + c_4 x_k X_{ik},$$

$$t_i = \frac{c_3(x_i(-3\lambda^2 + 5\lambda\mu + 18\mu^2) x_p x_q X_{pq} - 2\mu r^2(7\lambda + 9\mu) x_m X_{im})}{2r(\lambda + 3\mu)}$$

$$+ c_1 \left(\frac{2\mu x_m X_{im}}{r^5} - \frac{x_i(3\lambda + 10\mu) x_p x_q X_{pq}}{2r^7}\right)$$

$$+ c_2 \left(\frac{\mu x_m X_{im}}{r^3} - \frac{x_i(3\lambda + 10\mu) x_p x_q X_{pq}}{2r^5}\right) + \frac{2 c_4 \mu x_m X_{im}}{r}.$$

Unknown coefficients, c_1–c_4, are determined for each phase to satisfy the continuity condition of the displacement and traction.

3.3 Thermal Stress

Thermal stress analysis for composites (e.g., Rosen and Hashin (1970)) is an important subject for the integrity of aerospace structures as material failure is often caused by the existence of thermal stress due to a mismatch of thermal expansion coefficients or thermal conductivities between the matrix and inclusion phases. From the analysis standpoint of view, thermal stress problems are special cases of elasticity problems where the temperature effect is identified as a fictitious body force, and hence, if one can solve the elasticity equilibrium equation with a body force, thermal stress problems are automatically solved using the same procedure.

In this section, thermal stress analysis is carried out for an infinitely extended medium that has a spherical inclusion with a different thermal conductivity, thermal expansion coefficient, and elastic modulus from those of the matrix. First, thermal stress analysis is carried out when the source of thermal stress is a heat source that is present in the inclusion. Next, thermal stress analysis in which the thermal stress is due to a uniform heat flow at the far field is carried out. It is possible to derive analytical solutions for both cases with *Mathematica*, and both problems are good examples of how *Mathematica* can be used to handle all necessary algebra, which was not possible before.

3.3.1 Thermal Stress Due to Heat Source

In this subsection, a spherical inclusion with a radius, a, is surrounded by an infinitely extended matrix. The elastic moduli, thermal conductivities, and thermal expansion coefficients differ in the inclusion and in the matrix. A constant heat source, q, exists inside the spherical inclusion. The steady-state temperature distribution is derived first followed by the derivation of the thermal stress.

3.3.1.1 Temperature Distribution

The 3-D temperature field when there is a heat source, q, can be derived by solving the Poisson equation expressed as

$$k\, T_{,ii} + q = 0, \qquad (3.35)$$

where q is a heat source that exists inside the inclusion, but otherwise 0 and k is the thermal conductivity. As the heat source is a scalar and there is no other directional component in Equation (3.35), the temperature field, T, can be assumed as a function of $r = \sqrt{x_i x_i}$ alone as

$$T = T(r). \qquad (3.36)$$

Substituting Equation (3.36) into Equation (3.35) yields

$$k \left(T''(r) + \frac{2T'(r)}{r} \right) + q = 0,$$

which can be solved as

$$T(r) = -\frac{q}{6k} r^2 + d_1 + \frac{d_2}{r}, \qquad (3.37)$$

where d_1 and d_2 are integral constants yet to be determined. The following *Mathematica* code can derive Equation (3.37).

```
In[2]:= SetAttributes[δ, Orderless];
    δ[i_Integer, j_Integer] := If[i == j, 1, 0];
    δ[i_Symbol, i_Symbol] := 3;

    Unprotect[Times];
    Times[x[j_Symbol], δ[i_, j_Symbol]] := x[i];
    Times[δ[i_Symbol, j_], δ[i_Symbol, k_]] := δ[j, k];
    Times[δ[i_Symbol, j_], h_[i_Symbol, m_, n_]] := h[j, m, n];
    Times[δ[i_Symbol, j_], h_[m, i_Symbol, n_]] := h[m, j, n];
    Times[δ[i_Symbol, j_], h_[m, n, i_Symbol]] := h[m, n, j];
    □Times[δ[i_Symbol, j_], h_[i_Symbol]] := h[j];
    Protect[Times];

    Unprotect[Power];
    Power[x[i_Symbol], 2] := r^2;
    Protect[Power];

In[16]:= Unprotect[D];
    D[a_. r^n_. + b_., x[i_]] :=
      D[a, x[i]] r^n + n r^(n - 1) x[i] / r a + D[b, x[i]];
    D[a_. x[i_] + b_., x[j_]] := D[a, x[j]] x[i] + a δ[i, j] + D[b, x[j]];
    D[a_. x[i_Integer]^n_. + b_., x[j_]] :=
      a n x[i]^(n - 1) δ[i, j] + D[b, x[j]];
    D[a_. f_[r]^n_. + b_., x[i_]] :=
      D[a, x[i]] f[r]^n + a n f[r]^(n - 1) f'[r] x[i] / r + D[b, x[i]];
    D[a_List, b_] := Map[D[#, b] &, a]
    D[Times[a_, b_], x[i_]] := D[a, x[i]] b + a D[b, x[i]];
    Protect[D];
```

In[24]:= eq1 = D[D[f[r], x[i]], x[i]] // Simplify

Out[24]= $\dfrac{2 f'[r]}{r} + f''[r]$

In[25]:= sol = DSolve[eq1 + q == 0, f[r], r][[1]]

Out[25]= $\left\{ f[r] \to -\dfrac{q r^2}{6} - \dfrac{C[1]}{r} + C[2] \right\}$

Implicit differentiation can be implemented by adding rules to the D function.

Noting that the temperature must remain finite at $r = 0$ and tends to 0 as $r \to \infty$, the temperature for the inclusion phase and the matrix phase can be expressed as

1. Inside the inclusion:
$$T_1 = -\dfrac{q}{6k} r^2 + C_1.$$

2. In the matrix:
$$T_2 = \frac{C_2}{r}.$$

The temperature and heat flux must be continuous at the interface, $r = a$, as
$$T_1 = T_2,$$
$$k_1 \frac{dT_1}{dr} = k_2 \frac{dT_2}{dr}.$$

By imposing the aforementioned conditions, the unknown constants, C_1 and C_2, can be determined and the temperature for each phase can be obtained as

$$T_1 = -\frac{q}{6k_1}r^2 + \frac{a^2(2k_1 + k_2)}{6k_1 k_2}q, \quad 0 \le r < a, \tag{3.38}$$

$$T_2 = \frac{a^3 q}{3k_2}\frac{1}{r}, \quad a \le r < \infty. \tag{3.39}$$

The following *Mathematica* code derives the result automatically.

```
In[26]:= temp1 = f[r] /. sol /. C → C1
```
$$\text{Out[26]}= -\frac{q\,r^2}{6} - \frac{C1[1]}{r} + C1[2]$$

```
In[27]:= temp2 = f[r] /. sol /. C → C2
```
$$\text{Out[27]}= -\frac{q\,r^2}{6} - \frac{C2[1]}{r} + C2[2]$$

```
In[28]:= temp1 = temp1 /. C1[1] → 0
```
$$\text{Out[28]}= -\frac{q\,r^2}{6} + C1[2]$$

```
In[29]:= temp2 = temp2 /. {C2[2] → 0, q → 0}
```
$$\text{Out[29]}= -\frac{C2[1]}{r}$$

```
In[30]:= sol2 = Solve[{((temp1 - temp2) /. r → a) == 0,
         ((k1 D[temp1, r] - k2 D[temp2, r]) /. r → a) == 0}, {C1[2], C2[1]}][[1]]
```
$$\text{Out[30]}= \left\{C1[2] \to -\frac{-2\,a^2\,k1\,q - a^2\,k2\,q}{6\,k2},\ C2[1] \to -\frac{a^3\,k1\,q}{3\,k2}\right\}$$

```
In[31]:= temp1 = temp1 /. sol2
```
$$\text{Out[31]}= -\frac{-2\,a^2\,k1\,q - a^2\,k2\,q}{6\,k2} - \frac{q\,r^2}{6}$$

```
In[32]:= temp2 = temp2 /. sol2
```
$$\text{Out[32]}= \frac{a^3\,k1\,q}{3\,k2\,r}$$

3.3.1.2 Thermal Stress Field

Based on the temperature field obtained as Equations (3.38) and (3.39), the stress field induced by the heat source in the inclusion can be derived.

The general elasticity equilibrium equation for the displacement when a temperature distribution exists is expressed as

$$\mu \Delta u_m + (\mu + \lambda)u_{i,im} - (2\mu + 3\lambda)\alpha T_{,m} = 0, \qquad (3.40)$$

where u_m is the displacement, μ and λ are the Lamé constants, and α is the thermal expansion coefficient. If the temperature, T, is a function of r alone as in Equations (3.38) and (3.39), it follows that

$$T_{,m} = \frac{T'(r)}{r} x_m.$$

Therefore, the solution for Equation (3.40) can also be assumed as

$$u_m = \frac{h(r)}{r} x_m, \qquad (3.41)$$

where $h(r)$ is an unknown function of r alone yet to be determined. By substituting Equation (3.41) into Equation (3.40), the following differential equation for $h(r)$ can be obtained as

$$(2\mu + \lambda)\left(\frac{h''(r)}{r} + \frac{2h'(r)}{r^2} - \frac{2h(r)}{r^3}\right)x_m - \alpha(2\mu + 3\lambda)T_{,m} = 0,$$

or equivalently,

$$\left(\frac{1}{r^2}(r^2 h(r))'\right)' = \alpha \frac{2\mu + 3\lambda}{2\mu + \lambda} T'(r). \qquad (3.42)$$

The general solution to Equation (3.42) can be obtained as

$$r^2 h(r) = \alpha \frac{2\mu + 3\lambda}{2\mu + \lambda} \int^r r^2 T(r)dr + d_1 r^3 + d_2,$$

where d_1 and d_2 are integral constants. Therefore, the general expression of the displacement can be written as

$$u_m = \left(\alpha \frac{2\mu + 3\lambda}{2\mu + \lambda} \frac{1}{r^3}\int^r r^2 T(r)dr + d_1 + \frac{d_2}{r^3}\right)x_m.$$

The displacement inside the inclusion must be continuous and hence assumed as

$$u_m^1 = \left(\alpha \frac{2\mu_1 + 3\lambda_1}{2\mu_1 + \lambda_1} \frac{1}{r^3}\int^r r^2 T_1(r)dr + C_1\right)x_m,$$

and the displacement in the matrix must vanish as $r \to \infty$ and hence assumed as

$$u_m^2 = \left(\alpha \frac{2\mu_2 + 3\lambda_2}{2\mu_2 + \lambda_2} \frac{1}{r^3}\int^r r^2 T_2(r)dr + \frac{C_2}{r^3}\right)x_m,$$

where T_1 is the temperature inside the inclusion and T_2 is the temperature inside the matrix given by Equations (3.38) and (3.39), respectively. The unknowns, C_1 and C_2, can be determined from the continuity conditions of the displacement and the traction force at the interface, $r = a$ as

$$u_m^1 = u_m^2 \quad r = a$$
$$t_m^1 = t_m^2 \quad r = a.$$

By using *Mathematica*, the unknowns, C_1 and C_2, can be solved exactly as

$$\begin{aligned}
C_1 = &-((-(a^2 q \alpha_1 \lambda_1 (3\lambda_1 + 2\mu_1))/(15k_1(\lambda_1 + 2\mu_1)) \\
&+((10a^2 k_1 + 2a^2 k_2) q \alpha_1 \lambda_1 (3\lambda_1 + 2\mu_1))/(30k_1 k_2(\lambda_1 + 2\mu_1)) \\
&-(2a^2 q \alpha_1 \mu_1 (3\lambda_1 + 2\mu_1))/(15k_1(\lambda_1 + 2\mu_1)) \\
&+((10a^2 k_1 + 2a^2 k_2) q \alpha_1 \mu_1 (3\lambda_1 + 2\mu_1))/(45k_1 k_2(\lambda_1 + 2\mu_1)) \\
&+(2(10a^2 k_1 + 2a^2 k_2) q \alpha_1 (3\lambda_1 + 2\mu_1)\mu_2)/(45k_1 k_2(\lambda_1 + 2\mu_1)) \\
&-(a^2 q \alpha_2 \lambda_2 (3\lambda_2 + 2\mu_2))/(3k_2(\lambda_2 + 2\mu_2)) \\
&-(2a^2 q \alpha_2 \mu_2 (3\lambda_2 + 2\mu_2))/(3k_2(\lambda_2 + 2\mu_2)))/(3\lambda_1 + 2\mu_1 + 4\mu_2)), \\
C_2 = &-((-(a^5 q \alpha_1 \lambda_1 (3\lambda_1 + 2\mu_1))/(15k_1(\lambda_1 + 2\mu_1)) \\
&-(2a^5 q \alpha_1 \mu_1 (3\lambda_1 + 2\mu_1))/(15k_1(\lambda_1 + 2\mu_1)) \\
&+(a^5 q \alpha_2 \lambda_1 (3\lambda_2 + 2\mu_2))/(2k_2(\lambda_2 + 2\mu_2)) \\
&-(a^5 q \alpha_2 \lambda_2 (3\lambda_2 + 2\mu_2))/(3k_2(\lambda_2 + 2\mu_2)) \\
&+(a^5 q \alpha_2 \mu_1 (3\lambda_2 + 2\mu_2))/(3k_2(\lambda_2 + 2\mu_2)))/(3\lambda_1 + 2\mu_1 + 4\mu_2)).
\end{aligned}$$

Therefore, the displacement is expressed as

$$\begin{aligned}
u_i^1 = &\,(q(-(k_2 r^2 \alpha_1 (3\lambda_1 + 2\mu_1)(3\lambda_1 + 2\mu_1 + 4\mu_2)) + a^2(10k_1 \alpha_2(\lambda_1 + 2\mu_1) \\
&(3\lambda_2 + 2\mu_2) + k_2 \alpha_1 (3\lambda_1 + 2\mu_1)(5\lambda_1 + 6\mu_1 + 4\mu_2)))x_i) \\
&/(30k_1 k_2 (\lambda_1 + 2\mu_1)(3\lambda_1 + 2\mu_1 + 4\mu_2)), \\
u_i^2 = &\,(a^3 q(5k_1 r^2 \alpha_2 (3\lambda_2 + 2\mu_2)(3\lambda_1 + 2\mu_1 + 4\mu_2) \\
&+a^2 (2k_2 \alpha_1 (3\lambda_1 + 2\mu_1)(\lambda_2 + 2\mu_2) \\
&-5k_1 \alpha_2 (3\lambda_1 - 2\lambda_2 + 2\mu_1)(3\lambda_2 + 2\mu_2)))x_i) \\
&/(30k_1 k_2 r^3 (\lambda_2 + 2\mu_2)(3\lambda_1 + 2\mu_1 + 4\mu_2)).
\end{aligned}$$

The strain components are expressed as

$$\begin{aligned}
\epsilon_{im}^1 = &\,(q(-2k_2 \alpha_1 (3\lambda_1 + 2\mu_1)(3\lambda_1 + 2\mu_1 + 4\mu_2)x_i x_m \\
&+(-(k_2 r^2 \alpha_1 (3\lambda_1 + 2\mu_1)(3\lambda_1 + 2\mu_1 + 4\mu_2)) \\
&+a^2(10k_1 \alpha_2(\lambda_1 + 2\mu_1)(3\lambda_2 + 2\mu_2) + k_2 \alpha_1 (3\lambda_1 + 2\mu_1)(5\lambda_1 + 6\mu_1 + 4\mu_2))) \\
&\delta_{im}))/(30k_1 k_2 (\lambda_1 + 2\mu_1)(3\lambda_1 + 2\mu_1 + 4\mu_2)), \\
\epsilon_{im}^2 = &\,(a^3 q(-((5k_1 r^2 \alpha_2 (3\lambda_2 + 2\mu_2)(3\lambda_1 + 2\mu_1 + 4\mu_2)
\end{aligned}$$

$$+3a^2(2k_2\alpha_1(3\lambda_1+2\mu_1)(\lambda_2+2\mu_2)$$
$$-5k_1\alpha_2(3\lambda_1-2\lambda_2+2\mu_1)(3\lambda_2+2\mu_2)))x_ix_m)$$
$$+r^2(5k_1r^2\alpha_2(3\lambda_2+2\mu_2)(3\lambda_1+2\mu_1+4\mu_2)$$
$$+a^2(2k_2\alpha_1(3\lambda_1+2\mu_1)(\lambda_2+2\mu_2)-5k_1\alpha_2(3\lambda_1-2\lambda_2+2\mu_1)$$
$$(3\lambda_2+2\mu_2)))\delta_{im}))/(30k_1k_2r^5(\lambda_2+2\mu_2)(3\lambda_1+2\mu_1+4\mu_2)).$$

The stress components are expressed as

$$\sigma_{im}^1 = (q(3\lambda_1+2\mu_1)(-4k_2\alpha_1\mu_1(3\lambda_1+2\mu_1+4\mu_2)x_ix_m$$
$$+(-(k_2r^2\alpha_1(5\lambda_1+2\mu_1)(3\lambda_1+2\mu_1+4\mu_2))$$
$$+a^2(10k_1\alpha_2(\lambda_1+2\mu_1)(3\lambda_2+2\mu_2)+k_2\alpha_1(3\lambda_1+2\mu_1)$$
$$(5\lambda_1+6\mu_1+4\mu_2)))\delta_{im}))/(30k_1k_2(\lambda_1+2\mu_1)(3\lambda_1+2\mu_1+4\mu_2)),$$
$$\sigma_{im}^2 = (a^3q(-(\mu_2(5k_1r^2\alpha_2(3\lambda_2+2\mu_2)(3\lambda_1+2\mu_1+4\mu_2)$$
$$+3a^2(2k_2\alpha_1(3\lambda_1+2\mu_1)(\lambda_2+2\mu_2)-5k_1\alpha_2(3\lambda_1-2\lambda_2+2\mu_1)$$
$$(3\lambda_2+2\mu_2)))x_ix_m)+r^2(2a^2k_2\alpha_1(3\lambda_1+2\mu_1)\mu_2(\lambda_2+2\mu_2)$$
$$+5k_1\alpha_2(3\lambda_2+2\mu_2)(a^2(-3\lambda_1+2\lambda_2-2\mu_1)\mu_2$$
$$+r^2(\lambda_2+\mu_2)(3\lambda_1+2\mu_1+4\mu_2)))\delta_{im}))/$$
$$(15k_1k_2r^5(\lambda_2+2\mu_2)(3\lambda_1+2\mu_1+4\mu_2)).$$

Figure 3.13 is a plot for σ_{11} for two arbitrarily chosen material systems chosen from Table 3.1 below.

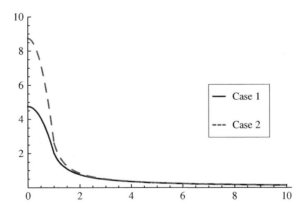

Figure 3.13 Thermal stress distribution due to heat source

Table 3.1 Material properties used in Figure 3.13

	α_1	α_2	k_1	k_2	q	a	λ_1	λ_2	μ_1	μ_2
Case 1	10	1	10	1	1	1	5	2	3	1
Case 2	10	1	100	1	1	1	100	2	100	1

The following is a *Mathematica* code for deriving the thermal stress.

```
SetAttributes[δ, Orderless];
SetAttributes[X, Orderless];
SetAttributes[σ∞, Orderless];
δ[i_Integer, j_Integer] := If[i = j, 1, 0];
δ[i_Symbol, i_Symbol] := 3;
X[i_Symbol, i_Symbol] := 0;
σ∞[i_Symbol, i_Symbol] := 0;

Unprotect[Times];
Times[x[j_Symbol], δ[i_, j_Symbol]] := x[i];
Times[X[i_Symbol, j_], δ[i_Symbol, k_]] := X[k, j];
Times[σ∞[i_Symbol, j_], δ[i_Symbol, k_]] := σ∞[k, j];
Times[δ[i_Symbol, j_], δ[i_Symbol, k_]] := δ[j, k];
Times[δ[i_Symbol, j_], h_[i_Symbol, m_, n_]] := h[j, m, n];
Times[δ[i_Symbol, j_], h_[m, i_Symbol, n_]] := h[m, j, n];
Times[δ[i_Symbol, j_], h_[m, n, i_Symbol]] := h[m, n, j];
Times[δ[i_Symbol, j_], h_[i_Symbol]] := h[j];
Protect[Times];

Unprotect[Power];
Power[x[i_Symbol], 2] := r^2;
Power[δ[i_Symbol, j_Symbol], 2] := 3;
Protect[Power];

In[73]:= Unprotect[D];
D[a_.r^n_.+b_., x[i_]] :=
   D[a, x[i]] r^n+n r^(n-1) x[i] /r a + D[b, x[i]];
D[a_.x[i_]+b_., x[j_]] := D[a, x[j]] x[i] + a δ[i, j] + D[b, x[j]];
D[a_.x[i_Integer]^n_.+b_., x[j_]] := a n x[i]^(n-1) δ[i, j] + D[b, x[j]];
D[a_.f_[r]^n_.+b_., x[i_]] :=
   D[a, x[i]] f[r]^n+a n f[r]^(n-1) f'[r] x[i] /r+D[b, x[i]];
D[a_List, b_] := Map[D[#, b] &, a]
D[Times[a_, b_], x[i_]] := D[a, x[i]] b + a D[b, x[i]];
Protect[D];
```

In[35]:= ui = (hh[r] + C[1] + C[2] / r^3) x[i]

Out[35]= $\left(C[1] + \dfrac{C[2]}{r^3} + hh[r]\right) x[i]$

```
In[37]:= sim = D[ui, x[m]];
         smm = sim /. m -> i // Expand;
         σim = 2 μ sim + λ δ[i, m] smm // Expand;
         ti = σim x[m] / r // Expand;
         phase1[f_] := f /. {hh -> hh1, μ -> μ1, λ -> λ1, C[2] -> 0, α -> α1};
         phase2[f_] := f /. {μ -> μ2, λ -> λ2, hh -> hh2, C -> C2, α -> α2} /. {C2[1] -> 0}
         ui1 = ui // phase1;
         ui2 = ui // phase2;
         ti1 = ti // phase1;
         ti2 = ti // phase2;
In[46]:=        (ui1 - ui2) /. r -> a // Expand // Factor

Out[46]=  (a³ C[1] - C2[2] + a³ hh1[a] - a³ hh2[a]) x[i]
         ─────────────────────────────────────────────────
                                a³

In[47]:= eq1 = (a³ C[1] - C2[2] + a³ hh1[a] - a³ hh2[a]) == 0;

In[48]:= (ti1 - ti2) /. r -> a // Expand // Factor

Out[48]=  1
         ── x[i] (3 a³ λ1 C[1] + 2 a³ μ1 C[1] + 4 μ2 C2[2] +
         a⁴
                3 a³ λ1 hh1[a] + 2 a³ μ1 hh1[a] - 3 a³ λ2 hh2[a] - 2 a³ μ2 hh2[a] +
                a⁴ λ1 hh1'[a] + 2 a⁴ μ1 hh1'[a] - a⁴ λ2 hh2'[a] - 2 a⁴ μ2 hh2'[a])

In[49]:= eq2 = (3 a³ λ1 C[1] + 2 a³ μ1 C[1] + 4 μ2 C2[2] +
                3 a³ λ1 hh1[a] + 2 a³ μ1 hh1[a] - 3 a³ λ2 hh2[a] - 2 a³ μ2 hh2[a] +
                a⁴ λ1 hh1'[a] + 2 a⁴ μ1 hh1'[a] - a⁴ λ2 hh2'[a] - 2 a⁴ μ2 hh2'[a]) == 0;

In[50]:= sol = Solve[{eq1, eq2}, {C[1], C2[2]}][[1]]

Out[50]= {C[1] -> -  1
                   ─────────────── (3 λ1 hh1[a] + 2 μ1 hh1[a] + 4 μ2 hh1[a] - 3 λ2 hh2[a] -
                   3 λ1 + 2 μ1 + 4 μ2
                6 μ2 hh2[a] + a λ1 hh1'[a] + 2 a μ1 hh1'[a] - a λ2 hh2'[a] - 2 a μ2 hh2'[a]),

          C2[2] -> -  1
                    ─────────────── (3 a³ λ1 hh2[a] - 3 a³ λ2 hh2[a] + 2 a³ μ1 hh2[a] -
                    3 λ1 + 2 μ1 + 4 μ2
                2 a³ μ2 hh2[a] + a⁴ λ1 hh1'[a] + 2 a⁴ μ1 hh1'[a] - a⁴ λ2 hh2'[a] - 2 a⁴ μ2 hh2'[a])}

In[51]:= Ui1 = ui1 /. sol
         Ui2 = ui2 /. sol

Out[51]= x[i] (hh1[r] -       1
                        ─────────────── (3 λ1 hh1[a] + 2 μ1 hh1[a] + 4 μ2 hh1[a] - 3 λ2 hh2[a] -
                        3 λ1 + 2 μ1 + 4 μ2
                6 μ2 hh2[a] + a λ1 hh1'[a] + 2 a μ1 hh1'[a] - a λ2 hh2'[a] - 2 a μ2 hh2'[a]))

Out[52]= x[i] (hh2[r] -
                (3 a³ λ1 hh2[a] - 3 a³ λ2 hh2[a] + 2 a³ μ1 hh2[a] - 2 a³ μ2 hh2[a] + a⁴ λ1 hh1'[a] +
                2 a⁴ μ1 hh1'[a] - a⁴ λ2 hh2'[a] - 2 a⁴ μ2 hh2'[a]) / (r³ (3 λ1 + 2 μ1 + 4 μ2)))

In[54]:= t1 = -q / 6 / k1 r^2 + a^2 (2 k1 + k2) / (6 k1 k2) q; t2 = a^3 q / (3 k2) / r;
         hh1[r_] := Evaluate[
             α1 (2 μ1 + 3 λ1) / (2 μ1 + λ1) Integrate[r^2 t1, r] / r^3 // Expand // Factor]
         hh2[r_] := Evaluate[
             α2 (2 μ2 + 3 λ2) / (2 μ2 + λ2) Integrate[r^2 t2, r] / r^3 // Expand // Factor]
In[57]:= Ui1 = Ui1 // FullSimplify

Out[57]= (q (-k2 r² α1 (3 λ1 + 2 μ1) (3 λ1 + 2 μ1 + 4 μ2) +
             a² (10 k1 α2 (λ1 + 2 μ1) (3 λ2 + 2 μ2) + k2 α1 (3 λ1 + 2 μ1) (5 λ1 + 6 μ1 + 4 μ2)))
         x[i]) / (30 k1 k2 (λ1 + 2 μ1) (3 λ1 + 2 μ1 + 4 μ2))
```

```
In[58]:= Ui2 = Ui2 // FullSimplify
```
Out[58]= $\big(a^3 q \big(5 k1 r^2 \alpha2 (3 \lambda2 + 2 \mu2) (3 \lambda1 + 2 \mu1 + 4 \mu2) +$
$\qquad a^2 (2 k2 \alpha1 (3 \lambda1 + 2 \mu1) (\lambda2 + 2 \mu2) - 5 k1 \alpha2 (3 \lambda1 - 2 \lambda2 + 2 \mu1) (3 \lambda2 + 2 \mu2))\big)$
$\qquad x[i]\big) / \big(30 k1 k2 r^3 (\lambda2 + 2 \mu2) (3 \lambda1 + 2 \mu1 + 4 \mu2)\big)$

```
In[59]:= σim1 = σim /. sol // phase1 // FullSimplify
```
Out[59]= $\big(q (3 \lambda1 + 2 \mu1) \big(-4 k2 \alpha1 \mu1 (3 \lambda1 + 2 \mu1 + 4 \mu2) x[i] x[m] +$
$\qquad \big(-k2 r^2 \alpha1 (5 \lambda1 + 2 \mu1) (3 \lambda1 + 2 \mu1 + 4 \mu2) +$
$\qquad a^2 (10 k1 \alpha2 (\lambda1 + 2 \mu1) (3 \lambda2 + 2 \mu2) + k2 \alpha1 (3 \lambda1 + 2 \mu1) (5 \lambda1 + 6 \mu1 + 4 \mu2))\big)$
$\qquad \delta[i, m]\big)\big) / \big(30 k1 k2 (\lambda1 + 2 \mu1) (3 \lambda1 + 2 \mu1 + 4 \mu2)\big)$

```
In[60]:= σim2 = (σim // phase2) /. sol // FullSimplify
```
Out[60]= $\big(a^3 q \big(-\mu2 \big(5 k1 r^2 \alpha2 (3 \lambda2 + 2 \mu2) (3 \lambda1 + 2 \mu1 + 4 \mu2) + 3 a^2$
$\qquad (2 k2 \alpha1 (3 \lambda1 + 2 \mu1) (\lambda2 + 2 \mu2) - 5 k1 \alpha2 (3 \lambda1 - 2 \lambda2 + 2 \mu1) (3 \lambda2 + 2 \mu2))\big)$
$\qquad x[i] x[m] + r^2 \big(2 a^2 k2 \alpha1 (3 \lambda1 + 2 \mu1) \mu2 (\lambda2 + 2 \mu2) + 5 k1 \alpha2 (3 \lambda2 + 2 \mu2)$
$\qquad \big(a^2 (-3 \lambda1 + 2 \lambda2 - 2 \mu1) \mu2 + r^2 (\lambda2 + \mu2) (3 \lambda1 + 2 \mu1 + 4 \mu2)\big)\big) \delta[i, m]\big)\big) /$
$\qquad \big(15 k1 k2 r^5 (\lambda2 + 2 \mu2) (3 \lambda1 + 2 \mu1 + 4 \mu2)\big)$

```
In[62]:= εim1 = (εim // phase1) /. sol // FullSimplify
```
Out[62]= $\big(q \big(-2 k2 \alpha1 (3 \lambda1 + 2 \mu1) (3 \lambda1 + 2 \mu1 + 4 \mu2) x[i] x[m] +$
$\qquad \big(-k2 r^2 \alpha1 (3 \lambda1 + 2 \mu1) (3 \lambda1 + 2 \mu1 + 4 \mu2) +$
$\qquad a^2 (10 k1 \alpha2 (\lambda1 + 2 \mu1) (3 \lambda2 + 2 \mu2) + k2 \alpha1 (3 \lambda1 + 2 \mu1) (5 \lambda1 + 6 \mu1 + 4 \mu2))\big)$
$\qquad \delta[i, m]\big)\big) / \big(30 k1 k2 (\lambda1 + 2 \mu1) (3 \lambda1 + 2 \mu1 + 4 \mu2)\big)$

```
In[63]:= εim2 = (εim // phase2) /. sol // FullSimplify
```
Out[63]= $\big(a^3 q \big(-\big(5 k1 r^2 \alpha2 (3 \lambda2 + 2 \mu2) (3 \lambda1 + 2 \mu1 + 4 \mu2) + 3 a^2$
$\qquad (2 k2 \alpha1 (3 \lambda1 + 2 \mu1) (\lambda2 + 2 \mu2) - 5 k1 \alpha2 (3 \lambda1 - 2 \lambda2 + 2 \mu1) (3 \lambda2 + 2 \mu2))\big)$
$\qquad x[i] x[m] + r^2 \big(5 k1 r^2 \alpha2 (3 \lambda2 + 2 \mu2) (3 \lambda1 + 2 \mu1 + 4 \mu2) +$
$\qquad a^2 (2 k2 \alpha1 (3 \lambda1 + 2 \mu1) (\lambda2 + 2 \mu2) - 5 k1 \alpha2 (3 \lambda1 - 2 \lambda2 + 2 \mu1) (3 \lambda2 + 2 \mu2))\big)$
$\qquad \delta[i, m]\big)\big) / \big(30 k1 k2 r^5 (\lambda2 + 2 \mu2) (3 \lambda1 + 2 \mu1 + 4 \mu2)\big)$

3.3.2 Thermal Stress Due to Heat Flow

If a composite is placed in a heat flow, thermal stresses result due to a mismatch of thermal expansion. In this subsection, the thermal stress field induced by a constant heat flux at the far field for an elastic medium that contains a spherical inclusion at the center is derived. Tauchert (1968) solved a thermal stress problem, by which this part of the book was motivated, using a method developed by Florence and Goodier (1960). However, Tauchert's solution is not complete as the heat flux is assumed to exist only in the z-direction, and the analytical method shown cannot be generalized. Other approaches include the use of complex variable theory such as Chao and Shen (1998), but they are limited to two-dimensional problems.

The formulation when the temperature distribution is generated due to heat flux is more involved than when the temperature is generated by a heat source because the heat source is a scalar (a zeroth-rank tensor) while the heat flux is a vector (a first-rank tensor).

3.3.2.1 Temperature Field Under Uniform Heat Flow at Infinity

Consider a 3-D body extended to infinity that contains a spherical inclusion at the center subject to a uniform heat flow at infinity. The temperature field is asymptotically expressed as

$$T \to \theta_k x_k \quad \text{as} \quad r \to \infty,$$

where θ_k is a constant temperature gradient. As the only source for the temperature is the temperature gradient, θ_k, at infinity, the temperature field must be proportional to θ_k. In addition, as the temperature is a scalar (a zeroth-rank tensor), the only possible formula for the temperature field is

$$T = h(r)x_k\theta_k, \quad (3.43)$$

where $h(r)$ is a function of the distance, $r = \sqrt{x_k x_k}$, alone.

As there is no heat source, the temperature field, T, is governed by the Laplace equation as

$$T_{,ii} = 0. \quad (3.44)$$

By substituting Equation (3.43) into Equation (3.44), the differential equation that $h(r)$ must satisfy is obtained as

$$T_{,ii} = \frac{rh''(r) + 4h'(r)}{r} x_m \theta_m = 0.$$

Therefore, by solving the differential equation, $rh''(r) + 4h'(r) = 0$, one obtains

$$h(r) = \frac{C_1}{r^3} + C_2,$$

where C_1 and C_2 are integral constants that need to be determined from the continuity conditions at the interface of the inclusion and the matrix for the temperature and heat flux.

The temperature field, T, is thus expressed as

$$T = \left(\frac{C_1}{r^3} + C_2\right) x_k \theta_k.$$

As the temperature must remain finite inside the inclusion and as $x \to \infty$, $T_{,i} \to \theta_i$, the temperature fields for the inclusion and the matrix phases must be expressed as

$$T^f = C_1 x_k \theta_k, \quad (3.45)$$

$$T^m = \left(\frac{C_2}{r^3} + 1\right) x_k \theta_k. \quad (3.46)$$

where the superscripts, f and m, refer to the inclusion phase and the matrix phase, respectively.

From Equations (3.45) and (3.46), the heat flux for each phase is expressed as

$$k_f \frac{\partial T^f}{\partial n} = k_f \frac{C_1}{r} x_k \theta_k,$$

$$k_m \frac{\partial T^m}{\partial n} = k_m \left(-2\frac{C_2}{r^4} + \frac{C_2}{r}\right) x_k \theta_k,$$

where k_f and k_m are the thermal conductivities for the inclusion phase and the matrix phase, respectively.

From the continuity conditions of the temperature and heat flux at $r = a$, the unknowns, C_1 and C_2, can be solved as

$$C_1 = -\frac{a^3(k_f - k_m)}{k_f + 2k_m},$$

$$C_2 = \frac{3k_m}{k_f + 2k_m}.$$

Thus, the temperature field for each phase is expressed as

$$T^f = \frac{3k_m}{k_f + 2k_m} x_k \theta_k, \qquad (3.47)$$

$$T^m = \left(-\frac{k_f - k_m}{k_f + 2k_m}\left(\frac{a}{r}\right)^3 + 1\right) x_k \theta_k. \qquad (3.48)$$

The following *Mathematica* code can derive the temperature field.

```
In[2]:= SetAttributes[δ, Orderless];
        δ[i_Integer, j_Integer] := If[i == j, 1, 0];
        δ[i_Symbol, i_Symbol] := 3;
        Unprotect[Times];
        Times[x_[j_Symbol], δ[i_, j_Symbol]] := x[i];
        Times[X_[i_Symbol, j_], δ[i_Symbol, k_]] := X[k, j];
        Times[δ[i_Symbol, j_], δ[i_Symbol, k_]] := δ[j, k];
        Times[δ[i_Symbol, j_], h_[i_Symbol]] := h[j];
        Protect[Times];
In[11]:= Unprotect[Power];
        Power[x[i_Symbol], 2] := r^2;
        Power[δ[i_Symbol, j_Symbol], 2] := 3;
        Protect[Power];

In[15]:= Unprotect[D];

D[a_. r^n_. + b_., x[i_]] :=
  D[a, x[i]] r^n + n r^(n-1) x[i]/r a + D[b, x[i]];
D[a_. x[i_] + b_., x[j_]] := D[a, x[j]] x[i] + a δ[i, j] + D[b, x[j]];
D[a_. x[i_Integer]^n_. + b_., x[j_]] :=
  a n x[i]^(n-1) δ[i, j] + D[b, x[j]];
D[a_. f_[r]^n_. + b_., x[i_]] :=
  D[a, x[i]] f[r]^n + a n f[r]^(n-1) f'[r] x[i]/r + D[b, x[i]];
D[a_List, b_] := Map[D[#, b] &, a]
D[Times[a_, b_], x[i_]] := D[a, x[i]] b + a D[b, x[i]];
Protect[D];
        In[23]:= temp = h[r] x[k] θ[k]
        Out[23]= h[r] x[k] θ[k]

        In[24]:= (D[D[temp, x[i]], x[i]] // Simplify) /.
                 x[i_] θ[i_] -> x[p] θ[p] // Simplify
        Out[24]= x[p] θ[p] (4 h'[r] + r h''[r])
                 ─────────────────────────────
                              r

        In[25]:= DSolve[ (4 h'[r] + r h''[r]) == 0, h[r], r][[1]]
        Out[25]= {h[r] -> - C[1]/(3 r^3) + C[2]}
```

```
In[26]:= tin = c[1] x[k] θ[k]; tout = (c[2]/r^3 + 1) x[k] θ[k];

In[27]:= eq1 =
   (((kin D[tin, x[i]] x[i]/r - kout D[tout, x[i]] x[i]/r) // Simplify) /.
      x[i_] θ[i_] → x[p] θ[p] // Simplify) /. r → a
```

$$\text{Out[27]=} \quad \frac{\left(-a^3 \text{ kout} + a^3 \text{ kin } c[1] + 2 \text{ kout } c[2]\right) x[p] \theta[p]}{a^4}$$

```
In[28]:= eq2 = (tin - tout) /. r → a // Simplify
```

$$\text{Out[28]=} \quad \left(-1 + c[1] - \frac{c[2]}{a^3}\right) x[k] \theta[k]$$

```
In[29]:= sol = Solve[{eq1 == 0, eq2 == 0}, {c[1], c[2]}][[1]]
```

$$\text{Out[29]=} \quad \left\{c[1] \to \frac{3 \text{ kout}}{\text{kin} + 2 \text{ kout}}, \; c[2] \to \frac{a^3 (-\text{kin} + \text{kout})}{\text{kin} + 2 \text{ kout}}\right\}$$

```
In[30]:= tin /. sol
```

$$\text{Out[30]=} \quad \frac{3 \text{ kout } x[k] \theta[k]}{\text{kin} + 2 \text{ kout}}$$

```
In[31]:= tout /. sol
```

$$\text{Out[31]=} \quad \left(1 + \frac{a^3 (-\text{kin} + \text{kout})}{(\text{kin} + 2 \text{ kout}) r^3}\right) x[k] \theta[k]$$

3.3.2.2 Thermal Stress Field

A 3-D elastic body is assumed to contain a spherical inclusion with a radius, a, of different elastic and thermal properties from the surrounding matrix subject to a contact temperature gradient, θ_m, at infinity. All the properties are assumed to be isotropic.

From Equations (3.47) and (3.48), the temperature gradient for each phase is expressed as

$$T_{,i}^f = \frac{3k_m}{k_f + 2k_m} \theta_i, \tag{3.49}$$

$$T_{,i}^m = \left(1 + \frac{k_m - k_f}{k_f + 2k_m}\left(\frac{a}{r}\right)^3\right)\theta_i - \frac{3(k_m - k_f)}{k_f + 2k_m}\left(\frac{a}{r}\right)^3 \frac{\theta_k x_k x_i}{r^2}. \tag{3.50}$$

The equilibrium equation for the displacements, u_i, with the thermal effect is expressed as

$$\mu \Delta u_i + (\mu + \lambda)u_{j,ji} - (2\mu + 3\lambda)\alpha T_{,i} = 0, \tag{3.51}$$

where μ and λ are the Lamé constants and α is the thermal expansion coefficient.

It is noted that the displacement and, hence, the stress are induced due to the presence of the $T_{,i}$ term in Equation (3.51). Therefore, with reference to Equations (3.49) and (3.50), the only possible way that the displacement is expressed is to assume that

$$u_i = f(r)\theta_i + \frac{g(r)}{r^2}\theta_k x_k x_i, \tag{3.52}$$

where $f(r)$ and $g(r)$ are unknown functions of r alone yet to be determined. Note that the term $1/r^2$ is introduced so that both $f(r)$ and $g(r)$ have the same dimension.

Substitution of Equation (3.52) into Equation (3.51) yields

$$\left(\mu f''(r) + \frac{(\lambda+\mu)f'(r)}{r} + \frac{2\mu f'(r)}{r} + \frac{(\lambda+\mu)g'(r)}{r} + \frac{2g(r)(\lambda+\mu)}{r^2} + \frac{2\mu g(r)}{r^2}\right)\theta_i$$

$$+ \left(\frac{(\lambda+\mu)f''(r)}{r^2} - \frac{(\lambda+\mu)f'(r)}{r^3} + \frac{(\lambda+\mu)g''(r)}{r^2} + \frac{\mu g''(r)}{r^2} + \frac{(\lambda+\mu)g'(r)}{r^3}\right.$$

$$\left. + \frac{2\mu g'(r)}{r^3} - \frac{4g(r)(\lambda+\mu)}{r^4} - \frac{6\mu g(r)}{r^4}\right)x_i\theta_k x_k$$

$$-(2\mu + 3\lambda)\alpha T_{,i} = 0.$$

Therefore, the differential equations that $f(r)$ and $g(r)$ must satisfy are expressed as follows:

1. Inside the inclusion:

$$\mu_f f''(r) + \frac{(\lambda_f + \mu_f)f'(r)}{r} + \frac{2\mu_f f'(r)}{r} + \frac{(\lambda_f + \mu_f)g'(r)}{r}$$

$$+ \frac{2g(r)(\lambda_f + \mu_f)}{r^2} + \frac{2\mu_f g(r)}{r^2} = \alpha_f(2\mu_f + 3\lambda_f)\frac{3k_m}{k_f + 2k_m},$$

$$\frac{(\lambda_f + \mu_f)f''(r)}{r^2} - \frac{(\lambda_f + \mu_f)f'(r)}{r^3}$$

$$+ \frac{(\lambda_f + \mu_f)g''(r)}{r^2} + \frac{\mu_f g''(r)}{r^2} + \frac{(\lambda_f + \mu_f)g'(r)}{r^3}$$

$$+ \frac{2\mu_f g'(r)}{r^3} - \frac{4g(r)(\lambda_f + \mu_f)}{r^4} - \frac{6\mu_f g(r)}{r^4} = 0.$$

2. In the matrix:

$$\mu_m f''(r) + \frac{(\lambda_m+\mu_m)f'(r)}{r} + \frac{2\mu_m f'(r)}{r} + \frac{(\lambda_m+\mu_m)g'(r)}{r}$$

$$+ \frac{2g(r)(\lambda_m+\mu_m)}{r^2} + \frac{2\mu_m g(r)}{r^2} = \alpha_m(2\mu_m + 3\lambda_m)\left(1 + \frac{k_m-k_f}{k_f+2k_m}\left(\frac{a}{r}\right)^3\right),$$

$$\frac{(\lambda_m+\mu_m)f''(r)}{r^2} - \frac{(\lambda_m+\mu_m)f'(r)}{r^3} + \frac{(\lambda_m+\mu_m)g''(r)}{r^2} + \frac{\mu_m g''(r)}{r^2} + \frac{(\lambda_m+\mu_m)g'(r)}{r^3}$$

$$+ \frac{2\mu_m g'(r)}{r^3} - \frac{4g(r)(\lambda_m+\mu_m)}{r^4} - \frac{6\mu_m g(r)}{r^4} = -\frac{3\alpha_m(2\mu_m+3\lambda_m)(k_m-k_f)}{k_f+2k_m}\frac{a^3}{r^5}.$$

By solving the aforementioned set of simultaneous differential equations for $f(r)$ and $g(r)$, one obtains the following:

1. Inside the inclusion:

$$f(r) = c_1 - \frac{c_2}{3r^3} - \frac{c_3}{r} + \frac{r^2}{2}c_4 + \frac{k_m r^2 \alpha_f(12\lambda_f^2 + 35\lambda_f\mu_f + 18\mu_f^2)}{10(k_f + 2k_m)\mu_f(\lambda_f + 2\mu_f)},$$

$$g(r) = \frac{c_2}{r^3} - \frac{\lambda_f + \mu_f}{r(\lambda_f + 3\mu_f)}c_3 - \frac{r^2(\lambda_f + 4\mu_f)}{2(2\lambda_f + 3\mu_f)}c_4$$

$$+\frac{\alpha_f k_m r^2 (12\lambda_f^2 + 35\lambda_f \mu_f + 18\mu_f^2)}{10\mu_f (k_f + 2k_m)(\lambda_f + 2\mu_f)}. \quad (3.53)$$

2. In the matrix:

$$f(r) = d_1 - \frac{d_2}{3r^3} - \frac{d_3}{r} + \frac{r^2}{2}d_4$$

$$+ \frac{\alpha_m(3\lambda_m + 2\mu_m)(5a^3\lambda_m(k_m - k_f) + r^3(k_f + 2k_m)(4\lambda_m + 9\mu_m))}{30\mu_m r(k_f + 2k_m)(\lambda_m + 2\mu_m)},$$

$$g(r) = \frac{d_2}{r^3} + \frac{(-\lambda_m - \mu_m)}{r(\lambda_m + 3\mu_m)}d_3 - \frac{r^2(\lambda_m + 4\mu_m)}{2(2\lambda_m + 3\mu_m)}d_4$$

$$- \frac{\alpha_m(3\lambda_m + 2\mu_m)(5a^3(k_f - k_m)(\lambda_m + 4\mu_m) + 2r^3(k_f + 2k_m)(\lambda_m + \mu_m))}{30\mu_m r(k_f + 2k_m)(\lambda_m + 2\mu_m)}, \quad (3.54)$$

where c_1–c_4 and d_1–d_4 are integral constants to be determined from the boundary and continuity conditions. As the displacement must remain finite inside the inclusion, it follows from Equation (3.53) that

$$c_2 = c_3 = 0.$$

The determination of d_1 and d_4 in Equation (3.54) comes from the condition that the stress vanishes at the far field, which requires the expression of the stress field. The rest of the integral constants can be determined from the conditions that both the displacement and the surface traction are continuous at $r = a$.

The stress components are expressed from the displacement as

$$\sigma_{ij} = C_{ijkl}(\epsilon_{kl} - \alpha\delta_{kl}T)$$
$$= 2\mu\epsilon_{ij} + \lambda\delta_{ij}\epsilon_{kk} - (2\mu + 3\lambda)\alpha T\delta_{ij}$$
$$= \mu u_{i,j} + \mu u_{j,i} + \lambda\delta_{ij}u_{k,k} - (2\mu + 3\lambda)\alpha T\delta_{ij}.$$

Therefore, the stress in the matrix is expressed from Equations (3.52) and (3.54) as

$$\sigma_{ij}^m = \left(\frac{A}{r^5} + \frac{B}{r^3} + C\right)(x_i\theta_j + x_j\theta_i)$$

$$+ \left(\frac{D}{r^7} + \frac{E}{r^5}\right)x_i x_j x_k \theta_k$$

$$+ \left(\frac{F}{r^5} + \frac{G}{r^3} + 4C\right)x_k x_k \delta_{ij}. \quad (3.55)$$

The constants, A–F, can be computed by *Mathematica* although they are not shown here because of their length. From Equation (3.55), it is seen that the condition that the stress vanishes as $r \to \infty$ is that C is 0. The explicit form of C is given as

$$C = \frac{(3\lambda_m + 2\mu_f)(2\alpha_m(6\lambda_m^2 + 25\lambda_m\mu_f + 24\mu_f^2) + 15d_4\mu_f(\lambda_m + 2\mu_f))}{30(\lambda_m + 2\mu_f)(2\lambda_m + 3\mu_f)}.$$

Therefore, by solving $C = 0$ for d_4, one of the integral constants, d_4, is now determined as

$$d_4 = -\frac{2\alpha_m(2\lambda_m + 3\mu_m)(3\lambda_m + 8\mu_m)}{15\mu_m(\lambda_m + 2\mu_m)}.$$

For the determination of the rest of the integral constants, a set of four simultaneous equations are derived from the condition that both the displacement and the traction are continuous at $r = a$. Note that the traction, t_i, is expressed as

$$t_i = \sigma_{ij}n_j$$
$$= \mu u_{i,j}n_j + \mu u_{j,i}n_j + \lambda u_{k,k}n_i - (2\mu + 3\lambda)\alpha T n_i.$$

3.3.2.3 Results

All the computations shown can be carried out by *Mathematica*. The stress inside the inclusion is expressed as

$$\sigma_{ij}^f = A'(x_i\theta_j + x_j\theta_i - 4x_k\theta_k\delta_{ij}),$$

where

$$A' = -\frac{2\mu_m\mu_f(3\lambda_f + 2\mu_m)(\alpha_m k_f - 3\alpha_f k_m + 2\alpha_m k_m)}{(k_f + 2k_m)(\lambda_f(3\mu_m + 2\mu_f) + 2\mu_m(\mu_m + 4\mu_f))}.$$

The stress in the matrix is expressed as

$$\sigma_{ij}^m = \left(\frac{A}{r^5} + \frac{B}{r^3}\right)(x_i\theta_j + x_j\theta_i)$$
$$+ \left(-\frac{5A}{r^7} - \frac{3B}{r^5}\right)x_ix_jx_k\theta_k$$
$$+ \left(\frac{A}{r^5} - \frac{B}{r^3}\right)x_kx_k\delta_{ij},$$

where

$$A = \frac{\begin{array}{c}a^5\mu_f(\alpha_m k_f(\lambda_f(3\lambda_m\mu_m + 6\lambda_m\mu_f - 6\mu_m\mu_f + 4\mu_f^2) + 2\mu_m(\lambda_m\mu_m + 12\lambda_m\mu_f - 2\mu_m\mu_f + 8\mu_f^2))\\ + k_m(6\alpha_f\mu_m(3\lambda_f + 2\mu_m)(\lambda_m + 2\mu_f) - \alpha_m(\lambda_f(21\lambda_m\mu_m + 6\lambda_m\mu_f + 30\mu_m\mu_f + 4\mu_f^2)\\ + 2\mu_m(7\lambda_m\mu_m + 12\lambda_m\mu_f + 10\mu_m\mu_f + 8\mu_f^2))))\end{array}}{((k_f + 2k_m)(\lambda_m + 2\mu_f)(\lambda_f(3\mu_m + 2\mu_f) + 2\mu_m(\mu_m + 4\mu_f)))},$$

$$B = \frac{a^3\alpha_m\mu_f(k_m - k_f)(3\lambda_m + 2\mu_f)}{(k_f + 2k_m)(\lambda_m + 2\mu_f)}.$$

Example

Figure 3.14 is a plot for σ_{11} for an arbitrarily chosen material system.

Inclusions in Infinite Media

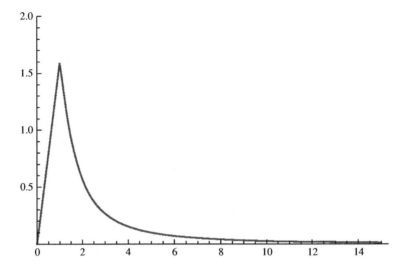

Figure 3.14 Thermal stress distribution due to heat flow

The following values were arbitrarily chosen for this plotting:

$$\mu^m = 1, \quad \mu^f = 12,$$
$$\lambda^m = 1, \quad \lambda^f = 3,$$
$$k^m = 1, \quad k^f = 10,$$
$$\alpha^m = 1, \quad \alpha^f = 2,$$
$$a = 1, \quad \theta = 1.$$

A stress concentration is seen at the interface, and the stress goes to 0 as $x \to \infty$. The following *Mathematica* code can do the computation mentioned:

```
In[2]:= SetAttributes[δ, Orderless];
   δ[i_Integer, j_Integer] := If[i == j, 1, 0];
   δ[i_Symbol, i_Symbol] := 3;
   Unprotect[Times];
   Times[x_[j_Symbol], δ[i_, j_Symbol]] := x[i];
   Times[X_[i_Symbol, j_], δ[i_Symbol, k_]] := X[k, j];
   Times[δ[i_Symbol, j_], δ[i_Symbol, k_]] := δ[j, k];
   Times[δ[i_Symbol, j_], h_[i_Symbol]] := h[j]; Protect[Times];

In[10]:= Unprotect[Power];
   Power[x[i_Symbol], 2] := r^2;
   Power[δ[i_Symbol, j_Symbol], 2] := 3;
   Protect[Power];
```

In[14]:= Unprotect[D];
    D[a_ . r^n_ . + b_ ., x[i_]] :=
        D[a, x[i]] r^n + n r^(n-1) x[i] / r a + D[b, x[i]];
    D[a_ . x[i_] + b_ ., x[j_]] := D[a, x[j]] x[i] + a δ[i, j] + D[b, x[j]];
    D[a_ . x[i_Integer]^n_ . + b_ ., x[j_]] :=
        a n x[i]^(n-1) δ[i, j] + D[b, x[j]];
    D[a_ . f_[r]^n_ . + b_ ., x[i_]] :=
        D[a, x[i]] f[r]^n + a n f[r]^(n-1) f'[r] x[i] / r + D[b, x[i]];
    D[Times[a_, b_], x[i_]] := D[a, x[i]] b + a D[b, x[i]];
    Protect[D];
In[21]:= mat = {km → 1, kf → 10, μm → 1, μf → 12, λm → 1, λf → 3, αm → 1, αf → 2, a → 1};
    mat = {};
    rule1 = {μm → em / 2 / (1 + νm), μf → ef / 2 / (1 + νf),
        λm → νm em / (1 + νm) / (1 - 2 νm), λf → νf ef / (1 + νf) / (1 - 2 νf)};

In[23]:= um = f[r] θ[m] + g[r] / r^2 x[p] θ[p] x[m];
    (*Δu_m*)
    j1 = (D[D[um, x[i]] // Expand, x[i]] // Expand) /. x[i] θ[i] → x[p] θ[p];
    (*u_{i,im}*)
    j2 =
        (D[(D[um, x[m]] // Expand) /. x[m] θ[m] -> x[p] θ[p], x[m]]) // Expand;
    (*μΔu_m + (μ+λ) u_{i,im} *)
    j3 = μ j1 + (μ + λ) j2;
    fsol = Coefficient[j3, θ[m]];
    gsol = Coefficient[j3, x[p] θ[p] x[m]];
    (* inside *)
    a1 = 3 km / (kf + 2 km);
    sol1 = DSolve[{(fsol /. {μ → μf, λ → λf}) == a1 αf (2 μf + 3 λf),
        (gsol /. {μ → μf, λ → λf}) == 0}, {f[r], g[r]}, r][[1]];
    (* outside *)
    a2 = (km - kf) / (kf + 2 km);
    sol2 =
        DSolve[{(fsol /. {μ → μm, λ → λm}) == αm (2 μm + 3 λm) (1 + a2 a^3 / r^3),
            (gsol /. {μ → μm, λ → λm}) == -3 αm (2 μm + 3 λm) a2 a^3 / r^5},
            {f[r], g[r]}, r][[1]];
In[33]:= (*---------------General solution----------------*)
    uin = ((f[r] θ[m] + g[r] / r^2 x[p] θ[p] x[m] /. sol1) /. C → C1) /.
        {C1[3] → 0, C1[2] → 0};
    uout = (f[r] θ[m] + g[r] / r^2 x[p] θ[p] x[m] /. sol2) /. C → C2;
In[35]:= (* ------------------ Stress in matrix ------------------ *)
    j1 = D[uout, x[n]]; j2 = D[uout /. m → n, x[m]];
    emnout = (j1 + j2) / 2;
    emnout = ((emnout /. n → m) // Expand) /. x[m] θ[m] -> x[p] θ[p];
    σmnout = 2 μm emnout + λm δ[m, n] emnout;
In[39]:= tempout = ((km - kf) / (kf + 2 km) (a / r)^3 + 1) x[p] θ[p];
    thermalstressout = (2 μm + 3 λm) αm tempout ;

In[41]:= `(* Stress in matrix *)`
```
stressoutpre = omnout - thermalstressout δ[m, n];
Collect[Coefficient[stressoutpre, x[n] θ[m]], r, Simplify];
solc24 =
  Solve[(2 αm (6 λm² + 25 λm μm + 24 μm²) + 15 μm (λm + 2 μm) C2[4]) == 0, C2[4]][[
    1]] // Factor;
stressout = stressoutpre /. solc24;
tracout =
  Collect[(stressout x[n] / r // Expand) /. x[n] θ[n] -> x[p] θ[p],
    {θ[m], x[m] x[p] θ[p]}, FullSimplify];
```

In[47]:=
```
j10 = D[uin, x[n]];
j11 = D[uin /. m -> n, x[m]];
εmnin = (j10 + j11) / 2;
εmnin =
  ((εmnin /. n -> m) // Expand) /. x[m] θ[m] -> x[p] θ[p] // FullSimplify;
σmnin = 2 μf εmnin + λf δ[m, n] εmnin;
tempin = a1 x[p] θ[p];
thermalstressin = (2 μf + 3 λf) αf tempin;
stressin = σmnin - thermalstressin δ[m, n];
tracin = Collect[Expand[stressin x[n] / r] /. x[n] θ[n] -> x[p] θ[p],
    {θ[m], x[m] x[p] θ[p]}, FullSimplify]
```
In[58]:=
```
j20 = Collect[(uin - uout) /. r -> a, {θ[m], x[m] x[p] θ[p]}];
j21 =
  Collect[(tracout - tracin) /. r -> a, {θ[m], x[m] x[p] θ[p]}, Simplify];
eq1 = Coefficient[j20, θ[m]] == 0;
eq2 = Coefficient[j20, x[m] x[p] θ[p]] == 0;
eq3 = Coefficient[j21, θ[m]] == 0;
eq4 = Coefficient[j21, x[m] x[p] θ[p]] == 0;
solall =
  (Solve[{eq1, eq2, eq3, eq4}, {C2[2], C2[3], C1[4], C1[1]}] /. {C2[4] ->
```
$$-\frac{2\,\alpha m\,(2\,\lambda m+3\,\mu m)\,(3\,\lambda m+8\,\mu m)}{15\,\mu m\,(\lambda m+2\,\mu m)},\ C2[1]\to 0\})[[1]]\ //\ \text{FullSimplify}$$

In[83]:= `stressinFinal = Collect[stressin /. solall,`
  `{x[n] θ[m] + x[m] θ[n], x[p] δ[m, n] θ[p]}, FullSimplify]`
In[84]:= `stressoutFinal = Collect[Expand[stressout /. solall],`
  `{x[n] θ[m], x[m] θ[n], x[p] δ[m, n] θ[p]}, Simplify]`

3.4 Airy's Stress Function Approach

For a 2-D elasticity problem (plane stress and plain strain), solving the stress components under a given boundary condition and geometry can be facilitated by using the Airy stress function (Sadd 2009), which is a mathematically elegant approach as the Airy stress function is a solution to the biharmonic differential equation in complex variable theory. This is a classical approach that has been in existence for the past 60 years, and most of the major textbooks on elasticity have one chapter dedicated to this topic with ample examples. However, it should be noted that there has not been prior work available for the Airy stress function applied to heterogeneous materials due to the complexity of algebra involved in manipulating complex variable calculus.

Mathematica provides an ideal tool to revitalize the capability of the Airy stress function applied to 2-D inclusion problems. Similar to the approach presented in the previous section, the Airy stress function approach can solve problems in which Eshelby's method fails to address. In this section, the Airy stress function approach is shown to be able to derive the elastic field for a three-phase medium that consists of two concentric circular inclusions in an infinitely extended matrix.

3.4.1 Airy's Stress Function

In state of plane stress (thin plates) where $\sigma_{zz} = 0$, the 2-D stress equilibrium equations without body force can be expressed as

$$\sigma_{xx,x} + \sigma_{xy,y} = 0, \quad \sigma_{yx,x} + \sigma_{yy,y} = 0. \tag{3.56}$$

Equation (3.56) is automatically satisfied if σ_{xx}, σ_{xy}, and σ_{yy} are derived from a single function, $\phi(x, y)$, as

$$\sigma_{xx} = \phi_{,yy}, \quad \sigma_{yy} = \phi_{,xx}, \quad \sigma_{xy} = -\phi_{,xy}. \tag{3.57}$$

The function $\phi(x, y)$ is called the Airy stress function (Sadd 2009).

As there are only two equations for the three unknowns in Equations (3.56), an additional equation is needed to solve for all three unknowns, σ_{xx}, σ_{xy}, and σ_{yy}. In Chapter 2, the compatibility condition for 2-D isotropic bodies is expressed in terms of the strain tensor as

$$\epsilon_{[i[j,k]l]} = 0,$$

which yields only one equation in 2-D as

$$\epsilon_{xx,yy} + \epsilon_{yy,xx} = 2\epsilon_{xy,xy}. \tag{3.58}$$

Using the 2-D stress–strain relation for an isotropic body,

$$\epsilon_{xx} = \frac{1}{E}(\sigma_{xx} - \nu\sigma_{yy}), \quad \epsilon_{yy} = \frac{1}{E}(\sigma_{yy} - \nu\sigma_{xx}), \quad \epsilon_{xy} = \frac{1+\nu}{E}\sigma_{xy},$$

Equation (3.58) is rewritten as

$$2(1+\nu)\sigma_{xy,xy} = (\sigma_{xx,yy} + \sigma_{yy,xx}) - \nu(\sigma_{xx,xx} + \sigma_{yy,xx}). \tag{3.59}$$

Equation (3.59) is further simplified (eliminating the σ_{xy} term). By differentiating the 2-D stress equilibrium equation, $\sigma_{ij,j} = 0$, with respect to x_i, one obtains

$$\sigma_{ij,ji} = 0,$$

or

$$\sigma_{xx,xx} + \sigma_{yy,yy} = -2\sigma_{xy,xy}. \tag{3.60}$$

By substituting Equation (3.60) into Equation (3.59), one obtains

$$2(1+v)\left(\frac{-1}{2}\right)(\sigma_{xx,xx}+\sigma_{yy,yy}) = (\sigma_{xx,yy}+\sigma_{yy,xx}) - v(\sigma_{xx,xx}+\sigma_{yy,xx}),$$

which is simplified to

$$\sigma_{xx,yy}+\sigma_{yy,xx}+\sigma_{xx,xx}+\sigma_{yy,yy} = 0,$$

or

$$\Delta(\sigma_{xx}+\sigma_{yy}) = 0. \tag{3.61}$$

Substitution of Equation (3.57) into Equation (3.61) yields

$$\Delta\Delta\phi(x,y) = 0. \tag{3.62}$$

Equation (3.62) is known as a biharmonic equation and has a general solution expressed in terms of complex variable functions as

$$\phi(x,y) = \Re(\bar{z}\gamma(z) + \chi(z)), \tag{3.63}$$

where $\Re(z)$ is the real part of a complex number, z, $\bar{z}$ is the complex conjugate of z, and $\gamma(z)$, and $\chi(z)$ are arbitrary complex analytic functions of z alone.

Proof. First, from the complex variable theory, it is noted that a general solution to the 2-D Laplace equation,

$$\Delta\phi = 0,$$

is expressed as

$$\phi(x,y) = \Re(f(z)),$$

where $f(z)$ is a complex analytic function (a function of z alone).
From the relationship,

$$z = x+iy, \quad \bar{z} = x-iy, \quad x = \frac{1}{2}(z+\bar{z}), \quad y = \frac{1}{2i}(z-\bar{z}),$$

and

$$\frac{\partial}{\partial z} = \frac{1}{2}\frac{\partial}{\partial x} + \frac{1}{2i}\frac{\partial}{\partial y}, \quad \frac{\partial}{\partial \bar{z}} = \frac{1}{2}\frac{\partial}{\partial x} - \frac{1}{2i}\frac{\partial}{\partial y}.$$

it follows that

$$\frac{\partial^2}{\partial z \partial \bar{z}} = \frac{1}{4}\left(\frac{\partial^2}{\partial x^2} + \frac{\partial^2}{\partial y^2}\right) = \frac{1}{4}\Delta. \tag{3.64}$$

Therefore, the equation

$$\Delta\Delta\phi = 0,$$

has a general solution,

$$\Delta\phi = \Re(f(z)), \tag{3.65}$$

where $f(z)$ is an arbitrary analytic function. Equation (3.65) with Equation (3.64) can be written as

$$\frac{\partial}{\partial z}\left(\frac{\partial \phi}{\partial \bar{z}}\right) = \Re(f(z)),$$

which can be integrated with respect to z as

$$\frac{\partial \phi}{\partial \bar{z}} = \Re\left(\int f(z)dz\right) \equiv \Re(\gamma(z)). \tag{3.66}$$

Integrating Equation (3.66) again but this time with respect to $\bar{z}$ yields

$$\phi = \Re(\bar{z}\gamma(z) + \chi(z)).$$

where $\chi(z)$ is what is equivalent to an integral constant, but in this case, it is a function of z alone, i.e., a complex analytic function.

Using the relationship,

$$\frac{\partial}{\partial x} = \frac{\partial}{\partial z} + \frac{\partial}{\partial \bar{z}}, \quad \frac{\partial}{\partial y} = i\frac{\partial}{\partial z} - i\frac{\partial}{\partial \bar{z}},$$

Equations (3.57) and (3.63) can be combined to yield the following expressions for the stress components as

$$\sigma_{xx} + \sigma_{yy} = 2(\gamma'(z) + \bar{\gamma}'(\bar{z})), \tag{3.67}$$

$$\sigma_{yy} - \sigma_{xx} + 2i\sigma_{xy} = 2(\bar{z}\gamma''(z) + \bar{\psi}'(\bar{z})), \tag{3.68}$$

$$u_x + iu_y = \frac{1}{2\mu}(\kappa\gamma(z) - z\bar{\gamma}'(\bar{z}) - \bar{\psi}(\bar{z})), \tag{3.69}$$

where

$$\chi'(z) \equiv \psi(z),$$

and $\bar{\gamma}'(\bar{z})$ means that $\bar{\gamma}(\bar{z})$, the complex conjugate of $\gamma(z)$, is differentiated with respect to $\bar{z}$. The quantity κ in Equation (3.69) is defined as

$$\kappa = \begin{cases} \frac{3-\nu}{1+\nu} & \text{plane stress} \\ 3 - 4\nu & \text{plane strain} \end{cases}$$

In the polar coordinate system, (r, θ), the stress components, σ_{rr}, $\sigma_{r\theta}$, and $\sigma_{\theta\theta}$, can be expressed as

$$\sigma_{rr} + \sigma_{\theta\theta} = 2(\gamma'(z) + \bar{\gamma}'(\bar{z}))e^{2i\theta}, \tag{3.70}$$

$$\sigma_{\theta\theta} - \sigma_{rr} + 2i\sigma_{r\theta} = 2(\bar{z}\gamma''(z) + \bar{\psi}'(\bar{z}))e^{2i\theta}, \tag{3.71}$$

$$u_r + iu_\theta = \frac{1}{2\mu}(\kappa\gamma(z) - z\bar{\gamma}'(\bar{z}) - \bar{\psi}(\bar{z}))e^{-i\theta}. \tag{3.72}$$

In general, $\gamma(z)$ and $\psi(z)$ take the form of the Laurent series about a particular point as

$$\gamma(z) = \sum_{n=-\infty}^{\infty} a_n z^n,$$

$$\psi(z) = \sum_{n=-\infty}^{\infty} b_n z^n,$$

where a_n and b_n are unknown complex coefficients that should be determined from the given boundary conditions. In a singly connected domain, the Taylor series should be used where n begins at 0 while in a multiply connected domain, the Laurent series should be used.

Example 3.1: Consider a rectangular bar defined as $(-a \leq x \leq a, -b \leq y \leq b)$ subject to a shear traction force $-W$, at $x = a$ and free at $y = \pm b$ as shown in Figure 3.15. The other end ($x = 0$) is fixed. This is a classical problem that is found in any textbook on strength of materials.

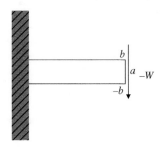

Figure 3.15 Bending of beam subject to shear

The boundary condition at $x = a$ is that $\sigma_{xx} = 0$ and $\int_{-b}^{b} \tau_{xy}\, dy = -W$. The boundary condition at $y = \pm b$ is that $\sigma_{yy} = 0$ and $\tau_{xy} = 0$. The functions $\gamma(z)$ and $\psi(z)$ are assumed to take the following form:

$$\gamma(z) = c_1 z^2 + c_2 z^3, \quad \psi(z) = d_1 z + d_2 z^2 + d_3 z^3,$$

where c_1–d_3 are complex coefficients.

The unknown coefficients of c_1–d_3 can be determined to satisfy the aforementioned boundary conditions and solved as

$$c_1 = 0, \quad c_2 = -\frac{3aW}{16b^3}i, \quad c_3 = \frac{W}{16b^3}i, \quad d_1 = -\frac{3W}{4b}i, \quad d_2 = \frac{3aW}{16b^3}i, \quad d_3 = -\frac{W}{8b^3}i.$$

Therefore, the stress functions, $f(z)$ and $h(z)$, are determined as

$$\gamma(z) = -\frac{3aW}{16b^3}iz^2 + \frac{W}{16b^3}iz^3, \quad \psi(z) = -\frac{3W}{4b}iz + \frac{3aW}{16b^3}iz^2 - \frac{W}{8b^3}iz^3.$$

Finally, the stress components are

$$\sigma_x = \frac{3W(a-x)y}{2b^3}, \quad \sigma_y = 0, \quad \tau_{xy} = \frac{3W(-b^2 + y^2)}{4b^3}.$$

This result agrees with those formulas found in structural mechanics textbooks. Even though this is a rather trivial example of the use of the Airy stress function, it illustrates the basic procedure to be followed.

Example 3.2: An infinitely extended body (isotropic) with a hole of a radius a is subject to the far-field stress shown in Figure 3.16.

The boundary conditions at infinity are

$$\sigma_x \to \sigma_x^\infty, \quad \sigma_y \to \sigma_y^\infty, \quad \tau \to \tau^\infty \quad \text{as} \quad x, y \to \infty.$$

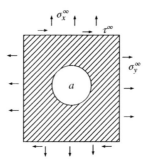

Figure 3.16 An infinitely extended body with a hole

Along the perimeter of the hole, there is no traction. Therefore, the following must be satisfied:
$$t_x = 0, t_y = 0 \quad \text{along} \quad x^2 + y^2 = a^2.^5$$

The two functions, $\gamma(z)$ and $\chi(z)$, can be chosen rather arbitrarily as

$$\gamma(z) = \frac{c_{-3}}{z^3} + \frac{c_{-2}}{z^2} + \frac{c_{-1}}{z} + c_0 + c_1 z,$$

$$\chi(z) = \frac{d_{-3}}{z^3} + \frac{d_{-2}}{z^2} + \frac{d_{-1}}{z} + d_0 + d_1 z.$$

The unknown coefficients, c_{-3}–d_1, are determined to satisfy the aforementioned boundary conditions. They can be solved as

$$c_{-3} = 0, \quad c_{-2} = 0, \quad c_{-1} = \frac{\sigma_x^\infty + \sigma_y^\infty}{4}, \quad c_0 = 0, \quad c_1 = \frac{\sigma_x^\infty - \sigma_y^\infty}{2} a^2 + a^2 \tau^\infty i.$$

$$d_{-3} = \frac{a^4}{2}(\sigma_x^\infty - \sigma_y^\infty) + i a^4 \tau^\infty, \quad d_{-2} = 0, \quad d_{-1} = -\frac{a^2}{2}(\sigma_x^\infty + \sigma_y^\infty),$$

$$d_0 = 0, \quad d_1 = \frac{\sigma_y^\infty - \sigma_x^\infty}{2} + i\tau^\infty.$$

The stress components are

$$\sigma_x = \frac{3a^4(\sigma_x^\infty - \sigma_y^\infty)(x^4 - 6x^2 y^2 + y^4)}{2(x^2 + y^2)^4} + \frac{6a^2 x^2 y^2(\sigma_x^\infty - \sigma_y^\infty)}{(x^2 + y^2)^3}$$
$$+ \frac{a^2 y^4(\sigma_x^\infty + \sigma_y^\infty)}{2(x^2 + y^2)^3} - \frac{a^2 x^4(5\sigma_x^\infty - 3\sigma_y^\infty)}{2(x^2 + y^2)^3}$$
$$- \frac{4a^2 \tau^\infty xy(-3a^2 x^2 + 3a^2 y^2 + 3x^4 + 2x^2 y^2 - y^4)}{(x^2 + y^2)^4} + \sigma_x^\infty,$$

---

5 $t_x = \sigma_x n_x + \tau n_y = \sigma_x \dfrac{x}{a} + \tau \dfrac{y}{a}, t_y = \tau n_x + \sigma_y n_y = \tau \dfrac{x}{a} + \sigma_y \dfrac{y}{a}.$

Inclusions in Infinite Media

$$\sigma_y = \frac{3a^4(\sigma_y^\infty - \sigma_x^\infty)(x^4 - 6x^2y^2 + y^4)}{2(x^2+y^2)^4} + \frac{6a^2x^2y^2(\sigma_y^\infty - \sigma_x^\infty)}{(x^2+y^2)^3} - \frac{a^2y^4(5\sigma_y^\infty - 3\sigma_x^\infty)}{2(x^2+y^2)^3}$$
$$+ \frac{a^2x^4(\sigma_x^\infty + \sigma_y^\infty)}{2(x^2+y^2)^3} + \frac{4a^2\tau^\infty xy(-3a^2x^2 + 3a^2y^2 + x^4 - 2x^2y^2 - 3y^4)}{(x^2+y^2)^4} + \sigma_y^\infty,$$

$$\tau = \frac{\tau^\infty(-3a^4x^4 + 18a^4x^2y^2 - 3a^4y^4 + 2a^2x^6 - 10a^2x^4y^2 - 10a^2x^2y^4 + 2a^2y^6 + x^8 + 4x^6y^2 + 6x^4y^4 + 4x^2y^6 + y^8)}{(x^2+y^2)^4}$$
$$- \frac{a^2xy(-6a^2\sigma_x^\infty x^2 + 6a^2\sigma_y^\infty x^2 + 6a^2\sigma_x^\infty y^2 - 6a^2\sigma_y^\infty y^2 + 5\sigma_x^\infty x^4 - 3\sigma_y^\infty x^4 + 2\sigma_x^\infty x^2y^2 + 2\sigma_y^\infty x^2y^2 - 3\sigma_x^\infty y^4 + 5\sigma_y^\infty y^4)}{(x^2+y^2)^4}.$$

When $\sigma_y^\infty = \tau^\infty = 0$, σ_x is reduced to

$$\sigma_x = \frac{3a^4\sigma_x^\infty(x^4 - 6x^2y^2 + y^4) - a^2(x^2+y^2)(5\sigma_x^\infty x^4 - 12\sigma_x^\infty x^2y^2 - \sigma_x^\infty y^4) + 2\sigma_x^\infty(x^2+y^2)^4}{2(x^2+y^2)^4}.$$

When $x = 0, y = a$, the aforementioned is reduced to

$$\sigma_x = 3\sigma_x^\infty,$$

which shows a stress concentration at the edge.

3.4.2 Mathematica Programming of Complex Variables

In *Mathematica*, the capital *I* is used to represent the imaginary number, $\sqrt{-1}$.[6]

In[1]:= `I^2`

Out[1]= -1

In[2]:= `I^101`

Out[2]= i

A complex number, $z = x + iy$, can be entered as

In[3]:= `z = x + I y`

Out[3]= $x + i y$

However, z^3 is not automatically expanded into the real part and the imaginary part as *Mathematica* does not know if x and y are complex or real.

In[4]:= `z^3`

Out[4]= $(x + i y)^3$

The `ComplexExpand` function assumes that all the symbols are real.

---

[6] One can also enter the imaginary number by pressing Esc, ii, Esc keys.

```
In[5]:= ComplexExpand[z^3]

Out[5]= x^3 - 3 x y^2 + i (3 x^2 y - y^3)
```

To extract the real and imaginary parts from a complex number, use the Re and Im functions followed by the ComplexExpand function as

```
In[5]:= ComplexExpand[z^3]

Out[5]= x^3 - 3 x y^2 + i (3 x^2 y - y^3)

In[6]:= Im[z^3]

Out[6]= Im[(x + i y)^3]

In[8]:= ComplexExpand[Re[z^3]]

Out[8]= x^3 - 3 x y^2

In[9]:= ComplexExpand[Im[z^3]]

Out[9]= 3 x^2 y - y^3
```

To implement the Airy stress function, first it is necessary to rewrite the partial derivative of a complex function with respect to z in terms of x and y as

$$\frac{\partial}{\partial z} = \frac{1}{2}\frac{\partial}{\partial x} - \frac{i}{2}\frac{\partial}{\partial y}. \quad (3.73)$$

For this purpose, a new function, dZ, is defined as

```
In[10]:= dZ[f_] := 1/2 D[f, x] - I/2 D[f, y];
         z = x + I y;  zbar = x - I y;

In[13]:= dZ[z^3 + Exp[z]]

Out[13]= -1/2 i (i e^(x+i y) + 3 i (x + i y)^2) + 1/2 (e^(x+i y) + 3 (x + i y)^2)

In[16]:= dZ[1/z^3] // ComplexExpand

Out[16]= -3 x^4/(x^2 + y^2)^4 + 18 x^2 y^2/(x^2 + y^2)^4 - 3 y^4/(x^2 + y^2)^4 + i (12 x^3 y/(x^2 + y^2)^4 - 12 x y^3/(x^2 + y^2)^4)
```

The solution procedure for Equations (3.67) and (3.68) can be implemented in *Mathematica* as

```
In[54]:= AiryStress[γ_, ψ_] := Module[{
         eq1 = (σx + σy == 2 ComplexExpand[dZ[γ] + Conjugate[dZ[γ]]]),
         eq2 = (σx - σy == ComplexExpand[Re[2 (zbar dZ[dZ[γ]] + dZ[ψ]) ]]),
         eq3 = (2 τxy == ComplexExpand[Im[2 (zbar dZ[dZ[γ]] + dZ[ψ]) ]])},
         sol = Solve[{eq1, eq2, eq3}, {σx, σy, τxy}];
         {sol[[1, 1, 2]], sol[[1, 2, 2]], sol[[1, 3, 2]]} // TrigReduce]
```

The function, `AiryStress`, takes two arguments, $\gamma(z)$ and $\psi(z)$,[7] and returns σ_{xx}, σ_{yy}, and σ_{xy} in the Cartesian coordinate system. For instance, if $\gamma(z) = 1/z + z^4$ and $\psi(z) = 1/z^3$, the function returns

In[60]:= `AiryStress[1/z + z^4, 1/z^3] // Simplify`

Out[60]= $\{(20 x^{11} + 68 x^9 y^2 + 72 x^7 y^4 + 8 x^5 y^6 - 28 x^3 y^8 -$
$12 x y^{10} + y^4 (-3 + 4 y^2) - 3 x^4 (1 + 4 y^2) + x^2 (18 y^2 - 8 y^4))/(x^2 + y^2)^4,$
$-(4 x^6 + 4 x^{11} + 52 x^9 y^2 - 3 y^4 + 168 x^7 y^4 + 232 x^5 y^6 + 148 x^3 y^8 + 36 x y^{10} - 6 x^2 y^2 (-3 + 2 y^2) -$
$x^4 (3 + 8 y^2))/(x^2 + y^2)^4, 4 y \left(3 x^2 + 3 y^2 + \frac{3 x^3}{(x^2+y^2)^4} - \frac{2 x^5}{(x^2+y^2)^4} + \frac{x y^2 (-3 + 2 y^2)}{(x^2+y^2)^4}\right)\}$

To obtain the displacement field, Equation (3.69) can be implemented by the function `AiryDisplacement` as

In[61]:= `AiryDisplacement[γ_, ψ_, κ_, μ_] := Module[{`
`  work = 1/(2 μ) (κ γ - z Conjugate[dZ[γ]] - Conjugate[ψ])},`
`  {ComplexExpand[Re[work]] // TrigReduce, ComplexExpand[Im[work]]}]`

The function `AiryDisplacement` takes four parameters, $\gamma(z)$, $\psi(z)$, κ, and μ, where κ is the bulk modulus and μ is the shear modulus. For instance, if $\gamma(z) = \frac{1}{z}$ and $\psi(z) = z^2$ are chosen, the displacement fields u_x and u_y are

In[65]:= `AiryDisplacement[1/z, z^2, k, μ] // Simplify`

Out[65]= $\{((1+k) x^3 - x^6 + (-3+k) x y^2 - x^4 y^2 + x^2 y^4 + y^6)/(2 (x^2+y^2)^2 \mu),$
$(y(-(-3+k) x^2 + 2 x^5 - (1+k) y^2 + 4 x^3 y^2 + 2 x y^4))/(2 (x^2+y^2)^2 \mu)\}$

3.4.3 Multiphase Inclusion Problems Using Airy's Stress Function

The Airy stress function combined with *Mathematica* provides a powerful tool for micromechanics. However, except for fracture mechanics applications (Shah 1995), there have been few applications of the Airy stress function to heterogeneous materials. The following example demonstrates that multiphase inclusion problems that were introduced in Section 3.2 can also be handled elegantly using the Airy stress function with *Mathematica* (Madhavan 2013). As in the second approach in Section 3.2, this approach has no limitation for the number of layers and can express the elastic field both inside and outside the inclusion to which the Eshelby method fails to address. The radii for the inner circle and the outer circle are denoted as a_1 and a_2, respectively.

The analysis begins by assuming that $\gamma(z)$ and $\psi(z)$ are expanded by the Laurent or Taylor series, depending on the region in question. In general, they are expressed as

$$\gamma(z) = \sum_{i=-\infty}^{\infty} c_i z^i, \quad \psi(z) = \sum_{i=-\infty}^{\infty} d_i z^i, \qquad (3.74)$$

---
[7] The Greek symbols, γ and ψ can be entered into *Mathematica* by pressing Esc g Esc keys and Esc psi Esc keys, respectively.

where c_i and d_i are unknown complex coefficients and take different values depending on the regions where $\gamma(z)$ and $\psi(z)$ are expanded.

Because of the symmetry of the problem, it is useful to employ the Airy stress and displacement functions that return the stress and displacement in polar form. The relationship between the stress and displacement components in polar form and the Airy stress function is cited here from the previous subsection as

$$\sigma_{rr} + \sigma_{\theta\theta} = 2(\gamma'(z) + \bar{\gamma}'(\bar{z}))e^{2i\theta}, \qquad (3.75)$$

$$\sigma_{\theta\theta} - \sigma_{rr} + 2i\sigma_{r\theta} = 2(\bar{z}\gamma''(z) + \bar{\psi}'(\bar{z}))e^{2i\theta}, \qquad (3.76)$$

$$u_r + iu_\theta = \frac{1}{2\mu}(\kappa\gamma(z) - z\bar{\gamma}'(\bar{z}) - \bar{\psi}(\bar{z}))e^{-i\theta}, \qquad (3.77)$$

The following *Mathematica* code implements Equations (3.75) and (3.76) and defines a function, `AiryStressPolar`, which returns the stress components in polar form, $(\sigma_r, \sigma_\theta, \tau_{r\theta})$ for given $\gamma(z)$ and $\psi(z)$.

```
In[79]:= polar = {x → r Cos[θ], y → r Sin[θ]};
AiryStressPolar[γ_, ψ_] := Module[{
    eq1 = (σr + σθ == 2 ComplexExpand[dZ[γ] + Conjugate[dZ[γ]]] /. polar),
    eq2 = (σθ - σr == ComplexExpand[Re[2 (zbar dZ[dZ[γ]] + dZ[ψ]) Exp[2 I θ]]] /. polar),
    eq3 = (2 τrθ == ComplexExpand[Im[2 (zbar dZ[dZ[γ]] + dZ[ψ]) Exp[2 I θ]]] /. polar)},
    sol = Solve[{eq1, eq2, eq3}, {σr, σθ, τrθ}];
    {sol[[1, 1, 2]], sol[[1, 2, 2]], sol[[1, 3, 2]]} // TrigReduce]
```

For instance, the stress field, $(\sigma_{rr}, \sigma_{\theta\theta}, \sigma_{r\theta})$, for $\gamma(z) = 1/z$ and $\psi(z) = 2z + 1/z^2$ is obtained as

```
In[81]:= AiryStressPolar[1/z, 2 z + 1/z^2]
```

$$\text{Out[81]}= \left\{-\frac{1}{r^3}2\left(-\text{Cos}[\theta] + 2r\,\text{Cos}[2\theta] + r^3\,\text{Cos}[2\theta]\right), \right.$$
$$\left(2\left(-\text{Cos}[\theta] + 2r\,\text{Cos}[2\theta] + r^3\,\text{Cos}[2\theta] - 2r\,\text{Cos}[\theta]^2\,\text{Cos}[2\theta] - 2r\,\text{Cos}[2\theta]\,\text{Sin}[\theta]^2\right)\right) /$$
$$\left(r^3\left(\text{Cos}[\theta]^2 + \text{Sin}[\theta]^2\right)^3\right), \left.\frac{2\,(\text{Sin}[\theta] - r\,\text{Sin}[2\theta] + r^3\,\text{Sin}[2\theta])}{r^3}\right\}$$

The following *Mathematica* code implements Equation (3.77) and defines a function, `AiryDisplacementPolar`, which returns the displacement components in polar form, (u_r, u_θ), for given $\gamma(z)$ and $\psi(z)$.

```
In[82]:= AiryDisplacementPolar[γ_, ψ_, κ_, μ_] := Module[{
    work = 1/(2 μ) (κ γ - z Conjugate[dZ[γ]] - Conjugate[ψ]) Exp[-I θ] /. polar},
    {ComplexExpand[Re[work]] // TrigReduce, ComplexExpand[Im[work]] // TrigReduce}]
```

For instance, the displacement field for $\gamma(z) = 1/z$ and $\psi(z) = 2z + 1/z^2$ can be obtained as

```
In[83]:= AiryDisplacementPolar[1/z, 2 z + 1/z^2, κ, μ] // Simplify
```

$$\text{Out[83]}= \left\{\frac{-\text{Cos}[\theta] + r\,(1 - 2r^2 + \kappa)\,\text{Cos}[2\theta]}{2\,r^2\,\mu}, \frac{(-1 + 2r\,(1 + 2r^2 - \kappa)\,\text{Cos}[\theta])\,\text{Sin}[\theta]}{2\,r^2\,\mu}\right\}$$

Inclusions in Infinite Media

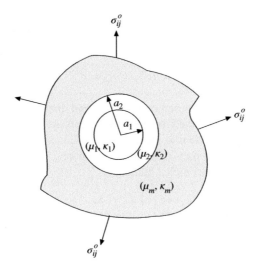

Figure 3.17 Three-phase composites

For the geometry of the three-phase material shown in Figure 3.17, it can be shown after some algebra that the expressions for Equation (3.74) for $\gamma(z)$ and $\psi(z)$ take the following form:

1. Inside $0 \le r \le a_1$

$$\gamma(z) = \sum_{i=1}^{3} c_i z^i, \quad \psi(z) = \sum_{i=1}^{1} d_i z^i. \tag{3.78}$$

2. Middle region $a_1 \le r \le a_2$

$$\gamma(z) = \sum_{i=-3}^{1} c_i z^i, \quad \psi(z) = \sum_{i=-3}^{1} d_i z^i. \tag{3.79}$$

3. Outside region $a_2 \le r \le \infty$

$$\gamma(z) = \left(\frac{\sigma_{xx}^\infty + \sigma_{yy}^\infty}{4} \right) z + \frac{c_{-1}}{z}, \quad \psi(z) = \left(\frac{\sigma_{xx}^\infty - \sigma_{yy}^\infty + 2i\sigma_{xy}^\infty}{2} \right) z + \sum_{i=1}^{3} \frac{d_i}{z^i}. \tag{3.80}$$

The first terms of $\gamma(z)$ and $\psi(z)$ in Equation (3.80) come from the boundary condition that $(\sigma_{xx}, \sigma_{yy}, \sigma_{xy}) \to (\sigma_{xx}^\infty, \sigma_{yy}^\infty, \sigma_{xy}^\infty)$ at the far field. These solutions must satisfy the continuity conditions of the displacement and the traction force across each interface. In the polar coordinate system, the continuity conditions of the displacement and the surface traction force can be expressed as

1. At $r = a_2$
$$u_r^{mid} = u_r^{out}, \quad u_\theta^{mid} = u_\theta^{out}, \quad \sigma_{\theta\theta}^{mid} = \sigma_{\theta\theta}^{out}, \quad \sigma_{r\theta}^{mid} = \sigma_{r\theta}^{out}.$$

2. At $r = a_1$
$$u_r^{in} = u_r^{mid}, \quad u_\theta^{in} = u_\theta^{mid}, \quad \sigma_{\theta\theta}^{in} = \sigma_{\theta\theta}^{mid}, \quad \sigma_{r\theta}^{in} = \sigma_{r\theta}^{mid}.$$

These continuity conditions are imposed on the stress and displacement components derived from the Airy stress function, and the unknown coefficients are solved analytically.

The following *Mathematica* code is implementation of Equation (3.78). The variables e and ee are the real and the imaginary parts of a complex number e. The variables f and ff are the real and the imaginary parts of a complex number f.

```
In[94]:= γ = Sum[(e[i] + ee[i] I) z^i, {i, 1, 3}];
        ψ = Sum[(f[i] + ff[i] I) z^i, {i, 1, 1}];
        InStress = AiryStressPolar[γ, ψ] // TrigReduce;
        InDis = AiryDisplacementPolar[γ, ψ, κin, μin];
        InTraction = {InStress[[1]], InStress[[3]]};
```

The following *Mathematica* code is implementation of Equation (3.79). The variables c and cc are the real and the imaginary parts of a complex number c. The variables d and dd are the real and the imaginary parts of a complex number d.

```
In[89]:= γ = Sum[(c[i] + cc[i] I) z^i, {i, -1, 3}];
        ψ = Sum[(d[i] + dd[i] I) z^i, {i, -3, 1}];
        MidStress = AiryStressPolar[γ, ψ] // Simplify // TrigReduce;
        MidDis = AiryDisplacementPolar[γ, ψ, κmid, μmid] // Simplify;
        MidTraction = {MidStress[[1]], MidStress[[3]]};
```

The following *Mathematica* code is implementation of Equation (3.80). The variables a and aa are the real and the imaginary parts of a complex number a. The variables b and bb are the real and the imaginary parts of a complex number b.

```
In[84]:= γ = (σx∞ + σy∞) / 4 z + Sum[(a[i] + aa[i] I) / z^i, {i, 1, 1}];
        ψ = (σx∞ - σy∞ + 2 I τ∞) / 2 z + Sum[(b[i] + bb[i] I) / z^i, {i, 1, 3}];
        OutStress = AiryStressPolar[γ, ψ] // Simplify // TrigReduce;
        OutDis = AiryDisplacementPolar[γ, ψ, κout, μout] // Simplify;
        OutTraction = {OutStress[[1]], OutStress[[3]]};
```

The displacement and traction continuity conditions at $r = r_0$ and $r = r_1$ can be entered as

```
In[38]:= DisEq1 = (OutDis - MidDis) /. r → r1
        DisEq2 = (MidDis - InDis) /. r → r0
        StressEq1 = (OutTraction - MidTraction) /. r → r1
        StressEq2 = (MidTraction - InTraction) /. r → r0
```

Inclusions in Infinite Media

$$\text{Out[38]=} \left\{ \frac{1}{8\, r1^3\, \mu_{out}} \Big(-r1^4\, \sigma x\infty + r1^4\, \chi_{out}\, \sigma x\infty - r1^4\, \sigma y\infty + r1^4\, \chi_{out}\, \sigma y\infty - 4\, r1^2\, b[1] - 4\, r1\, b[2]\, \cos[\theta] + \right.$$

$$\left(-2\, r1^4\, (\sigma x\infty - \sigma y\infty) + 4\, r1^2\, (1+\chi_{out})\, a[1] - 4\, b[3] \right)\, \cos[2\theta] - 4\, r1\, bb[2]\, \sin[\theta] +$$
$$4\, r1^4\, \tau\infty\, \sin[2\theta] + 4\, r1^2\, aa[1]\, \sin[2\theta] + 4\, r1^2\, \chi_{out}\, aa[1]\, \sin[2\theta] - 4\, bb[3]\, \sin[2\theta] \Big) -$$

$$\frac{1}{2\, r1^3\, \mu_{mid}} \Big(-r1^4\, c[1] + r1^4\, \chi_{mid}\, c[1] - r1^2\, d[-1] +$$

$$r1\, \cos[\theta] \left(r1^4\, (-2+\chi_{mid})\, c[2] - d[-2] + r1^2\, (\chi_{mid}\, c[0] - d[0]) \right) +$$
$$\cos[2\theta] \left(r1^2\, (1+\chi_{mid})\, c[-1] + r1^6\, (-3+\chi_{mid})\, c[3] - d[-3] - r1^4\, d[1] \right) -$$
$$r1^3\, \chi_{mid}\, cc[0]\, \sin[\theta] + 2\, r1^5\, cc[2]\, \sin[\theta] - r1^5\, \chi_{mid}\, cc[2]\, \sin[\theta] -$$
$$r1\, dd[-2]\, \sin[\theta] + r1^3\, dd[0]\, \sin[\theta] + r1^2\, cc[-1]\, \sin[2\theta] + r1^2\, \chi_{mid}\, cc[-1]\, \sin[2\theta] +$$
$$3\, r1^6\, cc[3]\, \sin[2\theta] - r1^6\, \chi_{mid}\, cc[3]\, \sin[2\theta] - dd[-3]\, \sin[2\theta] + r1^4\, dd[1]\, \sin[2\theta] \Big),$$

$$\frac{1}{4\, r1^3\, \mu_{out}} \Big(2\, r1^2\, bb[1] + 2\, r1\, bb[2]\, \cos[\theta] + 2\, \left(r1^4\, \tau\infty + r1^2\, (-1+\chi_{out})\, aa[1] + bb[3] \right)$$
$$\cos[2\theta] - 2\, r1\, b[2]\, \sin[\theta] + r1^4\, \sigma x\infty\, \sin[2\theta] - r1^4\, \sigma y\infty\, \sin[2\theta] +$$
$$2\, r1^2\, a[1]\, \sin[2\theta] - 2\, r1^2\, \chi_{out}\, a[1]\, \sin[2\theta] - 2\, b[3]\, \sin[2\theta] \Big) -$$

$$\frac{1}{2\, r1^3\, \mu_{mid}} \Big(r1^4\, cc[1] + r1^4\, \chi_{mid}\, cc[1] + r1^2\, dd[-1] +$$

$$r1\, \cos[\theta] \left(r1^4\, (2+\chi_{mid})\, cc[2] + dd[-2] + r1^2\, (\chi_{mid}\, cc[0] + dd[0]) \right) +$$
$$\cos[2\theta] \left(r1^2\, (-1+\chi_{mid})\, cc[-1] + r1^6\, (3+\chi_{mid})\, cc[3] + dd[-3] + r1^4\, dd[1] \right) -$$
$$r1^3\, \chi_{mid}\, c[0]\, \sin[\theta] + 2\, r1^5\, c[2]\, \sin[\theta] + r1^5\, \chi_{mid}\, c[2]\, \sin[\theta] -$$
$$r1\, d[-2]\, \sin[\theta] + r1^3\, d[0]\, \sin[\theta] + r1^2\, c[-1]\, \sin[2\theta] - r1^2\, \chi_{mid}\, c[-1]\, \sin[2\theta] +$$
$$3\, r1^6\, c[3]\, \sin[2\theta] + r1^6\, \chi_{mid}\, c[3]\, \sin[2\theta] - d[-3]\, \sin[2\theta] + r1^4\, d[1]\, \sin[2\theta] \Big) \Big\}$$

$$\text{Out[39]=} \left\{ \frac{1}{2\, r0^3\, \mu_{mid}} \Big(-r0^4\, c[1] + r0^4\, \chi_{mid}\, c[1] - r0^2\, d[-1] + \right.$$

$$r0\, \cos[\theta] \left(r0^4\, (-2+\chi_{mid})\, c[2] - d[-2] + r0^2\, (\chi_{mid}\, c[0] - d[0]) \right) +$$
$$\cos[2\theta] \left(r0^2\, (1+\chi_{mid})\, c[-1] + r0^6\, (-3+\chi_{mid})\, c[3] - d[-3] - r0^4\, d[1] \right) +$$
$$r0^3\, \chi_{mid}\, cc[0]\, \sin[\theta] + 2\, r0^5\, cc[2]\, \sin[\theta] - r0^5\, \chi_{mid}\, cc[2]\, \sin[\theta] -$$
$$r0\, dd[-2]\, \sin[\theta] + r0^3\, dd[0]\, \sin[\theta] + r0^2\, cc[-1]\, \sin[2\theta] +$$
$$r0^2\, \chi_{mid}\, cc[-1]\, \sin[2\theta] + 3\, r0^6\, cc[3]\, \sin[2\theta] -$$
$$r0^6\, \chi_{mid}\, cc[3]\, \sin[2\theta] - dd[-3]\, \sin[2\theta] + r0^4\, dd[1]\, \sin[2\theta] \Big) - \frac{1}{2\, \mu_{in}}$$

$$\big(-r0\, e[1] + r0\, \chi_{in}\, e[1] - 2\, r0^2\, \cos[\theta]\, e[2] + r0^2\, \chi_{in}\, \cos[\theta]\, e[2] - 3\, r0^3\, \cos[2\theta]\, e[3] +$$
$$r0^3\, \chi_{in}\, \cos[2\theta]\, e[3] - r0\, \cos[2\theta]\, f[1] + 2\, r0^2\, ee[2]\, \sin[\theta] - r0^2\, \chi_{in}\, ee[2]\, \sin[\theta] +$$
$$3\, r0^3\, ee[3]\, \sin[2\theta] - r0^3\, \chi_{in}\, ee[3]\, \sin[2\theta] + r0\, ff[1]\, \sin[2\theta] \big),$$

$$\frac{1}{2\, r0^3\, \mu_{mid}} \Big(r0^4\, cc[1] + r0^4\, \chi_{mid}\, cc[1] + r0^2\, dd[-1] +$$

$$r0\, \cos[\theta] \left(r0^4\, (2+\chi_{mid})\, cc[2] + dd[-2] + r0^2\, (\chi_{mid}\, cc[0] + dd[0]) \right) +$$
$$\cos[2\theta] \left(r0^2\, (-1+\chi_{mid})\, cc[-1] + r0^6\, (3+\chi_{mid})\, cc[3] + dd[-3] + r0^4\, dd[1] \right) -$$
$$r0^3\, \chi_{mid}\, c[0]\, \sin[\theta] + 2\, r0^5\, c[2]\, \sin[\theta] + r0^5\, \chi_{mid}\, c[2]\, \sin[\theta] -$$
$$r0\, d[-2]\, \sin[\theta] + r0^3\, d[0]\, \sin[\theta] + r0^2\, c[-1]\, \sin[2\theta] - r0^2\, \chi_{mid}\, c[-1]\, \sin[2\theta] +$$
$$3\, r0^6\, c[3]\, \sin[2\theta] + r0^6\, \chi_{mid}\, c[3]\, \sin[2\theta] - d[-3]\, \sin[2\theta] + r0^4\, d[1]\, \sin[2\theta] \Big) -$$

$$\frac{1}{2\, \mu_{in}} \big(r0\, ee[1] + r0\, \chi_{in}\, ee[1] + 2\, r0^2\, \cos[\theta]\, ee[2] + r0^2\, \chi_{in}\, \cos[\theta]\, ee[2] +$$
$$3\, r0^3\, \cos[2\theta]\, ee[3] + r0^3\, \chi_{in}\, \cos[2\theta]\, ee[3] + r0\, \cos[2\theta]\, ff[1] + 2\, r0^2\, e[2]\, \sin[\theta] +$$
$$r0^2\, \chi_{in}\, e[2]\, \sin[\theta] + 3\, r0^3\, e[3]\, \sin[2\theta] + r0^3\, \chi_{in}\, e[3]\, \sin[2\theta] + r0\, f[1]\, \sin[2\theta] \big) \Big\}$$

$$\text{Out[40]=} \left\{ \frac{1}{2\, r1^4} \big(r1^4\, \sigma x\infty + r1^4\, \sigma y\infty + 2\, r1^2\, b[1] + 4\, r1\, b[2]\, \cos[\theta] - r1^4\, \sigma x\infty\, \cos[2\theta] + \right.$$

$$r1^4\, \sigma y\infty\, \cos[2\theta] - 8\, r1^2\, a[1]\, \cos[2\theta] + 6\, b[3]\, \cos[2\theta] + 4\, r1\, bb[2]\, \sin[\theta] +$$
$$2\, r1^4\, \tau\infty\, \sin[2\theta] - 8\, r1^2\, aa[1]\, \sin[2\theta] + 6\, bb[3]\, \sin[2\theta] \big) - \frac{1}{r1^4}$$

$$\big(2\, r1^4\, c[1] + 2\, r1^5\, c[2]\, \cos[\theta] - 4\, r1^2\, c[-1]\, \cos[2\theta] + 3\, \cos[2\theta]\, d[-3] +$$
$$2\, r1\, \cos[\theta]\, d[-2] + r1^2\, d[-1] - r1^4\, \cos[\theta]\, d[1] - 2\, r1^5\, cc[2]\, \sin[\theta] +$$
$$2\, r1\, dd[-2]\, \sin[\theta] - 4\, r1^2\, cc[-1]\, \sin[2\theta] + 3\, dd[-3]\, \sin[2\theta] + r1^4\, dd[1]\, \sin[2\theta] \big),$$

$$\frac{1}{2\, r1^4} \big(-2\, r1^2\, bb[1] - 4\, r1\, bb[2]\, \cos[\theta] + 2\, r1^4\, \tau\infty\, \cos[2\theta] + 4\, r1^2\, aa[1]\, \cos[2\theta] -$$
$$6\, bb[3]\, \cos[2\theta] + 4\, r1\, b[2]\, \sin[\theta] + r1^4\, \sigma x\infty\, \sin[2\theta] -$$
$$r1^4\, \sigma y\infty\, \sin[2\theta] - 4\, r1^2\, a[1]\, \sin[2\theta] + 6\, b[3]\, \sin[2\theta] \big) - \frac{1}{r1^4}$$

$$\big(2\, r1^5\, cc[2]\, \cos[\theta] + 2\, r1^2\, cc[-1]\, \cos[2\theta] + 6\, r1^6\, cc[3]\, \cos[2\theta] -$$
$$3\, \cos[2\theta]\, dd[-3] - 2\, r1\, \cos[\theta]\, dd[-2] - r1^2\, dd[-1] + r1^4\, \cos[2\theta]\, dd[1] +$$
$$2\, r1^5\, c[2]\, \sin[\theta] + 2\, r1\, d[-2]\, \sin[\theta] - 2\, r1^2\, c[-1]\, \sin[2\theta] +$$
$$6\, r1^6\, c[3]\, \sin[2\theta] + 3\, d[-3]\, \sin[2\theta] + r1^4\, d[1]\, \sin[2\theta] \big) \Big\}$$

Out[47]= $\{-2\,e[1] - 2\,r0\,\text{Cos}[\theta]\,e[2] + \text{Cos}[2\,\theta]\,f[1] + 2\,r0\,ee[2]\,\text{Sin}[\theta] - ff[1]\,\text{Sin}[2\,\theta] +$

$\frac{1}{r0^4}\big(2\,r0^4\,c[1] + 2\,r0^5\,c[2]\,\text{Cos}[\theta] - 4\,r0^2\,c[-1]\,\text{Cos}[2\,\theta] + 3\,\text{Cos}[2\,\theta]\,d[-3] +$

$2\,r0\,\text{Cos}[\theta]\,d[-2] + r0^2\,d[-1] - r0^4\,\text{Cos}[2\,\theta]\,d[1] - 2\,r0^5\,cc[2]\,\text{Sin}[\theta] +$

$2\,r0\,dd[-2]\,\text{Sin}[\theta] - 4\,r0^2\,cc[-1]\,\text{Sin}[2\,\theta] + 3\,dd[-3]\,\text{Sin}[2\,\theta] + r0^4\,dd[1]\,\text{Sin}[2\,\theta]\big),$

$-2\,r0\,\text{Cos}[\theta]\,ee[2] - 6\,r0^2\,\text{Cos}[2\,\theta]\,ee[3] - \text{Cos}[2\,\theta]\,ff[1] -$

$2\,r0\,e[2]\,\text{Sin}[\theta] - 6\,r0^2\,e[3]\,\text{Sin}[2\,\theta] - f[1]\,\text{Sin}[2\,\theta] + \frac{1}{r0^4}$

$\big(2\,r0^5\,cc[2]\,\text{Cos}[\theta] + 2\,r0^2\,cc[-1]\,\text{Cos}[2\,\theta] + 6\,r0^6\,cc[3]\,\text{Cos}[2\,\theta] -$

$3\,\text{Cos}[2\,\theta]\,dd[-3] - 2\,r0\,\text{Cos}[\theta]\,dd[-2] - r0^2\,dd[-1] + r0^4\,\text{Cos}[2\,\theta]\,dd[1] +$

$2\,r0^5\,c[2]\,\text{Sin}[\theta] + 2\,r0\,d[-2]\,\text{Sin}[\theta] - 2\,r0^2\,c[-1]\,\text{Sin}[2\,\theta] +$

$6\,r0^6\,c[3]\,\text{Sin}[2\,\theta] + 3\,d[-3]\,\text{Sin}[2\,\theta] + r0^4\,d[1]\,\text{Sin}[2\,\theta]\big)\}$

To separate the aforementioned equation into independent equations, it is necessary to reduce the powers of sin θ and cos θ into the form of sin $n\theta$ and cos $n\theta$. The following function, EqMaker, automates this process.

In[103]:= **EqMaker[f_]** :=
  Module[{j1, j2, j3}, j1 = f //. {Cos[i_. * θ] → cosine^i, Sin[j_. * θ] → sine^j };
   j2 = CoefficientList[j1, {cosine, sine}]; j3 = Flatten[j2];
   j1 = DeleteCases[j3, 0]; Map[#1 == 0 &, j3]]

It is necessary to assemble the list of equations into eqlist and the list of unknown coefficients into varlist as

In[43]:= eqlist = {EqMaker[DisEq1], EqMaker[DisEq2],
       EqMaker[StressEq1], EqMaker[StressEq2]} // Flatten;
  eqlist = DeleteCases[eqlist, True]
  varlist = {Table[a[i], {i, 1, 1}], Table[aa[i], {i, 1, 1}],
       Table[b[i], {i, 1, 3}], Table[bb[i], {i, 1, 3}], Table[c[i], {i, -1, 3}],
       Table[cc[i], {i, -1, 3}], Table[d[i], {i, -3, 1}],
       Table[dd[i], {i, -3, 1}], Table[e[i], {i, 1, 3}], Table[ee[i], {i, 1, 3}],
       Table[f[i], {i, 1, 1}], Table[ff[i], {i, 1, 1}]} // Flatten

Inclusions in Infinite Media

Out[45]= $\{a[1], aa[1], b[1], b[2], b[3], bb[1], bb[2], bb[3], c[-1], c[0], c[1], c[2], c[3],$
$cc[-1], cc[0], cc[1], cc[2], cc[3], d[-3], d[-2], d[-1], d[0], d[1], dd[-3],$
$dd[-2], dd[-1], dd[0], dd[1], e[1], e[2], e[3], ee[1], ee[2], ee[3], f[1], ff[1]\}$

The first part of the aforementioned output is a list of simultaneous equations for the unknown coefficients and the second part of the output is a list of the variables to be solved. These equations are a set of underdetermined equations. The `Solve` function in *Mathematica* can handle underdetermined equations, and the output is a rule of solutions in terms of independent variables. As the output is lengthy, the `Short` command can display the output in short format in the number of lines specified (50 lines in this example).

In[52]:= `sol = Solve[eqlist, varlist] // Simplify;`
        `Short[sol, 50]`

Out[52]//Short= $\{\{a[1] \to$
$(r1^2 \{-6 r0^4 r1^4 (\mu in - \mu mid)(\mu in + \kappa in \mu mid)(\mu mid - \mu out)^2 + r0^5 r1^2 \{(3 + \kappa mid^2)$
$\mu in^2 - (-1 + \kappa in)(-3 + \kappa mid) \mu in \mu mid - 4 \kappa in \mu mid^2\}(\mu mid - \mu out)^2 +$
$r0^6 (\mu in - \mu mid)(\kappa mid \mu in - \kappa in \mu mid)(\mu mid - \mu out)(\mu mid + \kappa mid \mu out) +$
$r1^6 (\kappa mid \mu in + \mu mid)(\mu in + \kappa in \mu mid)(\mu mid - \mu out)(\mu mid + \kappa mid \mu out) +$
$r0^2 r1^4 (\mu in - \mu mid)(\mu in + \kappa in \mu mid)$
$(4 \mu mid^2 + 2(-3 + \kappa mid) \mu mid \mu out + (3 + \kappa mid^2) \mu out^2)\}(\alpha x \infty - \sigma y \infty)\} /$

```
(2 (-6 r0^4 r1^4 (μin - μmid) (μin + κin μmid) (μmid - μout) (κout μmid + μout) +
    r0^6 r1^2 ((3 + κmid^2) μin^2 - (-1 + κin) (-3 + κmid) μin μmid - 4 κin μmid^2)
     (μmid - μout) (κout μmid + μout) +
    r0^8 (μin - μmid) (κmid μin - κin μmid) (μmid - μout) (κout μmid - κmid μout) +
    r1^8 (κmid μin + μmid) (μin + κin μmid) (κout μmid + μout) (μmid + κmid μout) +
    r0^2 r1^6 (μin - μmid) (μin + κin μmid) (κout μmid (4 μmid + (-3 + κmid) μout) -
       μout ((-3 + κmid) μmid + (3 + κmid^2) μout)))),
aa[1] → -((r1^2 (-6 r0^4 r1^4 (μin - μmid) (μin + κin μmid) (μmid - μout)^2 +
        r0^6 r1^2 ((3 + κmid^2) μin^2 - (-1 + κin) (-3 + κmid) μin μmid - 4 κin μmid^2)
         (μmid - μout)^2 + r0^8 (μin - μmid) (κmid μin - κin μmid) (μmid - μout)
         (μmid + κmid μout) + r1^8 (κmid μin + μmid) (μin + κin μmid)
         (μmid - μout) (μmid + κmid μout) + r0^2 r1^6 (μin - μmid) (μin + κin μmid)
         (4 μmid^2 + 2 (-3 + κmid) μmid μout + (3 + κmid^2) μout^2)) τ∞) /
     (-6 r0^4 r1^4 (μin - μmid) (μin + κin μmid) (μmid - μout) (κout μmid + μout) +
      r0^6 r1^2 ((3 + κmid^2) μin^2 - (-1 + κin) (-3 + κmid) μin μmid - 4 κin μmid^2)
       (μmid - μout) (κout μmid + μout) +
      r0^8 (μin - μmid) (κmid μin - κin μmid) (μmid - μout) (κout μmid - κmid μout) +
      r1^8 (κmid μin + μmid) (μin + κin μmid) (κout μmid + μout) (μmid + κmid μout) +
      r0^2 r1^6 (μin - μmid) (μin + κin μmid) (κout μmid (4 μmid + (-3 + κmid) μout) -
         μout ((-3 + κmid) μmid + (3 + κmid^2) μout)))),
b[1] → 1/2 r1^2 (-1 + ((1 + κout) μmid (r1^2 (2 μin + (-1 + κin) μmid) +
          r0^2 ((-1 + κmid) μin + μmid - κin μmid))) /
     (2 r0^2 ((-1 + κmid) μin + μmid - κin μmid) (μmid - μout) +
       r1^2 (2 μin + (-1 + κin) μmid) (2 μmid + (-1 + κmid) μout))) (σx∞ + σy∞),
«29», f[1] → (r1^2 (1 + κmid) (1 + κout) μin μmid
    (-3 r0^4 r1^2 (μin - μmid) (μmid - μout) + r0^6 ((3 + κmid) μin - (3 + κin) μmid)
       (μmid - μout) + r1^6 (μin + κin μmid) (μmid + κmid μout)) (σx∞ - σy∞)) /
   (2 (-6 r0^4 r1^4 (μin - μmid) (μin + κin μmid) (μmid - μout) (κout μmid + μout) +
      r0^6 r1^2 ((3 + κmid^2) μin^2 - (-1 + κin) (-3 + κmid) μin μmid - 4 κin μmid^2)
       (μmid - μout) (κout μmid + μout) +
      r0^8 (μin - μmid) (κmid μin - κin μmid) (μmid - μout) (κout μmid - κmid μout) +
      r1^8 (κmid μin + μmid) (μin + κin μmid) (κout μmid + μout) (μmid + κmid μout) +
      r0^2 r1^6 (μin - μmid) (μin + κin μmid) (κout μmid (4 μmid + (-3 + κmid) μout) -
         μout ((-3 + κmid) μmid + (3 + κmid^2) μout)))),
ff[1] → (r1^2 (1 + κmid) (1 + κout) μin μmid (-3 r0^4 r1^2 (μin - μmid) (μmid - μout) +
      r0^6 ((3 + κmid) μin - (3 + κin) μmid) (μmid - μout) +
      r1^6 (μin + κin μmid) (μmid + κmid μout)) τ∞) /
   (-6 r0^4 r1^4 (μin - μmid) (μin + κin μmid) (μmid - μout) (κout μmid + μout) +
    r0^6 r1^2 ((3 + κmid^2) μin^2 - (-1 + κin) (-3 + κmid) μin μmid - 4 κin μmid^2)
     (μmid - μout) (κout μmid + μout) +
    r0^8 (μin - μmid) (κmid μin - κin μmid) (μmid - μout) (κout μmid - κmid μout) +
    r1^8 (κmid μin + μmid) (μin + κin μmid) (κout μmid + μout) (μmid + κmid μout) +
    r0^2 r1^6 (μin - μmid) (μin + κin μmid) (κout μmid (4 μmid + (-3 + κmid) μout) -
       μout ((-3 + κmid) μmid + (3 + κmid^2) μout)))}}
```

The stress components are now evaluated. For the innermost region ($0 < r < r_0$), σinpolor returns ($\sigma_{rr}, \sigma_{\theta\theta}, \sigma_{r\theta}$) and σinxy returns ($\sigma_{xx}, \sigma_{yy}, \sigma_{xy}$).

```
In[47]:= σinpolar = (InStress /. sol) [[1]];
         σinxy = { Cos[θ]^2 σinpolar[[1]] - 2 Cos[θ] Sin[θ] σinpolar[[3]] + Sin[θ]^2 σinpolar[[2]],
            Sin[θ]^2 σinpolar[[1]] + 2 Cos[θ] Sin[θ] σinpolar[[3]] + Cos[θ]^2 σinpolar[[2]],
            Cos[θ] Sin[θ] (σinpolar[[1]] - σinpolar[[2]]) + (Cos[θ]^2 - Sin[θ]^2) σinpolar[[3]]};
         uoutpolar = (OutDis /. sol) [[1]];
```

For instance, σ_{rr} in the innermost region is expressed as

```
In[50]:= σinpolar[[1]]

Out[50]= (r1^2 (1 + κmid) (1 + κout) μin μmid (σx∞ + σy∞)) /
          (2 (2 r0^2 ((-1 + κmid) μin + μmid - κin μmid) (μmid - μout) +
               r1^2 (2 μin + (-1 + κin) μmid) (2 μmid + (-1 + κmid) μout))) - 1/(2 (κout μmid + μout))
          (1 + κout) μmid (1 - (3 r1^2 (μin - μmid) (r0^4 (-μmid + μout) + r1^4 (μmid + κmid μout))) /
             (-3 r0^4 r1^2 (μin - μmid) (μmid - μout) + r0^6 ((3 + κmid) μin - (3 + κin) μmid) (μmid - μout) +
               r1^6 (μin + κin μmid) (μmid + κmid μout)) - ((μin - μmid)
             (r0^6 (κmid μin - κin μmid) (μmid - μout) + r1^6 (μin + κin μmid) (μmid + κmid μout))
```

$$\left(-4\,r0^6\,r1^2\,((3+\kappa mid)\,\mu in-(3+\kappa in)\,\mu mid)\,(\mu mid-\mu out)\,(\kappa out\,\mu mid+\mu out)+\right.$$
$$6\,r0^4\,r1^4\,((3+\kappa mid)\,\mu in-(3+\kappa in)\,\mu mid)\,(\mu mid-\mu out)\,(\kappa out\,\mu mid+\mu out)+$$
$$r0^8\,((3+\kappa mid)\,\mu in-(3+\kappa in)\,\mu mid)\,(\mu mid-\mu out)\,(\kappa out\,\mu mid-\kappa mid\,\mu out)-$$
$$r1^8\,(\mu in+3\,\kappa mid\,\mu in+(3+\kappa in)\,\mu mid)\,(\kappa out\,\mu mid+\mu out)\,(\mu mid+\kappa mid\,\mu out)-$$
$$r0^2\,r1^6\,(2\,\mu in-(3+\kappa in)\,\mu mid)$$
$$\left.\left(\kappa out\,\mu mid\,(4\,\mu mid+(-3+\kappa mid)\,\mu out)-\mu out\,((-3+\kappa mid)\,\mu mid+(3+\kappa mid^2)\,\mu out)\right)\right)\Big/$$
$$\left(\left(-3\,r0^4\,r1^2\,(\mu in-\mu mid)\,(\mu mid-\mu out)+r0^6\,((3+\kappa mid)\,\mu in-(3+\kappa in)\,\mu mid)\,(\mu mid-\mu out)+\right.\right.$$
$$r1^6\,(\mu in+\kappa in\,\mu mid)\,(\mu mid+\kappa mid\,\mu out)\Big)$$
$$\left(-6\,r0^4\,r1^4\,(\mu in-\mu mid)\,(\mu in+\kappa in\,\mu mid)\,(\mu mid-\mu out)\,(\kappa out\,\mu mid+\mu out)+\right.$$
$$r0^6\,r1^2\,\left((3+\kappa mid^2)\,\mu in^2-(-1+\kappa in)\,(-3+\kappa mid)\,\mu in\,\mu mid-4\,\kappa in\,\mu mid^2\right)$$
$$(\mu mid-\mu out)\,(\kappa out\,\mu mid+\mu out)+r0^8\,(\mu in-\mu mid)\,(\kappa mid\,\mu in-\kappa in\,\mu mid)$$
$$(\mu mid-\mu out)\,(\kappa out\,\mu mid-\kappa mid\,\mu out)+r1^8\,(\kappa mid\,\mu in+\mu mid)\,(\mu in+\kappa in\,\mu mid)$$
$$(\kappa out\,\mu mid+\mu out)\,(\mu mid+\kappa mid\,\mu out)+r0^2\,r1^6\,(\mu in-\mu mid)\,(\mu in+\kappa in\,\mu mid)$$
$$\left.\left.\left(\kappa out\,\mu mid\,(4\,\mu mid+(-3+\kappa mid)\,\mu out)-\mu out\,((-3+\kappa mid)\,\mu mid+(3+\kappa mid^2)\,\mu out)\right)\right)\right)$$

$(\sigma x\infty-\sigma y\infty)\,\text{Cos}[2\theta]+\dfrac{1}{\kappa out\,\mu mid+\mu out}\,(1+\kappa out)$

μmid

$\Big(1+$
$$\left(3\,r1^2\,(\mu in-\mu mid)\,\left(r0^4\,(-\mu mid+\mu out)+r1^4\,(\mu mid+\kappa mid\,\mu out)\right)\right)\Big/$$
$$\left(3\,r0^4\,r1^2\,(\mu in-\mu mid)\,(\mu mid-\mu out)+r0^6\,((-3+\kappa mid)\,\mu in-(-3+\kappa in)\,\mu mid)\,(\mu mid-\mu out)+\right.$$
$$r1^6\,(\mu in+\kappa in\,\mu mid)\,(\mu mid+\kappa mid\,\mu out)\Big)-\Big((\mu in-\mu mid)$$
$$\left(r0^6\,(\kappa mid\,\mu in-\kappa in\,\mu mid)\,(\mu mid-\mu out)+r1^6\,(\mu in+\kappa in\,\mu mid)\,(\mu mid+\kappa mid\,\mu out)\right)$$
$$\left(-4\,r0^6\,r1^2\,((-3+\kappa mid)\,\mu in-(-3+\kappa in)\,\mu mid)\,(\mu mid-\mu out)\,(\kappa out\,\mu mid+\mu out)-\right.$$
$$6\,r0^4\,r1^4\,(4\,\mu in+(-3+\kappa in)\,\mu mid)\,(\mu mid-\mu out)\,(\kappa out\,\mu mid+\mu out)+$$
$$r0^8\,((-3+\kappa mid)\,\mu in-(-3+\kappa in)\,\mu mid)\,(\mu mid-\mu out)\,(\kappa out\,\mu mid-\kappa mid\,\mu out)+$$
$$r1^8\,((-1+3\,\kappa mid)\,\mu in-(-3+\kappa in)\,\mu mid)\,(\kappa out\,\mu mid+\mu out)\,(\mu mid+\kappa mid\,\mu out)+$$
$$r0^2\,r1^6\,(4\,\mu in+(-3+\kappa in)\,\mu mid)$$
$$\left.\left(\kappa out\,\mu mid\,(4\,\mu mid+(-3+\kappa mid)\,\mu out)-\mu out\,((-3+\kappa mid)\,\mu mid+(3+\kappa mid^2)\,\mu out)\right)\right)\Big/$$
$$\left(\left(3\,r0^4\,r1^2\,(\mu in-\mu mid)\,(\mu mid-\mu out)+r0^6\,((-3+\kappa mid)\,\mu in-(-3+\kappa in)\,\mu mid)\,(\mu mid-\mu out)+\right.\right.$$
$$r1^6\,(\mu in+\kappa in\,\mu mid)\,(\mu mid+\kappa mid\,\mu out)\Big)$$
$$\left(-6\,r0^4\,r1^4\,(\mu in-\mu mid)\,(\mu in+\kappa in\,\mu mid)\,(\mu mid-\mu out)\,(\kappa out\,\mu mid+\mu out)+\right.$$
$$r0^6\,r1^2\,\left((3+\kappa mid^2)\,\mu in^2-(-1+\kappa in)\,(-3+\kappa mid)\,\mu in\,\mu mid-4\,\kappa in\,\mu mid^2\right)$$
$$(\mu mid-\mu out)\,(\kappa out\,\mu mid+\mu out)+$$
$$r0^8\,(\mu in-\mu mid)\,(\kappa mid\,\mu in-\kappa in\,\mu mid)\,(\mu mid-\mu out)\,(\kappa out\,\mu mid-\kappa mid\,\mu out)+$$
$$r1^8\,(\kappa mid\,\mu in+\mu mid)\,(\mu in+\kappa in\,\mu mid)\,(\kappa out\,\mu mid+\mu out)\,(\mu mid+\kappa mid\,\mu out)+$$
$$r0^2\,r1^6\,(\mu in-\mu mid)\,(\mu in+\kappa in\,\mu mid)\,\left(\kappa out\,\mu mid\,(4\,\mu mid+(-3+\kappa mid)\,\mu out)-\right.$$
$$\left.\left.\left.\mu out\,((-3+\kappa mid)\,\mu mid+(3+\kappa mid^2)\,\mu out)\right)\right)\right)\,\tau\infty\,\text{Sin}[2\theta]$$

The stress field in the intermediate region ($r_0 < r < r_1$) is stored in σmidpolar (in polar form) and in σmidxy (in the rectangular coordinate system).

```
In[51]:= σmidpolar = (MidStress /. sol)[[1]];
σmidxy =
 { Cos[θ]^2 σmidpolar[[1]] - 2 Cos[θ] Sin[θ] σmidpolar[[3]] + Sin[θ]^2 σmidpolar[[2]],
   Sin[θ]^2 σmidpolar[[1]] + 2 Cos[θ] Sin[θ] σmidpolar[[3]] + Cos[θ]^2 σmidpolar[[2]],
   Cos[θ] Sin[θ] (σmidpolar[[1]] - σmidpolar[[2]]) + (Cos[θ]^2 - Sin[θ]^2) σmidpolar[[3]] };
umidpolar = (MidDis /. sol)[[1]];
```

Similarly, the stress field in the outermost region $r_1 < r < \infty$ is stored in σoutpolar (in polar form) and in σoutxy (in the rectangular coordinate system).

```
In[54]:= σoutpolar = (OutStress /. sol)[[1]];
σoutxy =
 { Cos[θ]^2 σoutpolar[[1]] - 2 Cos[θ] Sin[θ] σoutpolar[[3]] + Sin[θ]^2 σoutpolar[[2]],
   Sin[θ]^2 σoutpolar[[1]] + 2 Cos[θ] Sin[θ] σoutpolar[[3]] + Cos[θ]^2 σoutpolar[[2]],
   Cos[θ] Sin[θ] (σoutpolar[[1]] - σoutpolar[[2]]) + (Cos[θ]^2 - Sin[θ]^2) σoutpolar[[3]] };
uoutpolar = (OutDis /. sol)[[1]];
```

By using the `RevolutionPlot3D` function, it is possible to plot functions in (r,θ).

```
In[70]:= in1 = in[[1]] /. σx∞ → 1
        mid1 = mid[[1]] /. σx∞ → 1
        out1 = out[[1]] /. σx∞ → 1
        sigmarr[r_, θ_] := Boole[r < 1] in1 + Boole[r > 1 && r < 2] mid1 + Boole[r > 2] out1
        RevolutionPlot3D[ sigmarr[r, θ], {r, 0, 5}, {θ, 0, 2 Pi}]
```

Out[70]= $\dfrac{40}{113} - \dfrac{152\,896\,\cos[2\,\theta]}{174\,719}$

Out[71]= $\dfrac{112}{339} + \dfrac{8}{339\,r^2} - \dfrac{137\,664\,\cos[2\,\theta]}{174\,719} + \dfrac{46\,272\,\cos[2\,\theta]}{174\,719\,r^4} - \dfrac{61\,504\,\cos[2\,\theta]}{174\,719\,r^2}$

Out[72]= $\dfrac{1}{2} - \dfrac{74}{113\,r^2} - \dfrac{1}{2}\cos[2\,\theta] + \dfrac{2\,645\,208\,\cos[2\,\theta]}{174\,719\,r^4} - \dfrac{912\,456\,\cos[2\,\theta]}{174\,719\,r^2}$

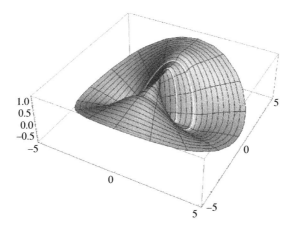

3.5 Effective Properties

One of the important subjects in micromechanics is the effective properties of heterogeneous materials. Finding the effective properties of heterogeneous materials has a long history dating back to Maxwell and others in the 1800s (Choy 1999). This problem is a so-called many-body problem, and it is known that there is no exact solution available for general geometry that can account for interactions among many particles. This subject became even more important over the last 40 years with the commercialization of composite materials in industry.

The majority of available approaches for the effective properties are based on the solution for a single inclusion problem, which this chapter mainly deals with. To define the overall properties for an infinitely extended medium, consider a material whose elastic moduli are functions of position. The effective elastic moduli, C^*, are defined by one of the following relationships:

$$<\sigma> = C^* <\epsilon> \qquad (3.81)$$

$$<\sigma\epsilon> = C^* <\epsilon><\epsilon>, \qquad (3.82)$$

where $<\cdot>$ denotes the spatial average defined as

$$<\cdot> \equiv \lim_{V \to \infty} \frac{1}{V} \int_V \cdot \, dx.$$

The definition of Equation (3.81) is based on the constitutive relationship while the definition of Equation (3.82) is based on the strain energy stored in the material. The equivalence of the aforementioned two definitions for C^* calls for the following relationship known as *Hill's Condition* (Hill 1963):

$$<\sigma\epsilon> = <\sigma><\epsilon>. \tag{3.83}$$

The conditions under which Equation (3.83) may be held have been discussed by many and somewhat controversial (see Buryachenko (2007), Qu and Cherkaoui (2006), and Nemat-Nasser and Hori (1999)). In a nutshell, Hill's condition, Equation (3.83), is held when the stress does not fluctuate much at the boundary of the composite.

For the overall elastic properties of heterogeneous materials, it is possible to derive the upper and lower bounds of the effective properties based on the variational principle. The minimum strain energy theorem gives the upper bound of the effective properties while the minimum complementary strain energy theorem gives the lower bound of the effective properties.

3.5.1 Upper and Lower Bounds of Effective Properties

The earliest known work is the upper and lower bounds known as Voigt (1910) and Reuss (1929) bounds based on the rule of mixtures, which is stated as

$$<C^{-1}>^{-1} \leq C^* \leq <C>. \tag{3.84}$$

In Equation (3.84), the inequality is understood as

$$A_{ijkl} \leq B_{ijkl} \iff A_{ijkl} v_{ij} v_{kl} \leq B_{ijkl} v_{ij} v_{kl},$$

where v_{ij} is an arbitrary, symmetric, second-rank tensor.

When applying Equation (3.84) to a two-phase material with isotropic components, the following inequalities are derived.

$$\frac{1}{\frac{v_1}{\mu_1} + \frac{v_2}{\mu_2}} \leq \mu^* \leq v_1 \mu_1 + v_2 \mu_2,$$

$$\frac{1}{\frac{v_1}{\kappa_1} + \frac{v_2}{\kappa_2}} \leq \kappa^* \leq v_1 \kappa_1 + v_2 \kappa_2,$$

where μ and κ are the shear and bulk moduli and v_1 and v_2 are the volume fractions of Phase 1 and Phase 2, respectively.

Voigt's bound is derived from the principle of minimum potential energy in which a uniform strain field is chosen as a permissible function, while Reuss' bound is derived from the principle of minimum complementary energy in which a uniform stress field is chosen as a permissible function. The upper and lower bounds by Reuss and Voigt are valid regardless of the constituents and their geometric distributions. However, the difference between the two bounds is too wide at intermediate volume fractions to be useful as a reasonable estimate of the effective moduli.

Hashin and Shtrikman (1963) derived tighter upper and lower bounds when the shape of inclusions is spherical using the concept of polarized stress. The upper and lower bounds for the bulk modulus, κ, and the shear modulus, μ, are expressed as

$$\kappa_1 + \frac{v_2}{\frac{1}{\kappa_2-\kappa_1} + \frac{3v_1}{3\kappa_1+4\mu_1}} \leq \kappa^* \leq \kappa_2 + \frac{v_1}{\frac{1}{\kappa_1-\kappa_2} + \frac{3v_2}{3\kappa_2+4\mu_2}}, \quad (3.85)$$

$$\mu_1 + \frac{v_2}{\frac{1}{\mu_2-\mu_1} + \frac{6v_1(\kappa_1+2\mu_1)}{5\mu_1(3\kappa_1+4\mu_1)}} \leq \mu^* \leq \mu_2 + \frac{v_1}{\frac{1}{\mu_1-\mu_2} + \frac{6v_2(\kappa_2+2\mu_2)}{5\mu_2(3\kappa_2+4\mu_2)}}. \quad (3.86)$$

where v_1 and v_2 are the volume fraction of Phase 1 and 2, respectively.

Hashin (1965) also derived the upper and lower bounds of fiber reinforced composite materials of arbitrary transverse-phase geometry based on the same principle. The upper and lower bounds for the bulk modulus, κ, and the shear modulus, μ, on the transverse plane are expressed as

$$\kappa_1 + \frac{v_2}{\frac{1}{\kappa_2-\kappa_1} + \frac{v_1}{\kappa_1+\mu_1}} \leq \kappa^* \leq \kappa_2 + \frac{v_1}{\frac{1}{\kappa_1-\kappa_2} + \frac{v_2}{\kappa_2+\mu_2}}, \quad (3.87)$$

$$\mu_1 + \frac{v_2}{\frac{1}{\mu_2-\mu_1} + \frac{v_1(\kappa_1+2\mu_1)}{2\mu_1(\kappa_1+\mu_1)}} \leq \mu^* \leq \mu_2 + \frac{v_1}{\frac{1}{\mu_1-\mu_2} + \frac{v_2(\kappa_2+2\mu_2)}{2\mu_2(\kappa_2+\mu_2)}}. \quad (3.88)$$

The bounds of Equations (3.85) and (3.86) are the best bounds when the only information about the phase geometry of the composite is isotropic and no extra information is given. Similarly, the bounds of Equations (3.87) and (3.88) are the best bounds when the only information about the phase geometry of the composite is transversely isotropic and no extra information is given.

Equation (3.85) is implemented in *Mathematica* with the following code:

```
In[1]:= κupper[κ1_, κ2_, μ1_, μ2_, v1_] := κ2 + v1 / (1 / (κ1 - κ2) + 3 (1 - v1) / (3 κ2 + 4 μ2))
        κlower[κ1_, κ2_, μ1_, μ2_, v1_] := κ1 + (1 - v1) / (1 / (κ2 - κ1) + 3 v1 / (3 κ1 + 4 μ1))

In[3]:= κ1 = 1; κ2 = 11; μ1 = 2; μ2 = 30;

In[4]:= Plot[{κupper[κ1, κ2, μ1, μ2, v1], κlower[κ1, κ2, μ1, μ2, v1]}, {v1, 0, 1}]
```

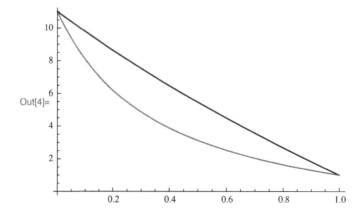

3.5.2 Self-Consistent Approximation

The upper and lower bounds derived by Hashin and Shtrikman are rigorous and give good estimates for the effective properties of composites. However, the difference between the upper and lower bounds at intermediate volume fractions is still wide, and it is desirable if alternative predictions that do not depend on the bound approach are available. The self-consistent method provides reasonable approximation to the effective properties of composites when the volume fractions are small by taking the effect of multiple inclusions into account. The self-consistent method also known as the effective medium theory has been in existence used mostly for conductivities or dielectric constants (Granqvist and Hunderi 1978). Hill (1965) applied the self-consistent method to elastic composites for spherical inclusions. It was later refined by Chou et al. (1980) and Laws and McLaughlin (1979) for composites with ellipsoidal inclusions. Oshima and Nomura (1985) applied the self-consistent method to three-phase materials discussed in the early part of this chapter.

To elucidate the concept of the self-consistent method, the linear strain–stress relationship without using index notation is expressed symbolically as

$$\sigma = C\epsilon. \qquad (3.89)$$

Each quantity in Equation (3.89) is a function of position. It should be noted that simple averaging of both sides of Equation (3.89) does not yield the relation of Equation (3.81) as $<C\epsilon>$ does not separate into $<C><\epsilon>$. However, if the strain at each point is known as a function of position and the average strains as

$$\epsilon = A <\epsilon>, \qquad (3.90)$$

then by substituting Equation (3.90) into Equation (3.89) and taking the spatial average, we obtain

$$C^* = <CA>. \qquad (3.91)$$

It is noted that in Equations (3.90) and (3.91), A depends on both the geometry and the material properties. As the elastic moduli, $C(x)$, are assumed to be piecewise constant across the phases, it follows that

$$C(x) = \sum_\alpha C_\alpha \psi_\alpha(x),$$

where C_α is the elastic modulus of the αth phase. We reserve $\alpha = 0$ for the matrix phase and $\alpha = 1, 2, \cdots, n$ for the inclusion phases. The function, $\psi_\alpha(x)$, is the characteristic function of the αth phase defined as

$$\psi_\alpha(x) = \begin{cases} 1 & x \in \Omega_\alpha \\ 0 & x \notin \Omega_\alpha, \end{cases}$$

where Ω_α represents the αth phase. By taking the spatial average of both sides of Equation (3.90), we have

$$<A> = I, \qquad (3.92)$$

where I is the identity tensor whose components are

$$I_{ijkl} = \frac{1}{2}(\delta_{ik}\delta_{jl} + \delta_{il}\delta_{jk}).$$

Rewriting Equation (3.92) and using the characteristic function yield

$$\begin{aligned} I &= <A> \\ &= <\sum_{\alpha=0}^{n} \psi_\alpha(x)A(x)> \\ &= <\psi_0(x)A(x)> + \sum_{\alpha=1}^{n} <\psi_i A(x)>. \end{aligned}$$

By using this relationship, Equation (3.91) can be rearranged as

$$\begin{aligned} C^* &= <CA> \\ &= C_0 <\psi_0(x)A(x)> + \sum_{i=1}^{n} C_i <\psi_i(x)A(x)> \\ &= C_0 + \sum_{i=1}^{n}(C_i - C_0)<\psi_i(x)A(x)>. \end{aligned} \quad (3.93)$$

Equation (3.93) implies that it is not necessary to know the strain field in the matrix phase to compute C^*. Note that Equation (3.93) is exact and has no approximating part.

The quantity $<\psi_i(x)A(x)>$ in Equation (3.93) can be rewritten as

$$\begin{aligned} <\psi_i(x)A(x)> &= \frac{1}{\Omega}\int_{\Omega_i} A(x)dx \\ &= \frac{\Omega_i}{\Omega}\bar{A}_i \\ &= v_i \bar{A}_i, \end{aligned}$$

where Ω, Ω_i, and v_i are the total volume, the volume for the ith phase, and the volume fraction for the ith phase, respectively. The quantity $\bar{A}_i$ is the average of A over the ith phase defined by

$$\bar{A}_i = \frac{1}{\Omega_i}\int_{\Omega_i} A(x)dx. \quad (3.94)$$

In the self-consistent method, $\bar{A}_i$ in Equation (3.94) is replaced by the proportionality factor for the strain field where a single inclusion is surrounded by a matrix whose properties are those of the composite. This approximation is easily visualized for the following scenario: If one of the inclusions in the composite is placed at the coordinate origin, the surrounding medium is seen as a matrix where the rest of the inclusions are mixed to form the matrix of the whole composite. This method is expected to give a reasonably accurate approximation when the distribution of inclusions is dilute and uniform. Therefore, the proportionality factor can be expressed according to the Eshelby method as

$$A = I + S\{(C^* - C_i)S - C^*\}^{-1}(C_i - C^*). \quad (3.95)$$

By using this, Equation (3.93) can be expressed as

$$C^* = C_0 + \sum_{i=1}^{n}(C_i - C_0)[I + S\{(C^* - C_i)S - C^*\}^{-1}(C_i - C^*)]. \tag{3.96}$$

Equation (3.96) is a set of nonlinear simultaneous equations for the unknown C^* that can be solved numerically.

Here is *Mathematica* implementation of the self-consistent method to estimate the effective properties of a composite consisting of spherical inclusions. First, a working directory (c:\tmp) is selected and the file, micromech.m, that contains all the necessary functions is loaded. The source list for micromech.m is found at the end of this section.

In[1]:= **SetDirectory["c:/tmp"]**

Out[1]= c:\tmp

In[2]:= **<< micromech.m**

The material properties for isotropic components can be entered by a pair of two independent components, i.e., $\{C_{1122}, C_{1212}\}$.

In[6]:= **cf = {10, 20}; cm = {1, 2};**

The effective properties of a composite with spherical inclusions are found to be $C^*_{1122} = 1.52929$ and $C^*_{1212} = 2.90589$.

In[7]:= **SphereEffectiveModulus[cf, cm, 0.2]**

Out[7]= {1.52929, 2.90589}

Here is a plot of C^*_{1122} versus the volume fractions for $v = 0$–1.0.

In[5]:= **Plot[SphereEffectiveModulus[cf, cm, vf][[1]], {vf, 0, 1}]**

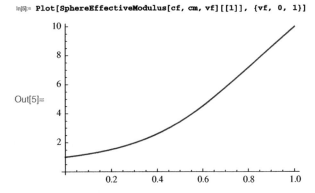

For composites reinforced by long fibers, the effective moduli are transversely isotropic and the material properties can be entered as

In[18]:= **cm = EngToModulus[1, 0.3]**

Out[18]= {1.34615, 1.34615, 0.576923, 0.576923, 0.576923, 0.384615}

In[19]:= **cf = EngToModulus[20, 0.3]**

Out[19]= {26.9231, 26.9231, 11.5385, 11.5385, 11.5385, 7.69231}

In[25]:= **FiberEffectiveModulus[cf, cm, 0.4]**

Out[25]= {9.32012, 2.85732, 1.20021, 1.20021, 1.14337, 1.13424}

In[29]:= **Plot[FiberEffectiveModulus[cf, cm, v][[2]], {v, 0, 1}]**

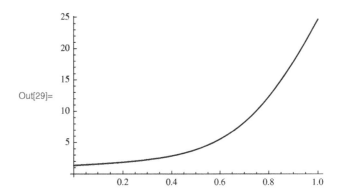

In[32]:= **Plot[FiberEffectiveModulus[cf, cm, v][[1]], {v, 0, 1}]**

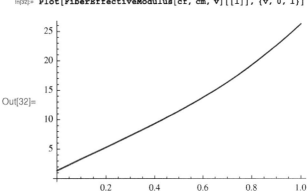

3.5.3 Source Code for micromech.m

This subsection shows a source code for micromech.m, which is a *Mathematica* package that contains various functions for manipulating second- and fourth-rank tensors as well as functions that compute the Eshelby tensors, the strain proportionality factor, and the effective properties using the self-consistent approximation.

To use this package, a working directory (c:\tmp) must be selected in which micromech.m is stored. After that, load the package using the « command as

In[1]:= **SetDirectory["c:/tmp"]**

Out[1]= c:\tmp

In[2]:= **<< micromech.m**

The code is also available from the author's website.

```
(* :Title:          micromech.m *)
(* :Author:             *)
(* :Summary:
```

This package contains functions that compute stress and strain
fields inside an inclusion of spherical and cylindrical shapes
embedded in an infinite matrix subject to uniform strains at
far fields. To support such functionality, several functions
manipulating second- and fourth-rank tensors are included. *)

(* BeginPackage["micromech`"] *)

Off[General::spell1];Off[General::spell];

TransverseInverse::usage="TransverseInverse[a] returns the
inverse of a transversely isotropic tensor, a(ijkl).
The transverse plane is the
x-y plane and the z axis is the axial direction.
Transversely isotropic tensors can be represented as
$\{c\_\{3333\}, c\_\{1111\}, c\_\{3311\},c\_\{1133\}, c\_\{1122\}, c\_\{1313\}\}$
 or
$\{c\_\{33\}, c\_\{11\}, c\_\{31\}, c\_\{13\},c\_\{12\}, c\_\{44\}\}$.";

TransverseProduct::usage="TransverseProduct[a,b] returns
a(ijkl)*b(klmn) where both a(ijkl) and b(klmn) are fourth-
rank tensors and the result is also a fourth-rank tensor.
The transverse plane is the x-y plane and the z axis is
the axial direction. Transversely isotropic tensors can
be input as
$\{c\_\{3333\}, c\_\{1111\}, c\_\{3311\},c\_\{1133\}, c\_\{1122\}, c\_\{1313\}\}$
 or
$\{c\_\{33\}, c\_\{11\}, c\_\{31\}, c\_\{13\},c\_\{12\}, c\_\{44\}\}$.";

Transverse24::usage="Transverse24[a,b] returns a(ij)*b(ijkl)
where a(ij) is a second-rank transversely isotropic tensor
and b(ijkl) is a fourth-rank transversely isotropic tensor.
The result is a second-rank tensor and is represented by
$\{b\_\{11\},b\_\{22\}, b\_\{33\},b\_\{23\},b\_\{13\},b\_\{12\}\}$.
The fourth-rank tensor can be input as
$\{c\_\{3333\}, c\_\{1111\}, c\_\{3311\},c\_\{1133\}, c\_\{1122\}, c\_\{1313\}\}$.";

Transverse42::usage="Transverse42[a,b] returns a(ijkl)*b(kl)
where a(ijkl) is a transversely isotropic fourth-rank tensor and
b(kl) is a second-rank tensor. The result is a second-rank
tensor and is represented by

```
{b_{11},b_{22}, b_{33},b_{23},b_{13},b_{12}}.
The fourth-rank tensor can be input as\n
{c_{3333}, c_{1111}, c_{3311},c_{1133}, c_{1122}, c_{1313}}.";

EngToModulus::usage="EngToModulus[E,v] returns c[ijkl]
from E and v where E and v are the Young modulus and the Poisson
ratio, respectively. The format of c[ijkl] is
{c_{3333}, c_{1111}, c_{3311}, c_{1133},c_{1122}, c_{1313}}
or\n
{c_{33}, c_{11}, c_{31}, c_{13}, c_{12},c_{44}}.";

CylinderStrainFactor::usage= "CylinderStrainFactor[cf,cm]
returns the proportionality factor of strain field inside
a cylindrical inclusion with the elastic modulus, cf,
embedded in the matrix with the elastic modulus, cm.
Transversely isotropic tensors can be input as
{c_{3333},c_{1111}, c_{3311}, c_{1133}, c_{1122}, c_{1313}}
or
{c_{33}, c_{11},c_{31}, c_{13}, c_{12}, c_{44}}.
CylinderStrainFactor x far_{s}train is the strain field
inside the inclusion subject to <epsilon_{ij}>
at far-field.";

CylinderStressFactor::usage= "CylinderStressFactor[cf,cm]
returns the proportionality factor of stress field inside
a cylindrical inclusion with the elastic modulus, cf,
embedded in the matrix with the elastic modulus, cm.
Transversely isotropic tensors can be input as
{c_{3333},c_{1111}, c_{3311}, c_{1133}, c_{1122}, c_{1313}}
or
{c_{33}, c_{11},c_{31}, c_{13}, c_{12}, c_{44}}
subject to <stress_{ij}> at far-field.";

FiberEffectiveModulus::usage="FiberEffectiveModulus[cf, cm, vf]
returns the effective modulus of long-fiber reinforced
composite materials with 'cf' as the fiber moduli, 'cm'
as the matrix moduli, and 'vf'as the fiber volume fraction
by the self-consistent method";

IsotropicProduct::usage= "IsotropicProduct[c1,c2] computes
product of two isotropic fourth-rank tensors, c1 \times c2.
Both c1 and c2 are represented in the format of {lambda,mu}
or {c_{1122},c_{1212}}.";

IsotropicInverse::usage= "IsotropicInverse[c] computes the
inverse of a fourth-rank tensor, c. The tensor_{c}_ is
```

represented by $\{c\_\{1122\},c\_\{1212\}\}$.";
Lame::usage= "Lame[e,nu] converts the Young modulus and the
Poisson ratio for isotropic elastic materials to lambda
(c_{1122}) and mu (c_{1212}).";

SphereStrainFactor::usage= "SphereStrainFactor[{lambda_{f},
mu_{f}},{lambda_{m}, mu_{m}}] returns the proportionality factor
of the strain field inside a spherical inclusion in the
format of {A_{1122},A_{1212}} subject to <epsilon_{ij}>
at far-field.";

SphereStressFactor::usage="SphereStressFactor[{lambda_{f}, mu_{f}},
{lambda_{m}, mu_{m}}] returns the proportionality factor the
stress field inside a spherical inclusion in the format of
{B_{1122}, B_{1212}} subject to <stress_{ij}> at far-field.";

SphereEffectiveModulus::usage="SphereEffectiveModulus[cf, cm,
vf] returns the effective modulus of composite reinforced by
spherical inclusions with 'cf' as the inclusion moduli, 'cm'
as the matrix moduli, and 'vf' as the inclusion volume fraction
by the self-consistent method";

IdentityTensor::usage="IdentityTensor is fourth rank identity
tensor.";

EshelbyIsotropic::usage="EshelbyIsotropic[t, nu] returns the
components of the Eshelby tensor for an isotropic medium for
the aspect ratio 't' and the Poisson ratio 'nu.' It returns
the result in transversely isotopic format as
{s_{3333},s_{1111}, s_{3311}, s_{1133}, s_{1122}, s_{1313}}
or
{s_{33}, s_{11},s_{31}, s_{13}, s_{12}, s_{44}}.";

IdentityTensor={1,1,0,0,0,1/2};

(* Begin["'Private'"]; *)

(*Tensor manipulation definitions

We represent fourth-rank tensors a[i,j,k,l] by arrays a[6],
where
a[1] <- a[[3,3,3,3]], a33,a[2] <- a[[1,1,1,1]]   a11
a[3] <- a[[3,3,1,1]] a31,a[4] <- a[[1,1,3,3]]   a13
a[5] <- a[[1,1,2,2]] a12,a[6] <- a[[1,3,1,3]]   a44 (=a55)

For a second-rank tensor, b[i,j], the following convention is

used:
  b[1]<-b[1,1] b[2]<-b[2,2] b[3]<-b[3,3]
  b[4]<-b[2,3] b[5]<-b[1,3] b[6]<-b[1,2].

*)

TransverseInverse[a_List]:=Module[
  {x=a[[1]]*(a[[2]]+a[[5]])-2a[[3]]*a[[4]]},
  {(a[[2]]+a[[5]])/x ,
  (a[[1]]a[[2]]-a[[3]]a[[4]])/(a[[2]]-a[[5]])/x,
  -a[[3]]/x,
  -a[[4]]/x,
  (a[[3]]a[[4]]-a[[1]]a[[5]])/(a[[2]]-a[[5]])/x,
  1/4/a[[6]]}];

TransverseProduct[a_List,b_List]:={
  a[[1]]*b[[1]] + 2*a[[3]]*b[[4]],
  a[[4]]*b[[3]] + a[[2]]*b[[2]] + a[[5]]*b[[5]],
  a[[1]]*b[[3]] + a[[3]]*b[[2]] + a[[3]]*b[[5]],
  a[[4]]*b[[1]] + a[[2]]*b[[4]] + a[[5]]*b[[4]],
  a[[4]]*b[[3]] + a[[2]]*b[[5]] + a[[5]]*b[[2]],
  2*a[[6]]*b[[6]]};

Transverse24[a_List,c_List]:={
  a[[1]] c[[2]]+a[[2]] c[[5]]+a[[3]] c[[3]],
  a[[1]] c[[5]]+a[[2]] c[[2]]+a[[3]] c[[3]],
  a[[1]] c[[4]]+a[[2]] c[[4]]+a[[3]] c[[1]],
  2 a[[4]] c[[6]],
  2 a[[5]] c[[6]],
  2 a[[6]] (c[[2]]-c[[5]])/2};

Transverse42[a_List,b_List]:={
  {a[[2]],a[[5]],a[[4]],0,0,0},
  {a[[5]],a[[2]],a[[4]],0,0,0},
  {a[[3]],a[[3]],a[[1]],0,0,0},
  {0,0,0,2a[[6]],0,0},
  {0,0,0,0,2a[[6]],0},
  {0,0,0,0,0,a[[2]]-a[[5]]}}.b;

EngToModulus[e_,nu_]:={
    e*(1-nu)/(1+nu)/(1-2*nu),
    e*(1-nu)/(1+nu)/(1-2*nu),
    nu*e/(1+nu)/(1-2*nu),
    nu*e/(1+nu)/(1-2*nu),
    nu*e/(1+nu)/(1-2*nu),
    e/2/(1+nu)};

```
EshelbyIsotropic[t_, nu_] := {(t*(((1 - 2*nu)*(Sqrt[-1
+ t^(-2)] -
ArcCos[t]))/(1 - t^{2})^(3/2) + (-1 + 5*t^{2} - 4*t^{4} +
3*t^{3}*Sqrt[-1 + t^{2}]*ArcCosh[t])/(t*(-1 + t^{2})^{3})))
/(2*(1 - nu)), (t*(t*
Sqrt[-1 + t^{2}]*(19 - 10*t^{2} + 8*nu*(-1 + t^{2})) +
(-13 + 4*t^{2} - 8*nu*(-1 + t^{2}))*ArcCosh[t]))/(16*(-1+
nu)*(-1 + t^{2})^(5/2)), ((2*(-1 + 2*nu)*
t*(Sqrt[-1 + t^(-2)] - ArcCos[t]))/(1 - t^{2})^(3/
2) + (Sqrt[-1 + t^{2}]*(2 + t^{2}) -
3*t*ArcCosh[t])/(-1 + t^{2})^(5/2))/(4*(1 -
nu)), (t*(t*
Sqrt[-1 + t^{2}]*(-3 - 2*nu*(-1 + t^{2})) + (1 + 2*t^{2} +
2*nu*(-1 + t^{2}))*ArcCosh[t]))/(4*(-1 +
nu)*(-1 + t^{2})^(5/2)), (t*(t*
Sqrt[-1 + t^{2}]*(9 - 6*t^{2} + 8*nu*(-1 + t^{2})) + (-7 + 4*t^{2} -
8*nu*(-1 + t^{2}))*ArcCosh[t]))/(16*(-1 +
nu)*(-1 + t^{2})^(5/
2)), ((2*(1 + t^{2})*(Sqrt[-1 + t^{2}]*(2 + t^{2}) -
3*t*ArcCosh[t]))/(-1 + t^{2})^(5/2) +
2*(1 - 2*nu)*
t*((2*(Sqrt[-1 + t^(-2)] - ArcCos[t]))/(1 - t^{2})^(3/2) + (t -
ArcCosh[t]/Sqrt[-1 + t^{2}])
/(-1 + t^{2})))/(16*(1 - nu))}//ComplexExpand//Chop;

(* ----------------------------------------------------
EshelbyCylinder::usage="EshelbyCylinder[Cm] returns the Eshelby
tensor for cylinder with Cm.";

EshelbyCylinder[c_List]:=
  Module[{c33=c[[1]],c11=c[[2]],c13=c[[3]],
  c12=c[[5]],c44=c[[6]]},
  {0,5/8+c12/(8*c11),0,c13/(2*c11),(-c11+3*c12)/(8*c11),
  1/4}];

(* The following routine computes CylindricalStrainFactor and
CylindricalStressFactor. *)

cm={cm33,cm11,cm31,cm31,cm12,cm44};
cf={cf33,cf11,cf31,cf31,cf12,cf44};

s=EshelbyCylinder[cm]//Simplify;

junk1=TransverseProduct[cm,s]//Simplify;
junk2=junk1-cm-TransverseProduct[cf,s]//Simplify;
junk1=TransverseInverse[junk2]//Simplify;
```

```
junk2=TransverseProduct[s,junk1]//Simplify;
junk1=TransverseProduct[junk2,cf-cm]//Simplify;
afactor=identity+junk1//Simplify;
stress=TransverseProduct[cf,afactor]//Simplify;

(*
a=IdentityTensor + TransverseProduct[ TransverseProduct[s,
TransverseInverse [ TransverseProduct[cm,s]//Simplify - cm +
TransverseProduct[cf,s]//Simplify]//Simplify]//Simplify,
cf-cm]//Simplify;
*)
Save["Afactor",afactor,stress]
------------------------------------------------------*)

CylinderStrainFactor[cf_List, cm_List]:=
Module[ {cf33=cf[[1]], cf11=cf[[2]], cf31=cf[[3]], cf13=cf[[4]],
cf12=cf[[5]], cf44=cf[[6]], cm33=cm[[1]], cm11=cm[[2]],
cm31=cm[[3]],
cm13=cm[[4]], cm12=cm[[5]], cm44=cm[[6]]},
Return[
{1, (cm11*(-5*cf11*cm11 + cf12*cm11 - 3*cm11^{2} + 3*cf11*cm12 +
      cf12*cm12 + 4*cm11*cm12 - cm12^{2}))/
    ((-cf11 - cf12 - cm11 + cm12)*
      (3*cf11*cm11 - 3*cf12*cm11 + cm11^{2} - cf11*cm12
      + cf12*cm12 - cm12^{2})),
  0, (cf31 - cm31)/(-cf11 - cf12 - cm11 + cm12),
  (cm11*(cf11*cm11 - 5*cf12*cm11 - cm11^{2} + cf11*cm12 +
  3*cf12*cm12 +
      4*cm11*cm12 - 3*cm12^{2}))/
    ((cf11 + cf12 + cm11 - cm12)*
      (3*cf11*cm11 - 3*cf12*cm11 + cm11^{2} - cf11*cm12
      + cf12*cm12 - cm12^{2})),
  cm44/(cf44 + cm44)}] ];

CylinderStressFactor[cf_List, cm_List]:=
Module[
{cf33=cf[[1]], cf11=cf[[2]], cf31=cf[[3]], cf13=cf[[4]],
cf12=cf[[5]], cf44=cf[[6]], cm33=cm[[1]], cm11=cm[[2]],
cm31=cm[[3]],
cm13=cm[[4]], cm12=cm[[5]], cm44=cm[[6]]},
Return[
{((-2*cf31^{2} + cf33*(cf11 + cf12 + cm11 - cm12))*
    (cm11 + cm12) + 2*cf31*(-cm11 + cm12)*cm31)/
  ((cf11 + cf12 + cm11 - cm12)*(-2*cm31^{2} +
    (cm11 + cm12)*cm33)),
 ((cm11*(3*cf11*(cm11 - cm12)*(cm11 + cm12) +
```

```
          cf11^{2}*(5*cm11 + cm12) -
         cf12*(5*cf12*cm11 + cm11^{2} + cf12*cm12 -
            cm12^{2})))/(3*cf11*cm11 - 3*cf12*cm11 +
         cm11^{2} - cf11*cm12 + cf12*cm12 - cm12^{2}) -
      ((cm11 - cm12)*cm31*(cf31*(cm11 + cm12) -
          (cf11 + cf12)*cm31))/(-2*cm31^{2} +
          (cm11 + cm12)*cm33))/
    ((cf11 + cf12 + cm11 - cm12)*(cm11 + cm12)),
    (-(cm31*(cf33*(cf11 + cf12 + cm11 - cm12) +
           2*cf31*(-cf31 + cm31))) + 2*cf31*cm11*cm33)/
      ((cf11 + cf12 + cm11 - cm12)*(-2*cm31^{2} +
          (cm11 + cm12)*cm33)),
    ((cm11 - cm12)*(cf31*(cm11 + cm12) -
         (cf11 + cf12)*cm31))/
      ((cf11 + cf12 + cm11 - cm12)*(-2*cm31^{2} +
          (cm11 + cm12)*cm33)),
    ((cm11*(cm11*(cf11^{2} - cf12*(cf12 - 3*cm11) -
            cf11*cm11) + 3*(-cf11^{2} + cf12^{2})*cm12 +
         (cf11 - 3*cf12)*cm12^{2}))/(3*cf11*cm11 -
         3*cf12*cm11 + cm11^{2} - cf11*cm12 + cf12*cm12 -
         cm12^{2}) - ((cm11 - cm12)*cm31*
          (cf31*(cm11 + cm12) - (cf11 + cf12)*cm31))/
         (-2*cm31^{2} + (cm11 + cm12)*cm33))/
      ((cf11 + cf12 + cm11 - cm12)*(cm11 + cm12)),
    cf44/(cf44 + cm44)}]];

FiberEffectiveModulus[cf_,cm_,vf_]:=
  Module[{ce=cm},
   f[ce_]:=cm+vf*TransverseProduct[(cf-cm),
   CylinderStrainFactor[cf,ce]];
   FixedPoint[f,ce,10]];

(* ---------------------------- *)
(* isotropic   spherical inclusion*)
(*
(* Various conversion table*)

lambda={2 mu nu/(1-2nu), mu (e-2 mu)/(3 mu-e),k-2/3 mu,
e nu/(1+nu)/(1-2nu),
3 k nu/(1+nu), 3k (3k-e)/(9k-e)};
mu={lambda (1-2nu)/2/nu,3/2 (k-lambda),e/2/(1+nu),
3k (1-2nu)/2/(1+nu),
3 k e /(9k-e)};
nu={lambda/2/(lambda+mu),lambda/(3k-lambda),e/2/mu-1,
(3k-2mu)/2/(3k+mu),
(3k-e)/6/k};
```

```
e={mu(3 lambda+2mu)/(mu+lambda),lambda (1+nu)(1-2nu)/nu,
9k(k-lambda)/(3k-lambda), 2mu (1+nu), 9k mu/(3k+mu),
3k(1-2nu)};
k={lambda+2/3 mu,lambda (1+nu)/3/nu, 3 mu (1+nu)/3/(1-2 nu),
mu e/3/(3mu-e), e/3/(1-2nu)};
*)

SetAttributes[IConv,Orderless];
IConv[e,k]={lambda->3k (3k-e)/(9k-e),mu->3 k e
/(9k-e),nu->(3k-e)/6/k};
IConv[e,mu]={lambda->mu (e-2 mu)/(3 mu-e),
 nu->e/2/mu-1,k->mu e/3/(3mu-e)};
IConv[e,nu]={lambda->e nu/(1+nu)/(1-2nu),mu->e/2/(1+nu),
 k->e/3/(1-2nu)};
IConv[e,lambda]={mu->(e - 3 lambda +(e^{2} + 2*e*lambda +
9*lambda^{2})^(1/2))/4,
nu->(2*lambda)/(e+lambda+(e^{2}+2*e*lambda+9*lambda^{2})^(1/2)),
 k->(e +3*lambda +(e^{2} + 2*e*lambda + 9*lambda^{2})^(1/2))/6};
IConv[k,mu]={lambda->k-2/3 mu,nu->(3k-2mu)/2/(3k+mu),
 e->9 k mu/(3k+mu)};
IConv[k,nu]={lambda->3k nu/(1+nu),mu->3 k(1-2nu)/2/(1+nu),
 e->3k (1-2nu)};
IConv[k,lambda]={mu->3/2 (k-lambda), nu-> lambda/(3k-lambda),
e->9k(k-lambda)/(3k-lambda)};
IConv[mu,nu]={lambda->2mu nu/(1-2nu),e->2 mu (1+nu),
 k->3mu(1+nu)/3/(1-2nu)};
IConv[mu,lambda]={nu->lambda/2/(lambda+mu),
e->mu(3lambda+2mu)/(mu+lambda), k->lambda+2/3 mu};
IConv[nu,lambda]={mu->lambda(1-2nu)/2/nu,
e->lambda(1+nu)(1-2nu)/nu,k->lambda(1+nu)/3/nu};

(*

A fourth-rank isotropic tensor, c_{ijkl}, is in general
expressed as

c_{ijkl} = alpha del_{ij} del_{kl} +
  beta (del_{ik} del_{jl} +del_{il} del_{jk})

and is represented as {alpha,beta} or {c_{1122}, c_{1212}}. If
c_{ijkl} is an elastic modulus, "alpha" is "lambda" and "beta"
is "mu").

*)

IsotropicProduct[c1_List,c2_List]:=
```

Inclusions in Infinite Media

```
{3c1[[1]]c2[[1]]+2(c1[[2]]c2[[1]]+c1[[1]]c2[[2]]),
 2 c1[[2]]c2[[2]]};

IsotropicInverse[c1_List]:={-c1[[1]]/2/c1[[2]]/
 (3c1[[1]]+2c1[[2]]),1/4/c1[[2]]};

Lame[e_,nu_]:={nu e/(1+nu)/(1-2nu),e/2/(1+nu)};

(* --------- *)

SphereStrainFactor[cf_List,cm_List]:=
Module[
{
kf=cf[[1]]+2/3cf[[2]],
muf=cf[[2]],
km=cm[[1]]+2/3cm[[2]],
mum=cm[[2]]
},
Return[{
((3*km + 4*mum)*(2*km*muf  -  5*kf*mum + 3*km*mum +
 4*muf*mum - 4*mum^{2}))/    ((3*kf + 4*mum)*(6*km*muf +
 9*km*mum + 12*muf*mum + 8*mum^{2})),
(5*mum*(3*km + 4*mum))/(12*km*muf + 18*km*mum +
24*muf*mum + 16*mum^{2})
}];

SphereStressFactor[cf_List,cm_List]:=
Module[
{
kf=cf[[1]]+2/3cf[[2]],
muf=cf[[2]],
km=cm[[1]]+2/3cm[[2]],
mum=cm[[2]]
},
Return[

   {(3*(km + (4*mum)/3)*(muf^{2}*(-6*(km - (2*mum)/3) +
  4*mum) + (kf -    (2*muf)/3)*mum*
     (9*(km - (2*mum)/3) + 14*mum) - muf*(9*(kf -
 (2*muf)/3)*(km -    (2*mum)/3) - 6*(kf - (2*muf)/3)*mum +
      14*(km - (2*mum)/3)*mum + 4*mum^{2})))/((3*(km -
 (2*mum)/3) + 2*mum)*    (3*(kf - (2*muf)/3) + 2*muf +
  4*mum)*(6*muf*(km - (2*mum)/3) + 16*muf*mum +
  9*(km - (2*mum)/3)*mum +
    14*mum^{2})), (15*muf*(km + (4*mum)/3))/
 (2*(6*muf*(km - (2*mum)/3) +
```

```
   16*muf*mum +
      9*(km - (2*mum)/3)*mum + 14*mum^{2}))
}]]

SphereEffectiveModulus[cf_,cm_,vf_]:=
   Module[{ce=cm},
   f[ce_]:=cm+vf*IsotropicProduct[(cf-cm),
   SphereStrainFactor[cf,ce]];
   FixedPoint[f,ce,10]];

(*
End[]
EndPackage[];
*)
```

3.6 Exercises

1. At the interface of an inclusion and a matrix, the displacement, u_i, and the traction, t_i, must be continuous, i.e.,

$$u_i^{in} = u_i^{out}, \quad t_i^{in} = t_i^{out}.$$

   By using the uniform stress field inside the inclusion, derive the stress component outside the inclusion.

2. Show that Green's function for a 2-D isotropic material is expressed as

$$g_{im}(\mathbf{r},\mathbf{r}') = \frac{1}{4\pi\mu(2\mu+\lambda)}\left\{(3\mu+\lambda)\delta_{ij}\ln\frac{1}{r} + (\mu+\lambda)\frac{x_i x_j}{r^2}\right\}.$$

3. Derive Equations (3.70)–(3.72).

4. Consider a steady-state temperature distribution problem when a cylinder is placed in a constant heat flux. The steady-state temperature distribution throughout must satisfy the Laplace equation,

$$\Delta T = 0.$$

   The boundary condition is such that as $x, y \to \infty$, the heat flux is constant, i.e.,

$$k_2\frac{\partial T}{\partial x} \to h_x, \quad k_2\frac{\partial T}{\partial y} \to h_y \quad \text{as} \quad x, y \to \infty.$$

   Derive the temperature field in both the cylinder and the surrounding medium using the complex variables.

5. Consider a rectangular bar ($-a \leq x \leq a, -b \leq y \leq b$) subject to a bending moment, M_x, at $x = \pm a$ and free at $y = \pm b$ shown in Figure 3.18.
   The boundary condition at $x = \pm a$ is $\int_{-b}^{b} y\,\sigma_{xx}\,dy = M_x$. The boundary condition at $x = \pm b$ is $\sigma_{yy} = 0, \sigma_{xy} = 0$. Assuming $\gamma(z) = c_1 z^2$ and $\chi(z) = d_1 z^2$, determine c_1 and d_1 so that the aforementioned boundary condition is satisfied.

Figure 3.18 Two-dimensional beam

6. Solve Equation (3.23) for plane stress using the approach in Section 3.2. It is necessary to use

$$X_{ij} = \hat{X}_{ij} + \frac{1}{2}X\delta_{ij}.$$

7. Under what condition is Equation (3.93) reduced to the Voigt and Reuss bounds?
Note: Refer www.wiley.com/go/nomura0615 for solutions for the Exercise section.

References

Bilby BA 1990 John Douglas Eshelby. 21 December 1916-10 December 1981. *Biographical Memoirs of Fellows of the Royal Society*, **36**, 127–150, JSTOR.

Buryachenko V 2007 *Micromechanics of Heterogeneous Materials*. Springer-Verlag.

Chao C and Shen M 1998 Thermal stresses in a generally anisotropic body with an elliptic inclusion subject to uniform heat flow. *Journal of Applied Mechanics* **65**(1), 51–58.

Chou T, Nomura S and Taya M 1980 A self-consistent approach to the elastic stiffness of short-fiber composites. *Journal of Composite Materials* **14**(3), 178.

Choy TC 1999 *Effective Medium Theory: Principles and Applications*. Clarendon Press, Oxford.

Christensen R and Lo K 1979 Solutions for effective shear properties in three phase sphere and cylinder models. *Journal of the Mechanics and Physics of Solids* **27**(4), 315–330.

Eshelby J 1957 The determination of the elastic field of an ellipsoidal inclusion, and related problems. *Proceedings of the Royal Society of London, Series A: Mathematical and Physical Sciences* **241**(1226), 376.

Eshelby J 1959 The elastic field outside an ellipsoidal inclusion. *Proceedings of the Royal Society of London, Series A: Mathematical and Physical Sciences* **252**, 561–569.

Florence A and Goodier J 1960 Thermal stresses due to disturbance of uniform heat flow by an insulated ovaloid hole. *Journal of Applied Mechanics* **27**(4), 635–639.

Granqvist C and Hunderi O 1978 Conductivity of inhomogeneous materials: effective-medium theory with dipole-dipole interaction. *Physical Review B* **18**(4), 1554.

Greenberg M 1971 *Application of Green's Functions in Science and Engineering*, vol. 30. Prentice-Hall.

Hashin Z 1965 On elastic behaviour of fibre reinforced materials of arbitrary transverse phase geometry. *Journal of the Mechanics and Physics of Solids* **13**(3), 119–134.

Hashin Z and Shtrikman S 1963 A variational approach to the theory of the elastic behaviour of multiphase materials. *Journal of the Mechanics and Physics of Solids* **11**(2), 127–140.

Herve E and Zaoui A 1993 N-layered inclusion-based micromechanical modelling. *International Journal of Engineering Science* **31**(1), 1–10.

Hill R 1963 Elastic properties of reinforced solids: some theoretical principles. *Journal of the Mechanics and Physics of Solids* **11**(5), 357–372.

Hill R 1965 A self-consistent mechanics of composite materials. *Journal of the Mechanics and Physics of Solids* **13**(4), 213–222.

Laws N and McLaughlin R 1979 The effect of fibre length on the overall moduli of composite materials. *Journal of the Mechanics and Physics of Solids* **27**(1), 1–13.

Lin S and Mura T 1973 Elastic fields of inclusions in anisotropic media (ii). *Physica Status Solidi A* **15**(1), 281–285.

Madhavan M 2013 *A Novel Approach for Two-Dimensional Inclusion/Hole Problems using the Airy Stress Function Method*. Master's thesis. University of Texas at Arlington, Arlington, TX.

Markenscoff X and Gupta A 2006 *Collected Works of JD Eshelby: The Mechanics of Defects and Inhomogeneities*. Springer-Verlag.

Mura T 1987 *Micromechanics of Defects in Solids*, vol. 3. Springer-Verlag.

Nemat-Nasser S and Hori M 1999 *Micromechanics: Overall Properties of Heterogeneous Materials*, vol. 2. Elsevier Amsterdam.

Nemat-Nasser S and Hori M 2013 *Micromechanics: Overall Properties of Heterogeneous Materials*. Elsevier.

Oshima N and Nomura S 1985 A method to calculate effective modulus of hybrid composite materials. *Journal of Composite Materials* **19**(3), 287.

Pan Y and Chou T 1979 Green's function solutions for semi-infinite transversely isotropic materials. *International Journal of Engineering Science* **17**(5), 545–551.

Qu J and Cherkaoui M 2006 *Fundamentals of Micromechanics of Solids*. Wiley Online Library.

Reuss A 1929 Berechnung der fließgrenze von mischkristallen auf grund der plastizitätsbedingung für einkristalle. *ZAMM-Journal of Applied Mathematics and Mechanics/Zeitschrift für Angewandte Mathematik und Mechanik* **9**(1), 49–58.

Rosen BW and Hashin Z 1970 Effective thermal expansion coefficients and specific heats of composite materials. *International Journal of Engineering Science* **8**(2), 157–173.

Sadd M 2009 *Elasticity: Theory, Applications, and Numerics*. Academic Press.

Shah SP 1995 *Fracture Mechanics of Concrete: Applications of Fracture Mechanics to Concrete, Rock and other Quasi-Brittle Materials*. John Wiley & Sons, Inc.

Sokolnikoff IS and Specht RD 1956 *Mathematical Theory of Elasticity*, vol. 83. McGraw-Hill, New York.

Stakgold I and Holst MJ 2011 *Green's Functions and Boundary Value Problems*, vol. 99. John Wiley & Sons, Inc.

Tauchert T 1968 Thermal stresses at spherical inclusions in uniform heat flow. *Journal of Composite Materials* **2**(4), 478–486.

Vinson JR and Sierakowski RL 2006 *The Behavior of Structures Composed of Composite Materials*, vol. 105. Springer Science & Business Media.

Voigt W 1910 *Lehrbuch der kristallphysik*, vol. 34. BG Teubner.

4

Inclusions in Finite Matrix

In Chapter 3, problems involving inclusions where the surrounding matrix medium is infinitely extended were addressed. It is noted that for elasticity problems with inclusions, analytical solutions for the elastic field are available only for a limited number of cases when a single ellipsoidal inclusion (or layered inclusions) is embedded in an infinitely extended medium. The matrix properties need to be either isotropic or transversely isotropic. Although this covers a wide range of practical material systems, in reality, all the existing material systems are of finite shape, and therefore, infinite media are an ideal scenario that are taken as a limit of finite media. Hence, all the analyses in the previous chapter are an approximation at best and should be taken as limiting cases of the existing material systems.

In this chapter, micromechanical approaches to derive thermal and mechanical fields in finite-sized bodies that contain an inclusion(s) are introduced. Contrary to intuition and common sense, this problem is, in fact, more difficult than the inclusion problems defined in infinite media in Chapter 3 and there is no closed/analytical solution available for any shape of inclusion if the matrix medium is finite. The common approach to deal with a finite-sized medium has been to use numerical methods represented by the finite element method. Available application software for the finite element method is mature, and many problems involving inclusions can be solved routinely. Nevertheless, it is still desirable if analytical or semianalytical solutions are available as such approaches can retain relevant material or geometrical parameters.

In this chapter, a semianalytical approach to derive approximate solutions to heat conduction and elasticity equations that involve inclusions is presented, which takes advantage of *Mathematica*. The approach is an implementation of the Galerkin method that has been used for many years and is also the basis for the finite element method (Finlayson 1972), but the availability of computer algebra systems makes the range of applicable problems extended significantly and permissible functions are no longer restricted to trivial geometrical shapes.

To elucidate the method used, this chapter begins with a brief introduction to the method of weighted residuals (MWR) and the Sturm–Liouville system. This lays the foundation for a paradigm to obtain approximate solutions to general boundary value problems in partial differential equations including elasticity equations and heat conduction equations.

After a brief introduction of MWR, steady-state heat conduction and elasticity equations are discussed with numerical examples. The major part of the elasticity solution is due to the PhD dissertation by Pathapalli (Pathapalli 2013).

Micromechanics with Mathematica, First Edition. Seiichi Nomura.
© 2016 John Wiley & Sons, Ltd. Published 2016 by John Wiley & Sons, Ltd.
Companion Website: www.wiley.com/go/nomura0615

4.1 General Approaches for Numerically Solving Boundary Value Problems

4.1.1 Method of Weighted Residuals

Among many approaches available for numerically solving boundary value problems in differential equations, MWR (Method of Weighted Residuals) (Finlayson 1972) is the most widely used method. The underlying idea of MWR is to approximate the solution to a differential equation by a linear combination of permissible functions each of which satisfies the homogeneous boundary condition with unknown coefficients. Depending on how the unknown coefficients for the permissible functions are determined, there are three major methods within MWR. They are (1) the collocation method, (2) the least squares method, and (3) the Galerkin method (Greenberg 1971). The finite element method is a special case of the Galerkin method in which the permissible functions are chosen to be piecewise continuous that take the value of 1 at their own nodes and 0 at all other nodes. This section briefly explains the concept of MWR and the Galerkin method in particular to be used for the analysis of heterogeneous materials.

The objective is to solve a system of general linear equations expressed symbolically as

$$Lu = c, \qquad (4.1)$$

where L is a linear operator, u is an unknown function to be solved, and c is a known function. Although MWR can be used for any type of linear operators (matrices, differential operators, or integral operators), it is limited to differential operators with the homogeneous boundary condition here.

An approximate solution, $\tilde{u}_N$, to Equation (4.1) is sought by a linear combination of N permissible functions in the function space as

$$\tilde{u}_N = \sum_{i=1}^{N} \tilde{u}_i \mathbf{e}_i,$$

where $\tilde{u}_i$ is an unknown coefficient yet to be determined and $\mathbf{e}_i$ is a permissible function in the linear space that satisfies the homogeneous boundary condition.

The residual (error), R, defined as the difference between the aforementioned approximate solution and the exact solution is expressed as

$$R \equiv L\tilde{u}_N - c$$

$$= L \sum_{i=1}^{N} \tilde{u}_i \mathbf{e}_i - c$$

$$= \sum_{i=1}^{N} \tilde{u}_i L\mathbf{e}_i - c.$$

The three representative methods to determine the unknown coefficients, $\tilde{u}_i$, based on R are discussed in the following:

1. Collocation method
   In the collocation method, the unknown coefficients, $\tilde{u}_i$, are chosen so that the residual (error) vanishes at N selected points, i.e.,

$$R(x_i) = 0, \quad i = 1, \ldots, N.$$

This leads to

$$Le_1|_{x=x_1}\tilde{u}_1 + Le_2|_{x=x_1}\tilde{u}_2 + Le_3|_{x=x_1}\tilde{u}_3 + \cdots + Le_N|_{x=x_1}\tilde{u}_N = c(x_1),$$
$$Le_1|_{x=x_2}\tilde{u}_1 + Le_2|_{x=x_2}\tilde{u}_2 + Le_3|_{x=x_2}\tilde{u}_3 + \cdots + Le_N|_{x=x_2}\tilde{u}_N = c(x_2),$$
$$\vdots$$
$$Le_1|_{x=x_N}\tilde{u}_1 + Le_2|_{x=x_N}\tilde{u}_2 + Le_3|_{x=x_N}\tilde{u}_3 + \cdots + Le_N|_{x=x_N}\tilde{u}_N = c(x_N),$$

or

$$\begin{pmatrix} Le_1|_{x=x_1} & Le_2|_{x=x_1} & \cdots & Le_N|_{x=x_1} \\ Le_1|_{x=x_2} & Le_2|_{x=x_2} & \cdots & Le_N|_{x=x_2} \\ \cdots & \cdots & \cdots & \cdots \\ Le_1|_{x=x_N} & Le_2|_{x=x_N} & \cdots & Le_N|_{x=x_N} \end{pmatrix} \begin{pmatrix} \tilde{u}_1 \\ \tilde{u}_2 \\ \cdots \\ \tilde{u}_N \end{pmatrix} = \begin{pmatrix} c(x_1) \\ c(x_2) \\ \cdots \\ c(x_N) \end{pmatrix}, \quad (4.2)$$

or

$$\tilde{L}\tilde{\mathbf{u}} = \mathbf{c},$$

where

$$(\tilde{L})_{ij} \equiv Le_j|_{x=x_i}, \quad c_i \equiv c|_{x=x_i}.$$

The unknown coefficients, $\tilde{u}_i$, can be determined by solving Equation (4.2). It is noted that the matrix on the left-hand side of Equation (4.2) is not symmetrical, which may pose stability problems in numerical inversion.

Although this method yields the exact values at the selected points, there is no guarantee that the approximation behaves nicely between the selected points.

2. Least squares method

In the least squares method, the unknown coefficients, $\tilde{u}_i$, are chosen so that the norm of residual (error) becomes the minimum, i.e.,

$$||R(x)||^2 \to \min.$$

The square of the norm of $R(x)$, $||R(x)^2||$, can be computed as

$$||R||^2 = (R, R)$$
$$= \left(\sum_{i=1}^{N} \tilde{u}_i Le_i - c, \sum_{j=1}^{N} \tilde{u}_j Le_j - c \right)$$
$$= \sum_{i=1}^{N} \sum_{j=1}^{N} \tilde{u}_i \tilde{u}_j (Le_i, Le_j) - 2 \sum_{i=1}^{N} \tilde{u}_i (Le_i, c) + c^2, \quad (4.3)$$

where (f, g) is an inner product in the function space defined as

$$(f, g) \equiv \int f(x)g(x)dx.$$

To minimize $||R(x)||^2$, Equation (4.3) is differentiated with respect to $\tilde{u}_i$ as

$$\frac{\partial ||R||^2}{\partial \tilde{u}_i} = 2 \sum_{j=1}^{N} \tilde{u}_j (Le_i, Le_j) - 2(Le_i, c) = 0,$$

which yields

$$\sum_{j=1}^{N}(Le_i, Le_j)\tilde{u}_j = (Le_i, c),$$

or

$$\begin{pmatrix} (Le_1, Le_1) & (Le_1, Le_2) & \cdots & (Le_1, Le_N) \\ (Le_2, Le_1) & (Le_2, Le_2) & \cdots & (Le_2, Le_N) \\ \cdots & \cdots & \cdots & \cdots \\ (Le_N, Le_1) & (Le_N, Le_2) & \cdots & (Le_N, Le_N) \end{pmatrix} \begin{pmatrix} \tilde{u}_1 \\ \tilde{u}_2 \\ \cdots \\ \tilde{u}_N \end{pmatrix} = \begin{pmatrix} (Le_1, c) \\ (Le_2, c) \\ \cdots \\ (Le_N, c) \end{pmatrix}. \quad (4.4)$$

Note that the matrix in Equation (4.4) is Hermitian (symmetrical). This method is expected to give an overall well-behaved approximation.

3. Galerkin's method

In the Galerkin method, the unknown coefficients, $\tilde{u}_i$, are chosen so that $R(x)$ is orthogonal to the N permissible functions (e_i), i.e.,

$$(R, e_i) = 0, \quad i = 1, \ldots, N. \quad (4.5)$$

The idea of the Galerkin method is that if e_i's span the entire function space, a function that is perpendicular to all the independent permissible functions must be a zero function as in Equation (4.5), i.e., $R \to 0$ as $N \to \infty$.

Equation (4.5) can be written as

$$\left(\sum_{j=1}^{N} \tilde{u}_j Le_j, e_i \right) = (c, e_i), \quad i = 1, \ldots, N,$$

or

$$\sum_{j=1}^{N}(Le_j, e_i)\tilde{u}_j = (c, e_i), \quad i = 1, \ldots, N,$$

or

$$A\tilde{u} = d, \quad (4.6)$$

where

$$A = \begin{pmatrix} (Le_1, e_1) & (Le_1, e_2) & \cdots & (Le_1, e_N) \\ (Le_2, e_1) & (Le_2, e_2) & \cdots & (Le_2, e_N) \\ \cdots & \cdots & \cdots & \cdots \\ (Le_N, e_1) & (Le_N, e_2) & \cdots & (Le_N, e_N) \end{pmatrix}, \quad d = \begin{pmatrix} (e_1, c) \\ (e_2, c) \\ \cdots \\ (e_N, c). \end{pmatrix}.$$

and $\tilde{u}$ is the column vector whose components are the unknown coefficients, $\tilde{u}_i$.

Note that the matrix, A, in Equation (4.6) is symmetrical when L is symmetrical (self-adjoint).[1]

---

[1] A linear operator, L, is symmetrical (or self-adjoint) if

$$(Lu, v) = (u, Lv),$$

for arbitrary functions, $u(x)$ and $v(x)$.

Examples

1. One-dimensional Ordinary Differential Equation

   The Galerkin method introduced is well suited for *Mathematica* implementations. As a demonstration, the following differential equation with the homogeneous boundary condition is considered.

   $$u''(x) - u(x) = x, \quad u(0) = u(1) = 0. \tag{4.7}$$

   The exact solution is easily obtained as

   $$u(x) = \frac{e^2 x - x + e^{1-x} - e^{x+1}}{1 - e^2}.$$

   In *Mathematica*, the DSolve function is available for solving differential equations as

   In[1]:= `DSolve[{u''[x] - u[x] == x, u[0] == 0, u[1] == 0}, u[x], x]`

   Out[1]= $\left\{\left\{u[x] \to \dfrac{e^{-x}\left(-e + e^{1+2x} + e^x x - e^{2+x} x\right)}{-1 + e^2}\right\}\right\}$

   In the DSolve function, the boundary conditions of $u(0) = u(1) = 0$ can be included as part of the definition of the differential equation. The output from DSolve is in list format because there may be multiple solutions, each of which is an element in the list in the format of *Mathematica*'s substitution rule. To plot a graph of the solution, the aforementioned substitution rule is used as

   In[2]:= `sol = DSolve[{u''[x] - u[x] == x, u[0] == 0, u[1] == 0}, u[x], x]`
   `Plot[u[x] /. sol, {x, 0, 1}]`

   Out[2]= $\left\{\left\{u[x] \to \dfrac{e^{-x}\left(-e + e^{1+2x} + e^x x - e^{2+x} x\right)}{-1 + e^2}\right\}\right\}$

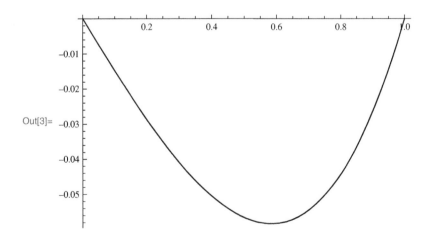

Although the exact solution is available, the Galerkin method is now employed to approximate the solution to Equation (4.7). First, a set of permissible functions, $e_i(x)$, that satisfy the homogeneous boundary conditions of $e_i(0) = e_i(1) = 0$ can be chosen as

$$e_i(x) = x^i(1-x), \quad i = 1, 2, \ldots, n.$$

The components of a_{ij} and d_i can be computed as

$$a_{ij} = \int_0^1 L(e_i)e_j dx$$

$$= \int_0^1 (e_i''(x) - e_i(x))e_j(x)dx$$

$$= \frac{2\left(-\frac{ij}{(i+j-1)(i+j)} - \frac{1}{(i+j+2)(i+j+3)}\right)}{i+j+1},$$

$$d_i = \int_0^1 e_i x \, dx$$

$$= \frac{1}{2+i} - \frac{1}{3+i}.$$

With $n = 3$, a_{ij} and d_i are computed as

$$a_{ij} = \begin{pmatrix} -\frac{11}{30} & -\frac{11}{60} & -\frac{23}{210} \\ -\frac{11}{60} & -\frac{1}{7} & -\frac{89}{840} \\ -\frac{23}{210} & -\frac{89}{840} & -\frac{113}{1260} \end{pmatrix}, \quad d_i = \begin{pmatrix} \frac{1}{12} \\ \frac{1}{20} \\ \frac{1}{30} \end{pmatrix}.$$

Therefore, the unknown coefficients, c_i, can be computed as

$$c_i = (a_{ij})^{-1} d_i = \begin{pmatrix} -\frac{14427}{96406} \\ -\frac{6944}{48203} \\ -\frac{21}{1121} \end{pmatrix}.$$

Finally, the Galerkin approximation by third-order polynomials ($n = 3$) yields

$$\tilde{u}_3 = \sum_{i=1}^{3} c_i e_i(x)$$

$$= -\frac{21(1-x)x^3}{1121} - \frac{6944(1-x)x^2}{48203} - \frac{14427(1-x)x}{96406}.$$

Inclusions in Finite Matrix

The aforementioned procedure is implemented in *Mathematica* as

In[2]:= `exact = y[x] /. DSolve[{y''[x] - y[x] == x, y[0] == 0, y[1] == 0}, y[x], x]`

Out[2]= $\left\{ \dfrac{e^{-x}\left(-e + e^{1+2x} + e^x x - e^{2+x} x\right)}{-1 + e^2} \right\}$

In[3]:=
```
order = 3;
myint[poly_] := Expand[poly] /. x^i_. -> 1/(i+1)
e[i_] := x^i (1-x);
l[f_] := D[f, {x, 2}] - f
aij = myint[l[e[i]] e[j]]
di = myint[e[i] x]
```

Out[7]= $-\dfrac{i}{-1+i+j} + \dfrac{i^2}{-1+i+j} - \dfrac{2i^2}{i+j} - \dfrac{1}{1+i+j} + \dfrac{i}{1+i+j} + \dfrac{i^2}{1+i+j} + \dfrac{2}{2+i+j} - \dfrac{1}{3+i+j}$

Out[8]= $\dfrac{1}{2+i} - \dfrac{1}{3+i}$

In[9]:=
```
aijmat = Table[aij, {i, order}, {j, order}]
divec = Table[di, {i, order}]
approximate = Inverse[aijmat].divec.Table[e[i], {i, order}]
Plot[{exact, approximate}, {x, 0, 1}]
```

Out[9]= $\left\{\left\{-\dfrac{11}{30}, -\dfrac{11}{60}, -\dfrac{23}{210}\right\}, \left\{-\dfrac{11}{60}, -\dfrac{1}{7}, -\dfrac{89}{840}\right\}, \left\{-\dfrac{23}{210}, -\dfrac{89}{840}, -\dfrac{113}{1260}\right\}\right\}$

Out[10]= $\left\{\dfrac{1}{12}, \dfrac{1}{20}, \dfrac{1}{30}\right\}$

Out[11]= $-\dfrac{14427\,(1-x)\,x}{96406} - \dfrac{6944\,(1-x)\,x^2}{48203} - \dfrac{21\,(1-x)\,x^3}{1121}$

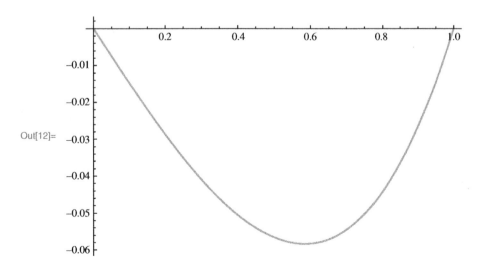

Out[12]=

The exact solution and the approximate solution by the Galerkin method by third-order polynomials are plotted together in the graph. As can be seen from the graph, there is hardly any difference between the exact solution and the Galerkin approximation with the third order polynomial, which suggests that the Galerkin method should work for other differential equations with good accuracy.

A note about the code is in order. It is known that, among many operations in *Mathematica*, integration of a function consumes a considerable amount of CPU, which impacts the performance of program execution. One way of getting around this issue is to replace the built-in Integrate function with user-defined private functions.

In the aforementioned code, the user-defined function, myint, is introduced for this purpose. Instead of integrating a polynomial directly using the built-in Integrate function, it replaces x^i by $\frac{1}{i+1}$ as

$$\int_0^1 x^i dx = \frac{1}{1+i},$$

with the code

```
myint[poly_]:=Expand[poly]/.x^i_. -> 1/(i+1)
```

This may not seem significant for a simple integration such as this example, but in the next example for a two-dimensional partial differential equation, it is shown that replacing the built-in Integrate function with a private function is shown to speed up the execution time significantly.

2. Two-dimensional Poisson Equation

As another example of the Galerkin method with *Mathematica*, a 2-D Poisson-type equation is chosen for this demonstration. A Poisson-type equation in this example is expressed as

$$\Delta u(x,y) + 1 = 0 \text{ in } D, \quad u = 0 \text{ on } \partial D, \tag{4.8}$$

where the domain, D, is a square given by $D = \{(x,y), 0 \leq x \leq 1, 0 \leq y \leq 1\}$ and Δ is the 2-D Laplacian. The boundary condition is the first type, i.e., u vanishes on the boundary, ∂D (the Dirichlet condition).

For this problem, the exact solution is available by the Fourier series expansion method as

$$u(x,y) = -\sum_{m,n=1}^{\infty} \frac{4((-1)^m - 1)((-1)^n - 1)}{(m^2 + n^2)mn\pi^4} \sin m\pi x \sin n\pi y. \tag{4.9}$$

As the exact solution is available, this problem can be used as a benchmark of how the Galerkin method with *Mathematica* compares with the exact solution.

In the Galerkin method, permissible functions that satisfy the homogeneous boundary condition must be chosen first. In general, it is difficult to select such functions for general boundary conditions with an arbitrary boundary shape. However, if the boundary shape is regular such as circular or rectangular and the boundary condition is the first type (Dirichlet type), it is straightforward to determine permissible functions. For the present problem, permissible functions, $e_m(x,y)$, can be chosen following Pascal's triangle as

$$e_m(x,y) = \{1, x, y, x^2, xy, y^2, x^3, x^2y, xy^2, y^3, x^4, x^3y, x^2y^2, xy^3, y^4, \ldots\} \times g(x,y),$$

where $g(x, y)$ is a function that vanishes on the boundary. For the square-shaped boundary, $g(x, y)$ is chosen as

$$g(x, y) = xy(1 - x)(1 - y).$$

An approximate solution to Equation (4.8) is sought as

$$\tilde{u} = \sum_{m=1}^{N} \tilde{u}_m e_m(x, y),$$

where $\tilde{u}_m$ is an unknown coefficient to be determined from the Galerkin method.

By using Equation (4.6), the matrix and the vector components to determine the unknown $\tilde{u}_m$ are expressed as

$$A\tilde{u} = d,$$

where

$$a_{mn} = -\int_0^1 \int_0^1 \nabla e_m \cdot \nabla e_n dx dy, \quad d_m = \int_0^1 \int_0^1 e_m dx dy.$$

The following *Mathematica* code implements this procedure.

```
In[1]:= myint[f_] :=
   Expand[f] /. {x^a_. y^b_. → 1/(a+1)/(b+1), x^a_. → 1/(a+1), y^b_. → 1/(b+1)};
  inner[f_, g_] := myint[f g];
  div[f_, g_] := D[f, x] D[g, x] + D[f, y] D[g, y];
  poly[n_, a_] := Sum[a[(i+j)(i+j+1)/2+i+1] x^j y^i, {i, 0, n}, {j, 0, n-i}];
  polyseq[m_] := Flatten[Table[Table[x^(j-i) y^i, {i, 0, j}], {j, 0, m}]]

In[5]:= basefunc = polyseq[4] x y (x-1) (y-1)

Out[6]= {(-1+x) x (-1+y) y, (-1+x) x^2 (-1+y) y, (-1+x) x (-1+y) y^2,
  (-1+x) x^3 (-1+y) y, (-1+x) x^2 (-1+y) y^2, (-1+x) x (-1+y) y^3, (-1+x) x^4 (-1+y) y,
  (-1+x) x^3 (-1+y) y^2, (-1+x) x^2 (-1+y) y^3, (-1+x) x (-1+y) y^4, (-1+x) x^5 (-1+y) y,
  (-1+x) x^4 (-1+y) y^2, (-1+x) x^3 (-1+y) y^3, (-1+x) x^2 (-1+y) y^4, (-1+x) x (-1+y) y^5}

In[7]:= terms = Length[basefunc]

Out[7]= 15

In[8]:= amat = Table[0, {i, terms}, {j, terms}];
  amat = Table[inner[div[basefunc[[i]], basefunc[[j]]], 1], {i, terms}, {j, terms}];
  Do[amat[[j, i]] = amat[[i, j]], {i, 1, terms}, {j, i+1, terms}]

In[11]:= bvec = Table[inner[ basefunc[[i]]], 1], {i, terms}]

Out[11]= {1/36, 1/72, 1/72, 1/120, 1/144, 1/120, 1/180, 1/240, 1/240, 1/180, 1/252, 1/360, 1/400, 1/360, 1/252}

In[12]:= approx = Inverse[amat].bvec.basefunc

Out[12]= 36051 (-1+x) x (-1+y) y / 17876  -  12243 (-1+x) x^2 (-1+y) y / 4469  +
  27489 (-1+x) x^3 (-1+y) y / 8938  -  3003 (-1+x) x^4 (-1+y) y / 4469  +  3003 (-1+x) x^5 (-1+y) y / 8938  -
  12243 (-1+x) x (-1+y) y^2 / 4469  +  70455 (-1+x) x^2 (-1+y) y^2 / 8938  -
  70455 (-1+x) x^3 (-1+y) y^2 / 8938  +  27489 (-1+x) x (-1+y) y^3 / 8938  -  70455 (-1+x) x^2 (-1+y) y^3 / 8938  +
  70455 (-1+x) x^3 (-1+y) y^3 / 8938  -  3003 (-1+x) x (-1+y) y^4 / 4469  +  3003 (-1+x) x (-1+y) y^5 / 8938

In[13]:= Plot3D[approx, {x, 0, 1}, {y, 0, 1}]
```

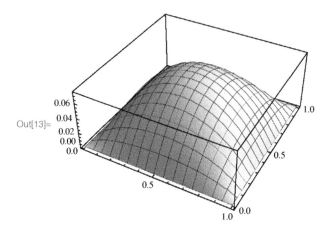

The user-defined function inner[f,g] computes the inner product of $f(x,y)$ and $g(x,y)$ as

$$(f,g) = \int_0^1 \int_0^1 f(x,y)g(x,y)dxdy.$$

The user-defined function div[f,g] computes the integration, $\int\int \Delta f g \, dxdy$, as

$$\int_0^1 \int_0^1 f_{,ii} g \, dxdy = -\int_0^1 \int_0^1 f_{,i} g_{,i} \, dxdy.$$

The user-defined function polyseq[m] generates a sequence of polynomials to be used as the permissible functions. The list amat stores the components of Equation (4.6). As the matrix is symmetrical, only a half of the components are evaluated. The variable approx is an approximation to the Poisson equation using fourth-order polynomials.

It is noted that the function myint[f] integrates a polynomial on x and y over the region $D = \{(x,y), 0 \le x \le 1, 0 \le y \le 1\}$ bypassing Mathematica's built-in function, Integrate. An example of how this function speeds up integration compared with the Integrate function is shown as follows:

In[53]:= `f = Sum[Random[Integer, {-15, 15}] x^i y^j, {i, 0, 100}, {j, 0, 100}];`

In[54]:= `Length[f]`

Out[54]= `9888`

In[59]:= `Short[f, 10]`

Out[59]//Short= $-7 - x - x^2 + 11 x^3 + 4 x^4 + 5 x^5 + 12 x^6 - 5 x^7 + 8 x^8 + x^9 + 15 x^{10} + 13 x^{11} - 8 x^{12} + 14 x^{13} - 7 x^{15} - 3 x^{16} + 8 x^{17} - 10 x^{19} + 3 x^{20} + 9 x^{21} - 3 x^{22} + 13 x^{23} + 7 x^{24} + 3 x^{25} - 9 x^{26} + 11 x^{27} - 8 x^{28} - 6 x^{29} + 10 x^{30} - 9 x^{31} + 5 x^{32} + 11 x^{33} + 2 x^{34} - 4 x^{35} - 14 x^{36} - 4 x^{37} + 7 x^{38} + x^{39} - 13 x^{40} + 7 x^{41} - 9 x^{42} + 5 x^{43} + \ll 14\,665 \gg + 5 x^{58} y^{100} + 13 x^{59} y^{100} - 9 x^{60} y^{100} + 3 x^{61} y^{100} - x^{62} y^{100} - 6 x^{63} y^{100} - 5 x^{64} y^{100} + 11 x^{65} y^{100} - x^{66} y^{100} - 3 x^{67} y^{100} + 2 x^{68} y^{100} + 2 x^{69} y^{100} - 7 x^{70} y^{100} - 14 x^{71} y^{100} - 4 x^{72} y^{100} - 3 x^{73} y^{100} - 4 x^{74} y^{100} - 7 x^{75} y^{100} + 11 x^{76} y^{100} - 13 x^{77} y^{100} + 3 x^{78} y^{100} + 4 x^{79} y^{100} - 11 x^{80} y^{100} - 6 x^{81} y^{100} - 5 x^{82} y^{100} - x^{83} y^{100} + 12 x^{84} y^{100} - 10 x^{85} y^{100} - 8 x^{86} y^{100} - 9 x^{87} y^{100} - 11 x^{88} y^{100} + 3 x^{89} y^{100} + 10 x^{90} y^{100} + 3 x^{91} y^{100} + 14 x^{92} y^{100} - 5 x^{93} y^{100} + 7 x^{94} y^{100} + 4 x^{95} y^{100} + 7 x^{96} y^{100} - x^{97} y^{100} + 7 x^{99} y^{100} - 14 x^{100} y^{100}$

In[56]:= **j1 = Timing[myint[f]]**

Out[56]= {0.03125,
219 474 646 791 770 145 500 269 437 166 493 627 947 182 631 341 305 220 332 256 208 828 \
331 706 172 075 573 871 /
28 302 713 639 769 180 630 836 475 513 805 447 184 920 984 395 949 222 067 424 458 728 \
170 630 641 757 120 000}

In[57]:= **j2 = Timing[Integrate[Integrate[f, {x, 0, 1}], {y, 0, 1}]]**

Out[57]= {6.21875,
219 474 646 791 770 145 500 269 437 166 493 627 947 182 631 341 305 220 332 256 208 828 \
331 706 172 075 573 871 /
28 302 713 639 769 180 630 836 475 513 805 447 184 920 984 395 949 222 067 424 458 728 \
170 630 641 757 120 000}

In[58]:= **j2[[1]] / j1[[1]]**

Out[58]= 199.

In the aforementioned example program, a polynomial, f, of the 200th order in x and y whose integer coefficients between -15 and 15 are randomly generated is integrated in two different ways. The user-defined function myint returns the integration of $x^a y^b$ over the square directly as

$$\int_0^1 \int_0^1 x^a y^b \, dy \, dx = \frac{1}{(a+1)(b+1)},$$

without calling the built-in Integrate function.

```
myint[f_]:=Expand[f]/.{x^a_. y^b_.->1/(a+1)/(b+1),
x^a.->1/(a+1), y^b_.->1/(b+1)};
```

The generated polynomial has 9988 terms as seen from the Length function and part of the polynomial can be shown using the Short function. The integration of f using the myint function takes 0.03125 seconds from the Timing function while the integration of f using the built-in Integrate function takes 6.21875 seconds. It is seen that using the myint function is faster than the built-in Integrate function by 200 times.

The exact series solution with 21 terms where the convergence is verified is shown as follows.

In[60]:= **exact = Sum[2 (-1 + (-1)^m) (-1 + (-1)^n) /m/n/π^2/ (m^2+n^2) /π^2
2 Sin[m π x] Sin[n π y], {n, 10}, {m, 10}]**

Out[60]= $\dfrac{8 \sin[\pi x] \sin[\pi y]}{\pi^4} + \dfrac{8 \sin[3\pi x] \sin[\pi y]}{15 \pi^4} +$
$\dfrac{8 \sin[5\pi x] \sin[\pi y]}{65 \pi^4} + \dfrac{8 \sin[7\pi x] \sin[\pi y]}{175 \pi^4} + \dfrac{8 \sin[9\pi x] \sin[\pi y]}{369 \pi^4} +$
$\dfrac{8 \sin[\pi x] \sin[3\pi y]}{15 \pi^4} + \dfrac{8 \sin[3\pi x] \sin[3\pi y]}{81 \pi^4} + \dfrac{8 \sin[5\pi x] \sin[3\pi y]}{255 \pi^4} +$
$\dfrac{8 \sin[7\pi x] \sin[3\pi y]}{609 \pi^4} + \dfrac{8 \sin[9\pi x] \sin[3\pi y]}{1215 \pi^4} +$
$\dfrac{8 \sin[\pi x] \sin[5\pi y]}{65 \pi^4} + \dfrac{8 \sin[3\pi x] \sin[5\pi y]}{255 \pi^4} + \dfrac{8 \sin[5\pi x] \sin[5\pi y]}{625 \pi^4} +$

$$\frac{8\,\text{Sin}[7\pi x]\,\text{Sin}[5\pi y]}{1295\,\pi^4} + \frac{8\,\text{Sin}[9\pi x]\,\text{Sin}[5\pi y]}{2385\,\pi^4} + \frac{8\,\text{Sin}[\pi x]\,\text{Sin}[7\pi y]}{175\,\pi^4} +$$
$$\frac{8\,\text{Sin}[3\pi x]\,\text{Sin}[7\pi y]}{609\,\pi^4} + \frac{8\,\text{Sin}[5\pi x]\,\text{Sin}[7\pi y]}{1295\,\pi^4} + \frac{8\,\text{Sin}[7\pi x]\,\text{Sin}[7\pi y]}{2401\,\pi^4} +$$
$$\frac{8\,\text{Sin}[9\pi x]\,\text{Sin}[7\pi y]}{4095\,\pi^4} + \frac{8\,\text{Sin}[\pi x]\,\text{Sin}[9\pi y]}{369\,\pi^4} + \frac{8\,\text{Sin}[3\pi x]\,\text{Sin}[9\pi y]}{1215\,\pi^4} +$$
$$\frac{8\,\text{Sin}[5\pi x]\,\text{Sin}[9\pi y]}{2385\,\pi^4} + \frac{8\,\text{Sin}[7\pi x]\,\text{Sin}[9\pi y]}{4095\,\pi^4} + \frac{8\,\text{Sin}[9\pi x]\,\text{Sin}[9\pi y]}{6561\,\pi^4}$$

In[81]:= `Plot3D[exact, {x, 0, 1}, {y, 0, 1}]`

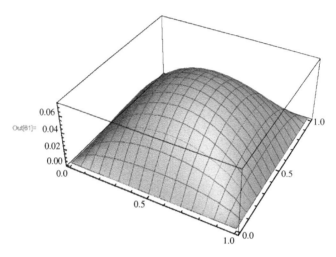

The cross-sectional profiles of the solution from both methods are compared by the following code:

In[82]:= `aprox2 = approx /. y → 1 / 2`

Out[82]= $-\dfrac{193\,467\,(-1+x)\,x}{572\,032} + \dfrac{27\,489\,(-1+x)\,x^2}{143\,008} - \dfrac{39\,501\,(-1+x)\,x^3}{143\,008} + \dfrac{3003\,(-1+x)\,x^4}{17\,876} - \dfrac{3003\,(-1+x)\,x^5}{35\,752}$

In[83]:= `exact2 = exact /. y → 1 / 2`

Out[83]= $\dfrac{6\,351\,224\,\text{Sin}[\pi x]}{839\,475\,\pi^4} + \dfrac{385\,240\,\text{Sin}[3\pi x]}{838\,593\,\pi^4} + \dfrac{1\,735\,124\,024\,\text{Sin}[5\pi x]}{17\,064\,376\,875\,\pi^4} + \dfrac{281\,660\,024\,\text{Sin}[7\pi x]}{7\,535\,598\,525\,\pi^4} + \dfrac{114\,922\,168\,\text{Sin}[9\pi x]}{6\,486\,959\,115\,\pi^4}$

In[84]:= `Plot[{approx2, exact2}, {x, 0, 1}]`

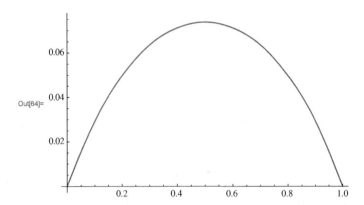

It is noted that virtually no difference is found between the exact solution and the approximation. This result is encouraging as the proposed method is shown to yield good agreement with the exact solution, which implies that the method works well where the exact solution is not available.

4.1.2 Rayleigh–Ritz Method

The Rayleigh–Ritz method is an another approach in addition to MWR for numerically approximating the solution to certain boundary value problems based on the variational principle. In the variational method, a functional, I, is defined as an integral of a function that contains an unknown function, $y(x)$, and its derivative, $y'(x)$, as

$$I \equiv \int_a^b f(x, y(x), y'(x))\, dx, \qquad (4.10)$$

where $y(x)$ is a twice-differentiable function that extremizes I.

Taking the variation on y in Equation (4.10) yields

$$\delta I = \int_a^b \left(\frac{\partial f}{\partial y'}(\delta y)' + \frac{\partial f}{\partial y}\delta y \right) dx$$

$$= \left[\frac{\partial f}{\partial y'}\delta y \right]_a^b + \int_a^b \left(\frac{\partial f}{\partial y} - \frac{d}{dx}\left(\frac{\partial f}{\partial y'} \right) \right) dx,$$

where an interchange of the variation and derivative was used as

$$(\delta y)' = \delta y'.$$

If the boundary condition of $y(x)$ is prescribed at $x = a$ and $x = b$, it follows $\delta y(a) = \delta y(b) = 0$. Therefore, $\delta I = 0$ is equivalent to

$$\frac{\partial f}{\partial y} - \frac{d}{dx}\left(\frac{\partial f}{\partial y'} \right) = 0, \quad y(a) \text{ and } y(b) \text{ are prescribed.} \qquad (4.11)$$

Equation (4.11) is known as the Euler equation.

If f is chosen to be

$$f \equiv \frac{1}{2}k(x)(y'(x))^2 + b(x)y(x), \tag{4.12}$$

where $k(x)$ and $b(x)$ are given functions, the Euler equation becomes

$$(k(x)y'(x))' = b(x). \tag{4.13}$$

Many governing equations in physics including the heat conduction equation and the elasticity equilibrium equation are similar to Equation (4.13) in 2-D or 3-D.

The variational method and its relationship to various differential equations have a long history, and it is not possible to exhaustively cover this topic in this book. One of the numerical method to extremize the functional, I, is the Ritz method that has been widely used. In the Ritz method, an approximation solution to Equation (4.13) is sought as

$$\tilde{y} = \sum_{i=1}^{N} c_i e_i(x), \tag{4.14}$$

where $e_i(x)$ is a permissible function satisfying the homogeneous boundary condition. Substituting Equation (4.14) into Equation (4.10) with Equation (4.12) yields

$$I = \int_a^b \left(\frac{1}{2} k \sum_{i,j}^{N} c_i c_j e'_i e'_j + b \sum_{i=1}^{N} c_i e_i \right) dx. \tag{4.15}$$

Taking the partial derivative of Equation (4.15) with respect to c_k yields

$$\frac{\partial I}{\partial c_k} = \sum_j c_j \int_a^b k e'_k e'_j dx + \int_a^b b e_k dx,$$

$$= 0,$$

which can be written in a matrix–vector equation as

$$\mathbf{A}\mathbf{c} + \mathbf{b} = 0, \tag{4.16}$$

where

$$a_{ij} = \int_a^b k e'_i e'_j dx, \quad b_i = \int_a^b b e_i dx.$$

Equation (4.16) is identical to the one obtained by the Galerkin method.

In elasticity, the strain energy is defined as

$$I \equiv \int_V \left(\frac{1}{2} C_{ikl} u_{i,j} u_{k,l} + b_i u_i \right) dV. \tag{4.17}$$

The Euler equation for Equation (4.17) is the elasticity equilibrium equation as

$$(C_{ijkl} u_{k,l})_{,j} + b_i = 0. \tag{4.18}$$

As will be shown in Section 4.3, the Galerkin method does not work properly for the elasticity equation in 2-D and 3-D when the permissible functions are chosen separately for each value of u_i. The Ritz method, however, yields proper equations for the unknown coefficients.

4.1.3 Sturm–Liouville System

As shown in the previous section, a linear differential equation can be numerically solved by one of the three methods in MWR, which requires solving a set of resulting linear equations whose matrix is not necessarily sparse nor symmetrical.

However, if the differential equation in question happens to be related to an eigenvalue problem known as the Sturm–Liouville system (Hassani 2013), numerical solutions to certain boundary value problems can be derived immediately once the corresponding eigenvalue and eigenfunction problem is solved without a lengthy process of solving a set of linear equations. The steady-state heat conduction equation and the elastic equilibrium equations in displacements belong to this type.

A Sturm–Liouville system is a boundary value problem defined by an differential equation expressed as

$$Ly = \lambda y, \quad (a < x < b), \tag{4.19}$$

where L is a symmetrical differential operator defined by

$$L \equiv -\frac{1}{w(x)}\left(\frac{d}{dx}\left(p(x)\frac{d}{dx}\right) + q(x)\right). \tag{4.20}$$

In Equation (4.20), $p(x)$ and $q(x)$ are given functions and $w(x)$ is a weight function used in the definition of the inner products as

$$(f(x), g(x)) \equiv \int_a^b f(x)g(x)w(x)dx.$$

The symbol, λ, is called an eigenvalue and takes discrete values to be determined by solving the boundary value problem.

Equation (4.19) is accompanied by the following homogeneous boundary conditions:

$$\begin{aligned} \alpha\, y(a) + \beta\, y'(a) &= 0, \\ \gamma\, y(b) + \delta\, y'(b) &= 0, \end{aligned} \tag{4.21}$$

where α, β, γ, and δ are constants. Equation (4.19) along with Equation (4.21) needs to be solved to obtain the eigenfunctions, $y(x)$, and their corresponding eigenvalues, λ.

It is noted that the Sturm–Liouville operator, L, of Equation (4.20) is symmetric (Hermitian), i.e.,

$$(Lu, v) = (u, Lv), \tag{4.22}$$

for arbitrary functions of u and v. Hence, the following important properties are held:

1. All the eigenvalues, λ_i, are real.
2. Eigenfunctions, y_i and y_j, are mutually orthogonal and can be normalized, i.e.,

$$(y_i, y_j) = \delta_{ij}.$$

This way, by solving the Sturm–Liouville system, one can automatically obtain an orthonormalized set of permissible functions, which can be used to express an arbitrary function in the function space by a linear combination of eigenfunctions.

The Sturm–Liouville system is useful as many physical systems are described by variations of S–L types of equations. In addition, most of the special functions (e.g., Bessel functions) are derived as the eigenfunctions of an S–L system (mother of all the special functions).

It is noted that both the steady-state heat conduction equation and static equilibrium equation in displacements are considered to be associated with the Sturm–Liouville system in 2-D or 3-D.

For the steady-state heat conduction, the differential operator, L, can be defined as

$$L \equiv -(k_{ij}T_{,j})_{,i}.$$

This is equivalent to choosing

$$w = 1, \quad p = k_{ij}, \quad q = 0.$$

For the static elasticity equilibrium equation in displacement, the differential operator, L, can be defined as

$$(Lu)_i \equiv -(C_{ijkl}u_{k,l})_{,j}.$$

This is equivalent to choosing

$$w = 1, \quad p = C_{ijkl}, \quad q = 0.$$

Solution to $Lu = c$

The Sturm–Liouville system can be used to solve linear equations of the type

$$Lu = c, \tag{4.23}$$

where L is a linear operator and c is a given function. It is assumed that the corresponding Sturm–Liouville equation, $Ly = \lambda y$, is already solved. By using the eigenfunction, y_i, the known function, c, can be expanded by

$$c = \sum_{i=1}^{\infty} c_i y_i, \tag{4.24}$$

where

$$c_i = (c, y_i).$$

The unknown function, u, is assumed to be expanded by y_is as

$$u = \sum_{i=1}^{\infty} u_i y_i, \tag{4.25}$$

where u_is are unknown coefficients. Substitution of Equations (4.24) and (4.25) into Equation (4.23) yields

$$L \sum_{i=1}^{\infty} u_i y_i = \sum_{i=1}^{\infty} c_i y_i,$$

or

$$\sum_{i=1}^{\infty} u_i \lambda_i y_i = \sum_{i=1}^{\infty} c_i y_i,$$

from which u_i can be solved as

$$u_i = \frac{c_i}{\lambda_i},$$

Finally, u can be expressed as

$$u = \sum_{i=1}^{\infty} \frac{c_i}{\lambda_i} y_i. \tag{4.26}$$

Example 4.1: The trigonometric functions (sine and cosine) are defined as the eigenfunctions for the following Strum–Liouville problem:

$$Ly = \lambda y, \quad L \equiv -\frac{d^2}{dx^2}, \quad y(0) = y(1) = 0. \tag{4.27}$$

Equation (4.27) can be solved analytically as

$$y_n = \sqrt{2} \sin \sqrt{\lambda_n} x, \quad \lambda_n = (n\pi)^2, \tag{4.28}$$

where λ_n is the nth eigenvalue. Note that the eigenfunctions, $y_n(x)$s, are orthogonal to each other and normalized as

$$(y_n, y_m) = \delta_{nm}, \tag{4.29}$$

where the inner product is defined as

$$(f, g) = \int_0^1 f(x)g(x)dx.$$

As the exact solution is known, the Galerkin method can be tried for this Strum–Liouville problem to determine whether it is suited for solving a general boundary value problem semi-analytically.

A set of permissible functions that satisfy the boundary condition in Equation (4.27) can be chosen as

$$e_i(x) = x^i(1-x), \quad i = 1, 2, 3, \ldots, n.$$

Note that $e_i(0) = e_i(1) = 0$. In the Galerkin method, an approximation to the eigenfunction, $y(x)$, $\tilde{y}(x)$, is sought by a linear combination of $e_i(x)$ as

$$\tilde{y}(x) = \sum_{i=1}^{n} c_i e_i(x). \tag{4.30}$$

By using Equation (4.30), the differential equation of Equation (4.27) is converted into an algebraic equation as

$$A\mathbf{c} = \lambda B\mathbf{c}, \tag{4.31}$$

where

$$a_{ij} = (Le_i, e_j), \quad b_{ij} = (e_i, e_j).$$

Equation (4.31) is a generalized eigenvalue problem in linear algebra and can be solved numerically for the eigenvectors, $\mathbf{c}$, and the eigenvalues, λ.

First few components of the matrices, A and B, are shown as

$$a_{ij} = \begin{pmatrix} \frac{1}{3} & \frac{1}{6} & \cdots \\ \frac{1}{6} & \frac{2}{15} & \cdots \\ \cdots & \cdots & \cdots \end{pmatrix}, \quad b_{ij} = \begin{pmatrix} \frac{1}{30} & \frac{1}{60} & \cdots \\ \frac{1}{60} & \frac{1}{105} & \cdots \\ \cdots & \cdots & \cdots \end{pmatrix}.$$

The number of the approximate eigenfunctions, $\tilde{y}_n(x)$, depends on the number of permissible functions used. For instance, using 10 permissible functions, the first of 10 approximate eigenfunctions is expressed as

$$\tilde{y}_1(x) = 1.72629 \times 10^{-6}(1-x)x^{10} + 0.0344002(1-x)x^9 - 0.137617(1-x)x^8$$
$$- 0.0827522(1-x)x^7 + 0.729932(1-x)x^6 + 0.744424(1-x)x^5$$
$$- 2.86596(1-x)x^4 - 2.86531(1-x)x^3$$
$$+ 4.44288(1-x)x^2 + 4.44288(1-x)x.$$

A *Mathematica* code to implement the aforementioned procedure is shown as follows:

```
In[1]:= inner[f_, g_] := Integrate[ f g , {x, 0, 1}]
        λ[n_] := (n Pi)^2
        exact[n_] := Sqrt[2] Sin[Sqrt[λ[n]] x]

In[4]:= cEigensystem[a_, b_] := Module[{u, λ, e1, e}, u = CholeskyDecomposition[b];
            {λ, e} = Eigensystem[Inverse[Transpose[u]].a.Inverse[u]];
            e1 = Transpose[Inverse[u].Transpose[e]];
            {λ, Map[# / Sqrt[#.b.#] &, e1]}];
        L[f_] := -D[f, {x, 2}]

In[6]:= e = Table[ x^i (1-x), {i, 1, 10}]; terms = Length[e];
        aij = Table[inner[L[e[[i]]], e[[j]]], {i, terms}, {j, terms}];
        bij = Table[ inner[ e[[i]], e[[j]]], {i, terms}, {j, terms}];
        {eval, evec} = cEigensystem[aij, N[bij]];
        approx = evec.e;
```

The function inner[f,g] defines the inner product between $f(x)$ and $g(x)$. The function cEgensystem[a, b] returns the eigenvalues and eigenvectors for a generalized eigenvalue problem of Equation (4.31). In the line,

{eval, evec} = cEigensystem[aij, N[bij]];,

the computed eigenvalues are stored in the variable, eval, and the computed eigenvectors are stored in the variable, evec. The obtained approximate eigenfunctions are stored in the variable, approx, which is a list. In the aforementioned example code, 10 permissible functions are chosen, which results in 10 approximate eigenfunctions.

Figure 4.1 is a comparison between the first approximate eigenfunction, $\tilde{y}_1(x)$, and the exact solution of $\sqrt{2} \sin \sqrt{\lambda_1} x$. As can be seen, two curves are indistinguishable.

Figure 4.2 is a comparison between the sixth approximate eigenfunction, $\tilde{y}_6(x)$, and the exact solution of $\sqrt{2} \sin \sqrt{\lambda_6} x$. Two curves are slightly distinguishable.

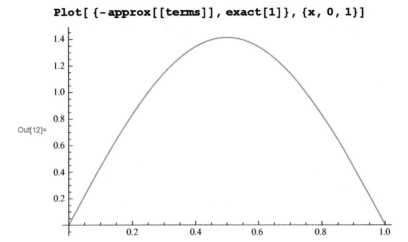

Figure 4.1 Comparison of the first eigenfunction and $\sqrt{2}\sin\sqrt{\lambda_1}x$

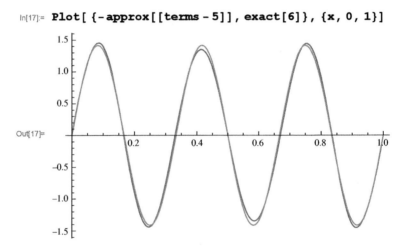

Figure 4.2 Comparison of the sixth eigenfunction and $\sqrt{2}\sin\sqrt{\lambda_6}x$

Example 4.2: The zeroth-order Bessel function of the first kind, $J_0(\lambda x)$, is defined as the eigenfunction for

$$Ly = \lambda y, \quad L \equiv -\left(\frac{d^2}{dx^2} + \frac{1}{x}\frac{d}{dx}\right), \quad y'(0) = y(1) = 0. \tag{4.32}$$

Equation (4.32) can be solved analytically as

$$y_n(x) = \frac{\sqrt{2}}{J_1(\lambda_n)}J_0(\lambda_n x), \tag{4.33}$$

where $J_1(x)$ is the first order Bessel function of the first kind and λ_n is the nth eigenvalue. Note that the eigenfunctions, $y_n(x)$s, are orthogonal to each other and normalized as

$$(y_n, y_m) = \delta_{nm}, \tag{4.34}$$

where the inner product is defined as

$$(f, g) = \int_0^1 xf(x)g(x)dx.$$

As the exact solution is available, the Galerkin method can be tried to determine the suitability of the present method as in Example 4.1.

A set of permissible functions that satisfy the boundary condition in Equation (4.32) can be chosen as

$$e_i(x) = x^{i+1} - 1, \quad i = 1, 2, 3, \ldots, n.$$

Note that $e_i'(0) = e_i(1) = 0$. In the Galerkin method, an approximation to $y(x)$, $\tilde{y}(x)$, is sought by a linear combination of $e_i(x)$ as

$$\tilde{y}(x) = \sum_{i=1}^{n} c_i e_i(x). \tag{4.35}$$

By using Equation (4.35), the differential equation of Equation (4.32) is converted into an algebraic equation as

$$Ac = \lambda Bc, \tag{4.36}$$

where

$$a_{ij} = (Le_i, e_j), \quad b_{ij} = (e_i, e_j).$$

Equation (4.36) is a generalized eigenvalue problem in linear algebra and can be solved numerically for the eigenvectors, c, and the eigenvalues, λ.

Parts of the matrices, A and B, are shown as

$$a_{ij} = \begin{pmatrix} 1 & \frac{6}{5} & \cdots \\ \frac{6}{5} & \frac{3}{2} & \cdots \\ \cdots & \cdots & \cdots \end{pmatrix}, \quad b_{ij} = \begin{pmatrix} \frac{1}{6} & \frac{27}{140} & \cdots \\ \frac{27}{140} & \frac{9}{40} & \cdots \\ \cdots & \cdots & \cdots \end{pmatrix}.$$

The number of the eigenfunctions, $y_n(x)$, depends on the number of permissible functions used. For instance, by using 10 permissible functions, the first of 10 eigenfunctions can be expressed as

$$\begin{aligned} y_1 = & 0.000755405(x^{11} - 1) - 0.00469829(x^{10} - 1) + 0.00798563(x^9 - 1) \\ & + 0.0100234(x^8 - 1) + 0.00891029(x^7 - 1) - 0.233494(x^6 - 1) \\ & + 0.00165695(x^5 - 1) + 1.42322(x^4 - 1) + 0.0000411283(x^3 - 1) \\ & - 3.93851(x^2 - 1) \end{aligned}$$

A *Mathematica* code to implement the aforementioned procedure is shown as follows:

```
In[19]:= inner[f_, g_] := Integrate[f g x, {x, 0, 1}]
         λ[n_] := BesselJZero[0, n]
         exact[n_] := Sqrt[2] / BesselJ[1, λ[n]] BesselJ[0, λ[n] x]
         (* inner[ exact[2], exact[2]]=1 *)

In[4]:= cEigensystem[a_, b_] := Module[{u, λ, e1, e}, u = CholeskyDecomposition[b];
         {λ, e} = Eigensystem[Inverse[Transpose[u]].a.Inverse[u]];
         e1 = Transpose[Inverse[u].Transpose[e]];
         {λ, Map[# / Sqrt[#.b.#] &, e1]}];
   L[f_] := D[f, {x, 2}] + 1 / x D[f, x]

In[6]:= e = Table[ x^(2 + i) - 1, {i, 0, 9}];
        terms = Length[e];
        aij = Table[inner[L[e[[i]]], e[[j]]], {i, terms}, {j, terms}];
        bij = Table[ inner[ e[[i]], e[[j]]], {i, terms}, {j, terms}];
        {eval, evec} = cEigensystem[aij, N[bij]];
        approx = evec.e;
```

Figure 4.3 is a comparison between the first approximate eigenfunction, $\tilde{y}_1(x)$, and the exact solution of $\frac{\sqrt{2}}{J_1(\lambda_1)} J_0(\lambda_1 x)$. As can be seen, two curves are indistinguishable.

Figure 4.4 is a comparison between the sixth approximate eigenfunction, $\tilde{y}_6(x)$, and the exact solution of $\frac{\sqrt{2}}{J_6(\lambda_1)} J_0(\lambda_6 x)$. Two curves are slightly distinguishable.

This example along with Example 4.1 shows that the Galerkin method combined with *Mathematica* offers an accurate tool for general boundary value problems.

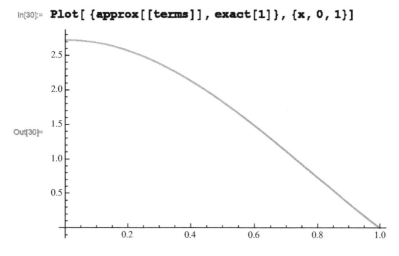

Figure 4.3 Comparison of the first approximate eigenfunction, $\tilde{y}_1(x)$, and $\frac{\sqrt{2}}{J_1(\lambda_1)} J_0(\lambda_1 x)$

In[27]:= `Plot[ {approx[[terms - 5]], exact[6]}, {x, 0, 1}]`

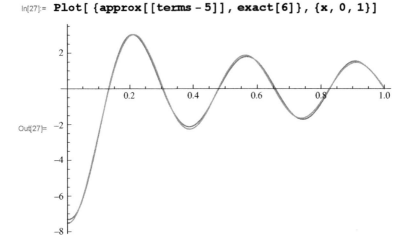

Out[27]=

Figure 4.4 Comparison of the sixth eigenfunction, $\tilde{y}_6(x)$, and $\frac{\sqrt{2}}{J_6(\lambda_1)} J_0(\lambda_6 x)$

Example 4.3: Solve
$$Ly = -x, \qquad (4.37)$$

where
$$L \equiv -\frac{d^2}{dx^2}, \quad y(0) = y(1) = 0.$$

The eigenfunctions and eigenvalues of the Sturm–Liouville system, $Le_n = \lambda e_n$, are
$$e_n = \sqrt{2} \sin \sqrt{\lambda_n} x, \quad \lambda_n = n^2 \pi^2.$$

From Equation (4.26), the coefficient for c is
$$c_n = (-x, e_n)$$
$$= \int_0^1 (-x)\sqrt{2} \sin \sqrt{\lambda_n} x \, dx$$
$$= \frac{\sqrt{2}(-1)^n}{n\pi}.$$

Therefore, the solution, y, is expressed as
$$y = \sum_{i=1}^{\infty} \frac{c_n}{\lambda_n} e_n(x)$$
$$= \sum_{i=1}^{\infty} \frac{\frac{\sqrt{2}(-1)^n}{n\pi}}{n^2 \pi^2} \sqrt{2} \sin (n\pi x)$$
$$= \sum_{i=1}^{\infty} \frac{2(-1)^n}{n^3 \pi^3} \sin (n\pi x). \qquad (4.38)$$

Note that Equation (4.38) is the Fourier series of the exact solution,

$$y = \frac{x^3}{6} - \frac{x}{6}.$$

This is no surprise as the Fourier series method is a special case of the eigenfunction expansion method.

4.2 Steady-State Heat Conduction Equations

Although numerical methods for boundary value problems introduced in the previous section are presented for 1-D problems, it is possible to apply the same approach to multidimensional boundary value problems.

To demonstrate the implementation of the semianalytical methods introduced in the previous section with *Mathematica*, the steady-state heat conduction equation for a finite medium that contains an inclusion is considered in this section. The partial differential equation for steady-state heat conduction is an elliptic-type differential equation expressed as

$$(k_{ij}T_{,i})_{,j} + b = 0,$$

where k_{ij} is the thermal conductivity, T is the steady-state temperature field, and b is the heat source.

If the medium is homogeneous and isotropic, the steady-state heat conduction equation is reduced to the Poisson equation expressed as

$$k\Delta T + b = 0, \tag{4.39}$$

where Δ is the Laplace operator. Analytical solutions to Equation (4.39) are available for only a limited number of cases of boundary conditions and boundary shapes. However, it is noted that if the material properties (k_{ij} for the heat conduction equation) are not uniform with the presence of an inclusion, no analytical solution is available except for infinitely extended media as discussed in Chapter 3. One of the challenging tasks in MWR is to find a set of permissible functions that satisfy the boundary condition for the given boundary shape. The finite element method, which is a special case of MWR, uses piecewise continuous functions that do not necessarily satisfy the continuity conditions or homogeneous boundary conditions.

In this section, a procedure to systematically derive permissible functions and to approximate the solution to the steady-state heat conduction equations is explained.

4.2.1 Derivation of Permissible Functions

In MWR, it is important to choose proper permissible functions that satisfy the homogeneous boundary condition as well as continuity conditions across different phases. In general, it is difficult to derive such functions as it involves a significant amount of algebra. Obviously, this issue can be alleviated with the use of *Mathematica*. This subsection describes the general procedure of how one can derive permissible functions for different boundary conditions with or without an inclusion using polynomials. As it is not possible to cover all possible geometric shapes and boundary conditions, those examples are shown to explain the general procedure and principle from which applications to different types of geometries and boundary conditions can be deduced. As examples, permissible functions for a homogeneous medium as well as a medium that contains a circular inclusion with the first (Dirichlet) and second (Neumann) types of boundary conditions are presented.

4.2.1.1 Homogeneous Medium (No Inclusion) with Dirichlet Boundary Condition

If the medium is homogeneous and contains no inclusion subject to the first type of boundary condition (the Dirichlet boundary condition) with the boundary consisting of piecewise curves expressed as $f_1(x, y) = 0, f_2(x, y) = 0, f_3(x, y) = 0 \ldots$ as shown in Figure 4.5, permissible functions can be expressed as a product of the equations that represent each of the segments of the boundary as

$$f_{ij}(x, y) = f_1(x, y) f_2(x, y) f_3(x, y) \ldots x^i y^j,$$

where i and j are integers.

For example, if the boundary of a medium is defined by a circle expressed as

$$x^2 + y^2 = a^2,$$

permissible functions can be chosen as

$$f_{ij}(x, y) = (x^2 + y^2 - a^2) x^i y^j.$$

If the boundary is rectangular shaped expressed as

$$\{(x, y), -a < x < a, -b < y < b\},$$

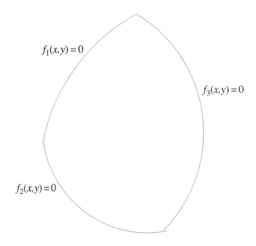

Figure 4.5 Dirichlet-type boundary condition

permissible functions can be chosen as

$$f_{ij}(x,y) = (x^2 - a^2)(y^2 - b^2)x^i y^j.$$

For a triangular boundary defined by $x = 0, y = 0, x + y = 1$, a permissible function can be chosen as

$$f_{ij}(x,y) = xy(x + y - a)x^i y^j.$$

4.2.1.2 Homogeneous Medium (No Inclusion) with the Neumann Boundary Condition

For the second type of boundary condition (the Neumann boundary condition) in which the heat flux (equivalent to the directional derivative of f) vanishes on the boundary, each permissible function, f, must satisfy

$$k_{ij} f_{,i} n_j = 0, \qquad (4.40)$$

where k_{ij} is the anisotropic thermal conductivity and n_j is the normal to the boundary. If k_{ij} is isotropic as

$$k_{ij} = k\delta_{ij},$$

Equation (4.40) is reduced to

$$\frac{\partial f}{\partial n} = f_{,i} n_i = 0.$$

The determination of permissible functions, f, that satisfy Equation (4.40) is more involved than the determination of permissible functions for the Dirichlet type (the first type) of boundary conditions. Unlike the permissible functions for the Dirichlet-type boundary condition in which the permissible functions can be obtained automatically by the product of polynomials each of which satisfies part of the boundary condition, the determination of the permissible functions for the Neumann-type boundary condition requires solving a set of simultaneous equations for the unknown coefficients of the polynomials.

A method to derive permissible functions for the Neumann-type boundary condition is illustrated in the following by a working example of elliptical boundary shown in Figure 4.6 in which permissible functions are sought that satisfy Equation (4.40). For other boundary shapes, a similar approach can be employed.

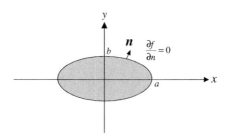

Figure 4.6 Neumann-type boundary condition

In this example, the boundary of the medium is chosen as an ellipse shown in Figure 4.6 expressed by

$$\left(\frac{x}{a}\right)^2 + \left(\frac{y}{b}\right)^2 = 1. \tag{4.41}$$

Permissible functions are sought in polynomials on x and y that satisfy Equation (4.40) on the boundary of Equation (4.41). A polynomial on x and y up to and including the nth order is defined as

$$p(n, x, y) = \sum_{l=0}^{n} \sum_{m=0}^{l} c_{lm} x^{l-m} y^m. \tag{4.42}$$

The normal, $\mathbf{n}$, to the ellipse of Equation (4.41) can be derived by taking the total derivative of Equation (4.41) as

$$\frac{2x}{a^2} dx + \frac{2y}{b^2} dy = 0,$$

implying that the vector, $(\frac{x}{a^2}, \frac{y}{b^2})$, is perpendicular to (dx, dy). Therefore, the normal, $\mathbf{n}$, can be obtained by normalizing $(\frac{x}{a^2}, \frac{y}{b^2})$ as

$$\mathbf{n} = \left(\frac{x}{a^2 \sqrt{\frac{x^2}{a^4} + \frac{y^2}{b^4}}}, \frac{y}{b^2 \sqrt{\frac{x^2}{a^4} + \frac{y^2}{b^4}}}\right). \tag{4.43}$$

Thus, Equation (4.40) can be expressed as

$$\frac{\partial f}{\partial n} = \nabla f \cdot \mathbf{n}$$

$$= \frac{\partial f}{\partial x}\left(\frac{x}{a^2 \sqrt{\frac{x^2}{a^4} + \frac{y^2}{b^4}}}\right) + \frac{\partial f}{\partial y}\left(\frac{y}{b^2 \sqrt{\frac{x^2}{a^4} + \frac{y^2}{b^4}}}\right). \tag{4.44}$$

As Equation (4.44) must vanish along the boundary of Equation (4.41), setting $x = a\cos\theta$ and $y = b\sin\theta$ automatically satisfies this condition. The common numerator of Equation (4.44) is

$$b^2 x \frac{\partial f}{\partial x} + a^2 y \frac{\partial f}{\partial y}, \tag{4.45}$$

which must vanish along the ellipse. Substituting $x = a\cos\theta$ and $y = b\sin\theta$ into Equation (4.45) results in a function of $\cos^n\theta$ and $\sin^m\theta$, which is then further reduced to a function of $\cos n\theta$ and $\sin m\theta$ in the format of

$$\sum_{m,n} d_{mn} \cos n\theta \sin m\theta. \tag{4.46}$$

Hence, by setting
$$d_{mn} = 0,$$
one can obtain a set of simultaneous equations for the unknown coefficient, c_{lm}, in Equation (4.42).

As a working example of the aforementioned algorithm, fifth-order polynomials that satisfy the Neumann boundary condition of Equation (4.40) are sought using *Mathematica*. A complete fifth-order polynomial consisting of all the necessary terms on x and y is expressed as

$$p(x,y) = c_1 + c_2 x + c_3 y + c_4 x^2 + c_5 xy + c_6 y^2 + c_7 x^3 + c_8 x^2 y + c_9 xy^2 + c_{10} y^3$$
$$+ c_{11} x^4 + c_{12} x^3 y + c_{13} x^2 y^2 + c_{14} xy^3 + c_{15} y^4 + c_{16} x^5 + c_{17} x^4 y$$
$$+ c_{18} x^3 y^2 + c_{19} x^2 y^3 + c_{20} xy^4 + c_{21} y^5,$$

which can be generated by poly[n, a] in *Mathematica* as

$$\text{poly}(n, a) = \sum_{i=0}^{n} \sum_{j=0}^{n-i} a \left[\frac{(i+j)(i+j+1)}{2} + i + 1 \right] x^j y^i.$$

In[1]:= order = 5; terms = (order + 1) (order + 2) / 2;
poly[n_, a_] := Sum[a[(i+j) (i+j+1) /2+i+1] x^j y^i, {i, 0, n}, {j, 0, n-i}]

In[3]:= j1 = poly[order, c]

Out[3]= c[1] + x c[2] + y c[3] + x² c[4] + x y c[5] + y² c[6] + x³ c[7] +
x² y c[8] + x y² c[9] + y³ c[10] + x⁴ c[11] + x³ y c[12] + x² y² c[13] + x y³ c[14] +
y⁴ c[15] + x⁵ c[16] + x⁴ y c[17] + x³ y² c[18] + x² y³ c[19] + x y⁴ c[20] + y⁵ c[21]

The variable terms stands for the number of terms of poly[order, c].

In[4]:= terms

Out[4]= 21

This fifth-order polynomial has 21 terms. It is necessary to determine 21 unknown coefficients, $c[1]$–$c[21]$. The normal vector, **n**, in Equation (4.43), to the ellipse is entered in *Mathematica* as

In[5]:= n = {x/a^2, y/b^2}/Sqrt[(x/a^2)^2 + (y/b^2)^2]

Out[5]= $\left\{ \dfrac{x}{a^2 \sqrt{\frac{x^2}{a^4} + \frac{y^2}{b^4}}}, \dfrac{y}{b^2 \sqrt{\frac{x^2}{a^4} + \frac{y^2}{b^4}}} \right\}$

The directional derivative in the normal direction, $\dfrac{\partial f}{\partial n}$, can be computed as

$$\frac{\partial f}{\partial n} = \nabla f \cdot \mathbf{n}.$$

The variable *j2* stores the directional derivative of *j1* as

In[6]:= **j2 = {D[j1, x], D[j1, y]}.n**

Out[6]= $\dfrac{1}{a^2\sqrt{\dfrac{x^2}{a^4}+\dfrac{y^2}{b^4}}}$ x $\big(c[2]+2\,x\,c[4]+y\,c[5]+3\,x^2\,c[7]+2\,x\,y\,c[8]+y^2\,c[9]+4\,x^3\,c[11]+3\,x^2\,y\,c[12]+2\,x\,y^2\,c[13]+$
$y^3\,c[14]+5\,x^4\,c[16]+4\,x^3\,y\,c[17]+3\,x^2\,y^2\,c[18]+2\,x\,y^3\,c[19]+y^4\,c[20]\big)+\dfrac{1}{b^2\sqrt{\dfrac{x^2}{a^4}+\dfrac{y^2}{b^4}}}$
y $\big(c[3]+x\,c[5]+2\,y\,c[6]+x^2\,c[8]+2\,x\,y\,c[9]+3\,y^2\,c[10]+x^3\,c[12]+2\,x^2\,y\,c[13]+$
$3\,x\,y^2\,c[14]+4\,y^3\,c[15]+x^4\,c[17]+2\,x^3\,y\,c[18]+3\,x^2\,y^2\,c[19]+4\,x\,y^3\,c[20]+5\,y^4\,c[21]\big)$

An intermediate variable j3 stores the numerator of $\dfrac{\partial f}{\partial n}$ after simplifying the expression using the Numerator function as

In[11]:= **j3 = Numerator[Simplify[j2]]**

Out[11]= b^2 x $\big(c[2]+5\,x^4\,c[16]+4\,x^3\,(c[11]+y\,c[17])+$
$3\,x^2\,(c[7]+y\,(c[12]+y\,c[18]))+2\,x\,(c[4]+y\,(c[8]+y\,(c[13]+y\,c[19])))+$
$y\,(c[5]+y\,(c[9]+y\,(c[14]+y\,c[20])))\big)+$
a^2 y $\big(c[3]+x^4\,c[17]+x^3\,(c[12]+2\,y\,c[18])+x^2\,(c[8]+2\,y\,c[13]+3\,y^2\,c[19])+$
x $\big(c[5]+2\,y\,c[9]+3\,y^2\,c[14]+4\,y^3\,c[20]\big)+$
y $\big(2\,c[6]+3\,y\,c[10]+4\,y^2\,c[15]+5\,y^3\,c[21]\big)\big)$

The 21 unknown coefficients, $c[i]$, in the aforementioned numerator expression must be determined so that *j3* satisfies the condition

$$\dfrac{\partial f}{\partial n}=0.$$

As the boundary of the ellipse is defined by $(x/a)^2+(y/b)^2=1$, this condition is automatically satisfied by setting $x=a\cos\theta$ and $y=b\sin\theta$ in *j3* as [2]

In[14]:= **j4 = j3 /. {x → a Cos[θ], y → b Sin[θ]}**

Out[14]= $a^2\,b\,\text{Sin}[\theta]\,\big(c[3]+a^4\,c[17]\,\text{Cos}[\theta]^4+a^3\,\text{Cos}[\theta]^3\,(c[12]+2\,b\,c[18]\,\text{Sin}[\theta])+$
$a^2\,\text{Cos}[\theta]^2\,\big(c[8]+2\,b\,c[13]\,\text{Sin}[\theta]+3\,b^2\,c[19]\,\text{Sin}[\theta]^2\big)+$
a $\text{Cos}[\theta]\,\big(c[5]+2\,b\,c[9]\,\text{Sin}[\theta]+3\,b^2\,c[14]\,\text{Sin}[\theta]^2+4\,b^3\,c[20]\,\text{Sin}[\theta]^3\big)+$
b $\text{Sin}[\theta]\,\big(2\,c[6]+3\,b\,c[10]\,\text{Sin}[\theta]+4\,b^2\,c[15]\,\text{Sin}[\theta]^2+5\,b^3\,c[21]\,\text{Sin}[\theta]^3\big)\big)$
$a\,b^2\,\text{Cos}[\theta]\,\big(c[2]+5\,a^4\,c[16]\,\text{Cos}[\theta]^4+4\,a^3\,\text{Cos}[\theta]^3\,(c[11]+b\,c[17]\,\text{Sin}[\theta])+$
$3\,a^2\,\text{Cos}[\theta]^2\,(c[7]+b\,\text{Sin}[\theta]\,(c[12]+b\,c[18]\,\text{Sin}[\theta]))+$
$2\,a\,\text{Cos}[\theta]\,(c[4]+b\,\text{Sin}[\theta]\,(c[8]+b\,\text{Sin}[\theta]\,(c[13]+b\,c[19]\,\text{Sin}[\theta])))+$
$b\,\text{Sin}[\theta]\,(c[5]+b\,\text{Sin}[\theta]\,(c[9]+b\,\text{Sin}[\theta]\,(c[14]+b\,c[20]\,\text{Sin}[\theta])))\big)$

It is necessary to rearrange the output *j4* into linearly independent terms, cos $m\theta$ and sin $n\theta$, on the variable θ so that each of the independent terms is set to 0 to generate a set of simultaneous

---

[2] The Greek letter, θ, can be entered in *Mathematica* by pressing Esc q Esc.

equations for the unknown coefficients, $c[i]$. The function `TrigReduce` can be used to convert the power of trigonometric functions to the first order terms. Note that cos $m\theta$ and sin $n\theta$ are linearly independent of one another. Here is the result of rewriting $j4$ in terms of cos $m\theta$ and sin $n\theta$.

In[15]:= `j5 = TrigReduce[j4]`

Out[15]= $a^2 b^2 c[4] + a^2 b^2 c[6] + \frac{3}{2} a^4 b^2 c[11] + \frac{1}{4} a^4 b^2 c[13] + \frac{1}{4} a^2 b^4 c[13] + \frac{3}{2} a^2 b^4 c[15] +$
$a b^2 c[2] \cos[\theta] + \frac{9}{4} a^3 b^2 c[7] \cos[\theta] + \frac{1}{2} a^3 b^2 c[9] \cos[\theta] + \frac{1}{4} a b^4 c[9] \cos[\theta] +$
$\frac{25}{8} a^5 b^2 c[16] \cos[\theta] + \frac{1}{4} a^5 b^2 c[18] \cos[\theta] + \frac{3}{8} a^3 b^4 c[18] \cos[\theta] +$
$\frac{1}{2} a^3 b^4 c[20] \cos[\theta] + \frac{1}{8} a b^6 c[20] \cos[\theta] + a^2 b^2 c[4] \cos[2\theta] - a^2 b^2 c[6] \cos[2\theta] +$
$2 a^4 b^2 c[11] \cos[2\theta] - 2 a^2 b^4 c[15] \cos[2\theta] + \frac{3}{4} a^3 b^2 c[7] \cos[3\theta] -$
$\frac{1}{2} a^3 b^2 c[9] \cos[3\theta] - \frac{1}{4} a b^4 c[9] \cos[3\theta] + \frac{25}{16} a^5 b^2 c[16] \cos[3\theta] -$
$\frac{1}{8} a^5 b^2 c[18] \cos[3\theta] - \frac{3}{16} a^3 b^4 c[18] \cos[3\theta] - \frac{3}{4} a^3 b^4 c[20] \cos[3\theta] -$
$\frac{3}{16} a b^6 c[20] \cos[3\theta] + \frac{1}{2} a^4 b^2 c[11] \cos[4\theta] - \frac{1}{4} a^4 b^2 c[13] \cos[4\theta] -$
$\frac{1}{4} a^2 b^4 c[13] \cos[4\theta] + \frac{1}{2} a^2 b^4 c[15] \cos[4\theta] + \frac{5}{16} a^5 b^2 c[16] \cos[5\theta] -$
$\frac{1}{8} a^5 b^2 c[18] \cos[5\theta] - \frac{3}{16} a^3 b^4 c[18] \cos[5\theta] + \frac{1}{4} a^3 b^4 c[20] \cos[5\theta] +$
$\frac{1}{16} a b^6 c[20] \cos[5\theta] + a^2 b c[3] \sin[\theta] + \frac{1}{4} a^4 b c[8] \sin[\theta] +$
$\frac{1}{2} a^2 b^3 c[8] \sin[\theta] + \frac{9}{4} a^2 b^3 c[10] \sin[\theta] + \frac{1}{8} a^6 b c[17] \sin[\theta] +$
$\frac{1}{2} a^4 b^3 c[17] \sin[\theta] + \frac{3}{8} a^4 b^3 c[19] \sin[\theta] + \frac{1}{4} a^2 b^5 c[19] \sin[\theta] +$
$\frac{25}{8} a^2 b^5 c[21] \sin[\theta] + \frac{1}{2} a^3 b c[5] \sin[2\theta] + \frac{1}{2} a b^3 c[5] \sin[2\theta] +$
$\frac{1}{4} a^5 b c[12] \sin[2\theta] + \frac{3}{4} a^3 b^3 c[12] \sin[2\theta] + \frac{3}{4} a^3 b^3 c[14] \sin[2\theta] +$
$\frac{1}{4} a b^5 c[14] \sin[2\theta] + \frac{1}{4} a^4 b c[8] \sin[3\theta] + \frac{1}{2} a^2 b^3 c[8] \sin[3\theta] -$
$\frac{3}{4} a^2 b^3 c[10] \sin[3\theta] + \frac{3}{16} a^6 b c[17] \sin[3\theta] + \frac{3}{4} a^4 b^3 c[17] \sin[3\theta] +$
$\frac{3}{16} a^4 b^3 c[19] \sin[3\theta] - \frac{1}{8} a^2 b^5 c[19] \sin[3\theta] - \frac{25}{16} a^2 b^5 c[21] \sin[3\theta] +$
$\frac{1}{8} a^5 b c[12] \sin[4\theta] + \frac{3}{8} a^3 b^3 c[12] \sin[4\theta] - \frac{3}{8} a^3 b^3 c[14] \sin[4\theta] -$
$\frac{1}{8} a b^5 c[14] \sin[4\theta] + \frac{1}{16} a^6 b c[17] \sin[5\theta] + \frac{1}{4} a^4 b^3 c[17] \sin[5\theta] -$
$\frac{3}{16} a^4 b^3 c[19] \sin[5\theta] - \frac{1}{8} a^2 b^5 c[19] \sin[5\theta] + \frac{5}{16} a^2 b^5 c[21] \sin[5\theta]$

It is necessary to collect the coefficients of (sin θ, cos θ, sin 2θ, cos 2θ, ...) so that independent terms can be isolated and separated. Such separations can be easily performed if the formula were a polynomial on x and y using the `CoefficientList` functions as

```
In[24]:= poly = 1 + a x^2 + b x y + c y^2
        xmax = Exponent[poly, x];
        ymax = Exponent[poly, y];
        temp1 = CoefficientList[ poly, {x, y}]
        Sum[temp1[[i, j]] x^(i - 1) y^(j - 1), {i, xmax + 1}, {j, ymax + 1}]
```

Out[24]= $1 + a x^2 + b x y + c y^2$

Out[27]= $\{\{1, 0, c\}, \{0, b, 0\}, \{a, 0, 0\}\}$

Out[28]= $1 + a x^2 + b x y + c y^2$

In the aforementioned example code, xmax returns the maximum power of poly on x and ymax returns the maximum power of poly on y. The CoefficientList function separates poly into a nested list of x and y and the last statement restores poly.

However, a *Mathematica* function that is equivalent to the CoefficientList function that works for sin $n\theta$ and cos $n\theta$ is not available. Therefore, a small trick needs to be devised that modifies j5 by replacing cos $m\theta$ and sin $n\theta$ with x^m and y^n so that the CoefficientList function can be used on them. In the following code, cos $n\theta$ and sin $n\theta$ are replaced by cosine ^{n} and sine ^{n}, respectively.

```
In[17]:= j7 = CoefficientList[j6, {cosine, sine}]
```

Out[17]= $\{\{a^2 b^2 c[4] + a^2 b^2 c[6] + \frac{3}{2} a^4 b^2 c[11] + \frac{1}{4} a^4 b^2 c[13] + \frac{1}{4} a^2 b^4 c[13] + \frac{3}{2} a^2 b^4 c[15],$
$a^2 b c[3] + \frac{1}{4} a^4 b c[8] + \frac{1}{2} a^2 b^3 c[8] + \frac{9}{4} a^2 b^3 c[10] + \frac{1}{8} a^6 b c[17] +$
$\frac{1}{2} a^4 b^3 c[17] + \frac{3}{8} a^4 b^3 c[19] + \frac{1}{4} a^2 b^5 c[19] + \frac{25}{8} a^2 b^5 c[21],$
$\frac{1}{2} a^3 b c[5] + \frac{1}{2} a b^3 c[5] + \frac{1}{4} a^5 b c[12] + \frac{3}{4} a^3 b^3 c[12] + \frac{3}{4} a^3 b^3 c[14] + \frac{1}{4} a b^5 c[14],$
$\frac{1}{4} a^4 b c[8] + \frac{1}{2} a^2 b^3 c[8] - \frac{3}{4} a^2 b^3 c[10] + \frac{3}{16} a^6 b c[17] +$
$\frac{3}{4} a^4 b^3 c[17] + \frac{3}{16} a^4 b^3 c[19] + \frac{1}{8} a^2 b^5 c[19] - \frac{25}{16} a^2 b^5 c[21],$
$\frac{1}{8} a^5 b c[12] + \frac{3}{8} a^3 b^3 c[12] - \frac{3}{8} a^3 b^3 c[14] - \frac{1}{8} a b^5 c[14],$
$\frac{1}{16} a^6 b c[17] + \frac{1}{4} a^4 b^3 c[17] - \frac{3}{16} a^4 b^3 c[19] - \frac{1}{8} a^2 b^5 c[19] + \frac{5}{16} a^2 b^5 c[21]\},$
$\{a b^2 c[2] + \frac{9}{4} a^3 b^2 c[7] + \frac{1}{2} a^3 b^2 c[9] + \frac{1}{4} a b^4 c[9] + \frac{25}{8} a^5 b^2 c[16] +$
$\frac{1}{4} a^5 b^2 c[18] + \frac{3}{8} a^3 b^4 c[18] + \frac{1}{2} a^3 b^4 c[20] + \frac{1}{8} a b^6 c[20], 0, 0, 0, 0\},$
$\{a^2 b^2 c[4] - a^2 b^2 c[6] + 2 a^4 b^2 c[11] - 2 a^2 b^4 c[15], 0, 0, 0, 0, 0\},$
$\{\frac{3}{4} a^3 b^2 c[7] - \frac{1}{2} a^3 b^2 c[9] - \frac{1}{4} a b^4 c[9] + \frac{25}{16} a^5 b^2 c[16] - \frac{1}{8} a^5 b^2 c[18] -$
$\frac{3}{16} a^3 b^4 c[18] - \frac{3}{4} a^3 b^4 c[20] - \frac{3}{16} a b^6 c[20], 0, 0, 0, 0, 0\},$
$\{\frac{1}{2} a^4 b^2 c[11] - \frac{1}{4} a^4 b^2 c[13] - \frac{1}{4} a^2 b^4 c[13] + \frac{1}{2} a^2 b^4 c[15], 0, 0, 0, 0, 0\},$
$\{\frac{5}{16} a^5 b^2 c[16] - \frac{1}{8} a^5 b^2 c[18] - \frac{3}{16} a^3 b^4 c[18] + \frac{1}{4} a^3 b^4 c[20] + \frac{1}{16} a b^6 c[20],$
$0, 0, 0, 0, 0\}\}$

As the variable j6 is a polynomial on sine and cosine, it is possible to collect like terms for each of the powers on cosine and sine by using the CoefficientList function as

Inclusions in Finite Matrix

`In[11]:= j7 = CoefficientList[j6, {cosine, sine}]`

$$Out[11]= \{\{a^2 b^2 c[4] + a^2 b^2 c[6] + \tfrac{3}{2}a^4 b^2 c[11] + \tfrac{1}{4}a^4 b^2 c[13] + \tfrac{1}{4}a^2 b^4 c[13] + \tfrac{3}{2}a^2 b^4 c[15],$$

$$a^2 b c[3] + \tfrac{1}{4}a^4 b c[8] + \tfrac{1}{2}a^2 b^3 c[8] + \tfrac{9}{4}a^2 b^3 c[10] + \tfrac{1}{8}a^6 b c[17] +$$

$$\tfrac{1}{2}a^4 b^3 c[17] + \tfrac{3}{8}a^4 b^3 c[19] + \tfrac{1}{4}a^2 b^5 c[19] + \tfrac{25}{8}a^2 b^5 c[21],$$

$$\tfrac{1}{2}a^3 b c[5] + \tfrac{1}{2}a b^3 c[5] + \tfrac{1}{4}a^5 b c[12] + \tfrac{3}{4}a^3 b^3 c[12] + \tfrac{3}{4}a^3 b^3 c[14] + \tfrac{1}{4}a b^5 c[14],$$

$$\tfrac{1}{4}a^4 b c[8] + \tfrac{1}{2}a^2 b^3 c[8] - \tfrac{3}{4}a^2 b^3 c[10] + \tfrac{3}{16}a^6 b c[17] + \tfrac{1}{4}a^4 b^3 c[17] + \tfrac{3}{16}a^4 b^3 c[19] +$$

$$\tfrac{1}{8}a^2 b^5 c[19] - \tfrac{25}{16}a^2 b^5 c[21], \tfrac{1}{8}a^5 b c[12] + \tfrac{3}{8}a^3 b^3 c[12] - \tfrac{3}{8}a^3 b^3 c[14] - \tfrac{1}{8}a b^5 c[14],$$

$$\tfrac{1}{16}a^6 b c[17] + \tfrac{1}{4}a^4 b^3 c[17] - \tfrac{3}{16}a^4 b^3 c[19] - \tfrac{1}{8}a^2 b^5 c[19] + \tfrac{5}{16}a^2 b^5 c[21]\},$$

$$\{a b^2 c[2] + \tfrac{9}{4}a^3 b^2 c[7] + \tfrac{1}{2}a^3 b^2 c[9] + \tfrac{1}{4}a b^4 c[9] + \tfrac{25}{8}a^5 b^2 c[16] +$$

$$\tfrac{1}{4}a^5 b^2 c[18] + \tfrac{3}{8}a^3 b^4 c[18] + \tfrac{1}{2}a^3 b^4 c[20] + \tfrac{1}{8}a b^6 c[20], 0, 0, 0, 0, 0\},$$

$$\{a^2 b^2 c[4] - a^2 b^2 c[6] + 2 a^4 b^2 c[11] - 2 a^2 b^4 c[15], 0, 0, 0, 0, 0\},$$

$$\{\tfrac{3}{4}a^3 b^2 c[7] - \tfrac{1}{2}a^3 b^2 c[9] - \tfrac{1}{4}a b^4 c[9] + \tfrac{25}{16}a^5 b^2 c[16] -$$

$$\tfrac{1}{8}a^5 b^2 c[18] - \tfrac{3}{16}a^3 b^4 c[18] - \tfrac{3}{4}a^3 b^4 c[20] - \tfrac{3}{16}a b^6 c[20], 0, 0, 0, 0, 0\},$$

$$\{\tfrac{1}{2}a^4 b^2 c[11] - \tfrac{1}{4}a^4 b^2 c[13] - \tfrac{1}{4}a^2 b^4 c[13] + \tfrac{1}{2}a^2 b^4 c[15], 0, 0, 0, 0, 0\},$$

$$\{\tfrac{5}{16}a^5 b^2 c[16] - \tfrac{1}{8}a^5 b^2 c[18] - \tfrac{3}{16}a^3 b^4 c[18] + \tfrac{1}{4}a^3 b^4 c[20] + \tfrac{1}{16}a b^6 c[20], 0, 0, 0, 0, 0\}\}$$

Each expression containing $c[i]$s separated by a comma represents the coefficient of the independent terms of cosine ^{m} and sine ^{n}, which are to be equated to 0 to generate a set of simultaneous equations for $c[i]$s. The "0" entry can be eliminated by using the `DeleteCases` function. As the aforementioned list is nested, the `Flatten` function can be used to remove nested list structures so that all the elements are arranged at the same level. With the following code, each entry in the list represents the left-hand side of an individual equation to be equated to 0.

`In[12]:= j8 = DeleteCases[j7 // Flatten, 0]`

$$Out[12]= \{a^2 b^2 c[4] + a^2 b^2 c[6] + \tfrac{3}{2}a^4 b^2 c[11] + \tfrac{1}{4}a^4 b^2 c[13] + \tfrac{1}{4}a^2 b^4 c[13] + \tfrac{3}{2}a^2 b^4 c[15],$$

$$a^2 b c[3] + \tfrac{1}{4}a^4 b c[8] + \tfrac{1}{2}a^2 b^3 c[8] + \tfrac{9}{4}a^2 b^3 c[10] + \tfrac{1}{8}a^6 b c[17] +$$

$$\tfrac{1}{2}a^4 b^3 c[17] + \tfrac{3}{8}a^4 b^3 c[19] + \tfrac{1}{4}a^2 b^5 c[19] + \tfrac{25}{8}a^2 b^5 c[21],$$

$$\tfrac{1}{2}a^3 b c[5] + \tfrac{1}{2}a b^3 c[5] + \tfrac{1}{4}a^5 b c[12] + \tfrac{3}{4}a^3 b^3 c[12] + \tfrac{3}{4}a^3 b^3 c[14] + \tfrac{1}{4}a b^5 c[14],$$

$$\tfrac{1}{4}a^4 b c[8] + \tfrac{1}{2}a^2 b^3 c[8] - \tfrac{3}{4}a^2 b^3 c[10] + \tfrac{3}{16}a^6 b c[17] + \tfrac{1}{4}a^4 b^3 c[17] + \tfrac{3}{16}a^4 b^3 c[19] +$$

$$\tfrac{1}{8}a^2 b^5 c[19] - \tfrac{25}{16}a^2 b^5 c[21], \tfrac{1}{8}a^5 b c[12] + \tfrac{3}{8}a^3 b^3 c[12] - \tfrac{3}{8}a^3 b^3 c[14] - \tfrac{1}{8}a b^5 c[14],$$

$$\tfrac{1}{16}a^6 b c[17] + \tfrac{1}{4}a^4 b^3 c[17] - \tfrac{3}{16}a^4 b^3 c[19] - \tfrac{1}{8}a^2 b^5 c[19] + \tfrac{5}{16}a^2 b^5 c[21],$$

$$a b^2 c[2] + \tfrac{9}{4}a^3 b^2 c[7] + \tfrac{1}{2}a^3 b^2 c[9] + \tfrac{1}{4}a b^4 c[9] + \tfrac{25}{8}a^5 b^2 c[16] + \tfrac{1}{4}a^5 b^2 c[18] + \tfrac{3}{8}a^3 b^4 c[18] +$$

$$\tfrac{1}{2}a^3 b^4 c[20] + \tfrac{1}{8}a b^6 c[20], a^2 b^2 c[4] - a^2 b^2 c[6] + 2 a^4 b^2 c[11] - 2 a^2 b^4 c[15],$$

$$\tfrac{3}{4}a^3 b^2 c[7] - \tfrac{1}{2}a^3 b^2 c[9] - \tfrac{1}{4}a b^4 c[9] + \tfrac{25}{16}a^5 b^2 c[16] - \tfrac{1}{8}a^5 b^2 c[18] - \tfrac{3}{16}a^3 b^4 c[18] -$$

$$\tfrac{3}{4}a^3 b^4 c[20] - \tfrac{3}{16}a b^6 c[20], \tfrac{1}{2}a^4 b^2 c[11] - \tfrac{1}{4}a^4 b^2 c[13] - \tfrac{1}{4}a^2 b^4 c[13] + \tfrac{1}{2}a^2 b^4 c[15],$$

$$\tfrac{5}{16}a^5 b^2 c[16] - \tfrac{1}{8}a^5 b^2 c[18] - \tfrac{3}{16}a^3 b^4 c[18] + \tfrac{1}{4}a^3 b^4 c[20] + \tfrac{1}{16}a b^6 c[20]\}$$

From the aforementioned list, a set of simultaneous equations for $c[i]$s satisfying the homogeneous second-type boundary conditions can be obtained by equating each of the elements in the list to 0. This can be achieved automatically using the Map function. The Map function can map a function onto each individual element in a list as

In[13]:= **eqs = Map[# == 0 &, j8]**

Out[13]= $\{a^2 b^2 c[4] + a^2 b^2 c[6] + \frac{3}{2} a^4 b^2 c[11] + \frac{1}{4} a^4 b^2 c[13] + \frac{1}{4} a^2 b^4 c[13] + \frac{3}{2} a^2 b^4 c[15] = 0,$

$a^2 b c[3] + \frac{1}{4} a^4 b c[8] + \frac{1}{2} a^2 b^3 c[8] + \frac{9}{4} a^2 b^3 c[10] + \frac{1}{8} a^6 b c[17] +$
$\frac{1}{2} a^4 b^3 c[17] + \frac{3}{8} a^4 b^3 c[19] + \frac{1}{4} a^2 b^5 c[19] + \frac{25}{8} a^2 b^3 c[21] = 0,$

$\frac{1}{2} a^3 b c[5] + \frac{1}{2} a b^3 c[5] + \frac{1}{4} a^5 b c[12] + \frac{3}{4} a^3 b^3 c[12] + \frac{3}{4} a^3 b^3 c[14] + \frac{1}{4} a b^5 c[14] = 0,$

$\frac{1}{4} a^4 b c[8] + \frac{1}{2} a^2 b^3 c[8] - \frac{3}{4} a^2 b^3 c[10] + \frac{3}{16} a^6 b c[17] + \frac{3}{4} a^4 b^3 c[17] + \frac{3}{16} a^4 b^3 c[19] +$
$\frac{1}{8} a^2 b^5 c[19] - \frac{25}{16} a^2 b^3 c[21] = 0, \frac{1}{8} a^5 b c[12] + \frac{3}{8} a^3 b^3 c[12] - \frac{3}{8} a^3 b^3 c[14] - \frac{1}{8} a b^5 c[14] = 0,$

$\frac{1}{16} a^6 b c[17] + \frac{1}{4} a^4 b^3 c[17] - \frac{3}{16} a^4 b^3 c[19] - \frac{1}{8} a^2 b^5 c[19] + \frac{5}{16} a^2 b^5 c[21] = 0,$

$a b^2 c[2] + \frac{9}{4} a^3 b^2 c[7] + \frac{1}{2} a^3 b^2 c[9] + \frac{1}{4} a b^4 c[9] + \frac{25}{8} a^5 b^2 c[16] +$
$\frac{1}{4} a^5 b^2 c[18] + \frac{3}{8} a^3 b^4 c[18] + \frac{1}{2} a^3 b^4 c[20] + \frac{1}{8} a b^6 c[20] = 0,$

$a^2 b^2 c[4] - a^2 b^2 c[6] + 2 a^4 b^2 c[11] - 2 a^2 b^4 c[15] == 0, \frac{3}{4} a^3 b^2 c[7] - \frac{1}{2} a^3 b^2 c[9] -$
$\frac{1}{4} a b^4 c[9] + \frac{25}{16} a^5 b^2 c[16] - \frac{1}{8} a^5 b^2 c[18] - \frac{3}{16} a^3 b^4 c[18] - \frac{3}{4} a^3 b^4 c[20] - \frac{3}{16} a b^6 c[20] = 0,$

$\frac{1}{2} a^4 b^3 c[11] - \frac{1}{4} a^4 b^3 c[13] - \frac{1}{4} a^2 b^4 c[13] + \frac{1}{2} a^2 b^4 c[15] = 0,$

$\frac{5}{16} a^5 b^2 c[16] - \frac{1}{8} a^5 b^2 c[18] - \frac{3}{16} a^3 b^4 c[18] + \frac{1}{4} a^3 b^4 c[20] + \frac{1}{16} a b^6 c[20] = 0\}$

The list created is a set of underdetermined simultaneous equations that need to be solved for $c[1]$–$c[21]$. The Solve function can be used to solve a set of underdetermined equations in terms of independent coefficients as a rule.

In[14]:= **sol = Solve[eqs, Table[c[i], {i, 1, terms}]][[1]]**

Out[14]= $\{c[2] \to -(2 a^2 + b^2) c[9] - (4 a^2 b^2 + b^4) c[20], c[3] \to -(a^2 + 2 b^2) c[8] - (a^4 + 4 a^2 b^2) c[17],$

$c[4] \to -2 a^2 c[11], c[5] \to -\frac{(a^4 + 3 a^2 b^2) c[12]}{a^2 + b^2}, c[6] \to 2 a^2 c[11] - (a^2 + b^2) c[13],$

$c[7] \to -\frac{(-2 a^2 - b^2) c[9]}{3 a^2} - \frac{1}{3}(2 a^2 + 3 b^2) c[18] + \frac{2 (4 a^2 b^2 + b^4) c[20]}{3 a^2},$

$c[10] \to -\frac{(-a^2 - 2 b^2) c[8]}{3 b^2} + \frac{2 (a^4 + 4 a^2 b^2) c[17]}{3 b^2} - \frac{1}{3}(3 a^2 + 2 b^2) c[19],$

$c[14] \to -\frac{(-a^4 - 3 a^2 b^2) c[12]}{b^2 (3 a^2 + b^2)}, c[15] \to -\frac{a^2 c[11]}{b^2} - \frac{(-a^2 - b^2) c[13]}{2 b^2},$

$c[16] \to -\frac{(-2 a^2 - 3 b^2) c[18]}{5 a^2} - \frac{(4 a^2 b^2 + b^4) c[20]}{5 a^4},$

$c[21] \to -\frac{(a^4 + 4 a^2 b^2) c[17]}{5 b^4} - \frac{(-3 a^2 - 2 b^2) c[19]}{5 b^2}\}$

In the given output, $c[8]$, $c[9]$, $c[11]$, $c[12]$, $c[13]$, $c[17]$, $c[18]$, $c[19]$, and $c[20]$ are independent coefficients and all other coefficients are expressed as a linear combination of them. Noting that the solution, sol, from the Solve function is a set of *Mathematica* rules (substitution), the polynomials that satisfy the homogeneous boundary condition can be obtained by applying these rules to the assumed polynomial, $j1$, with unknown coefficients as

In[15]:= **j9 = j1 /. sol**

Out[15]= $c[1] + x^2 y c[8] + xy^2 c[9] - 2a^2 x^2 c[11] + x^4 c[11] - \dfrac{(a^4+3a^2b^2) xy c[12]}{a^2+b^2} +$

$x^3 y c[12] - \dfrac{(-a^4-3a^2b^2) xy^3 c[12]}{b^2(3a^2+b^2)} + x^2 y^2 c[13] + y^4 \left(\dfrac{a^2 c[11]}{b^2} - \dfrac{(-a^2-b^2) c[13]}{2b^2} \right) +$

$y^2 \left(2a^2 c[11] - (a^2+b^2) c[13] \right) + x^4 y c[17] + y \left(-(a^2+2b^2) c[8] - (a^4+4a^2b^2) c[17] \right) +$

$x^3 y^2 c[18] + x^2 y^3 c[19] + y^5 \left(-\dfrac{(a^4+4a^2b^2) c[17]}{5b^4} - \dfrac{(-3a^2-2b^2) c[19]}{5b^2} \right) +$

$y^3 \left(-\dfrac{(-a^2-2b^2) c[8]}{3b^2} + \dfrac{2(a^4+4a^2b^2) c[17]}{3b^2} - \dfrac{1}{3}(3a^2+2b^2) c[19] \right) + xy^4 c[20] +$

$x(-(2a^2+b^2) c[9] - (4a^2b^2+b^4) c[20]) + x^5 \left(-\dfrac{(-2a^2-3b^2) c[18]}{5a^2} - \dfrac{(4a^2b^2+b^4) c[20]}{5a^4} \right) +$

$x^3 \left(-\dfrac{(-2a^2-b^2) c[9]}{3a^2} - \dfrac{1}{3}(2a^2+3b^2) c[18] + \dfrac{2(4a^2b^2+b^4) c[20]}{3a^2} \right)$

The expressions of $j9$ can be grouped into independent polynomials by collecting the coefficients of each $c[i]$ using the Coefficient function as

In[16]:= **j10 = DeleteCases[Table[Coefficient[j9, c[i]], {i, 1, terms}], 0]**

Out[16]= $\{ 1, -a^2 y - 2b^2 y + x^2 y + \dfrac{2y^3}{3} + \dfrac{a^2 y^3}{3b^2}, -2a^2 x - b^2 x + \dfrac{2x^3}{3} + \dfrac{b^2 x^3}{3a^2} + xy^2, -2a^2 x^2 + x^4 + 2a^2 y^2 - \dfrac{a^2 y^4}{b^2},$

$-\dfrac{(a^4+3a^2b^2) xy}{a^2+b^2} + x^3 y - \dfrac{(-a^4-3a^2b^2) xy^3}{b^2(3a^2+b^2)}, -a^2 y^2 - b^2 y^2 + x^2 y^2 + \dfrac{y^4}{2} + \dfrac{a^2 y^4}{2b^2},$

$-a^4 y - 4a^2 b^2 y + x^4 y + \dfrac{8a^2 y^3}{3} + \dfrac{2a^4 y^3}{3b^2} - \dfrac{a^4 y^5}{5b^4} - \dfrac{4a^2 y^5}{5b^2} - \dfrac{2}{3} a^2 x^2 - b^2 x^3 + \dfrac{2x^5}{5} + \dfrac{3b^2 x^5}{5a^2} + x^3 y^2,$

$-a^2 y^3 - \dfrac{2b^2 y^3}{3} + x^2 y^3 + \dfrac{2y^5}{5} + \dfrac{3a^2 y^5}{5b^2}, -4a^2 b^2 x - b^4 x + \dfrac{8b^2 x^3}{3} + \dfrac{2b^4 x^3}{3a^2} - \dfrac{4b^2 x^5}{5a^2} - \dfrac{b^4 x^5}{5a^4} + xy^4 \}$

It is seen that there are 10 independent permissible functions that satisfy the second-type boundary condition for the order of polynomial $n = 5$.

4.2.1.3 Medium with an Inclusion (the Dirichlet Boundary Condition)

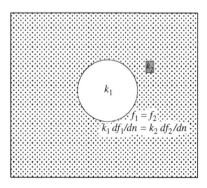

If a finite medium contains an inclusion inside, the derivation of the permissible functions for such a medium is obviously more involved than a medium with no inclusion. As there are two phases (the inclusion and the matrix), it is necessary to prepare two separate sets of permissible functions, one for the inclusion phase and the other for the matrix phase. The permissible function for the inclusion phase must satisfy the continuity condition at the inclusion–matrix interface with the permissible function defined for the matrix phase in which the two functions and their flux must be continuous as

$$f_1 = f_2 \quad \text{at the interface,} \tag{4.47}$$

$$k_1 \frac{\partial f_1}{\partial n} = k_2 \frac{\partial f_2}{\partial n} \quad \text{at the interface,} \tag{4.48}$$

where f_1 and f_2 are permissible functions for the inclusion and the matrix, respectively, and k_1 and k_2 are the thermal conductivities for the inclusion and the matrix, respectively. In addition, the permissible function, f_2, for the matrix must also satisfy the first-type homogeneous boundary condition along the boundary as

$$f_2 = 0 \quad \text{on the boundary.}$$

To illustrate this procedure, a working example of a rectangular medium defined by $\{(x, y), -a_1 < x < a_1, -a_2 < y < a_2\}$ that contains a circular inclusion with a radius of a at the center as shown in Figure 4.7 is used. The thermal conductivity for the inclusion is denoted by k_1 and the thermal conductivity for the matrix is denoted by k_2.

The following three steps are required to obtain the permissible functions:

1. Prepare two polynomials, one for the matrix phase and the other for the inclusion phase, with unknown coefficients.
2. Impose the homogeneous boundary condition on the boundary as well as the continuity conditions at the interface.
3. Solve a set of underdetermined simultaneous equations for the unknown coefficients.

For this working example, it can be shown that the minimum order of polynomials for the permissible functions that satisfy both the continuity conditions across the phases and the

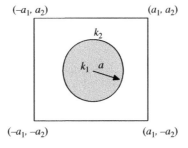

Figure 4.7 Circular inclusion in a finite medium

boundary condition is $n = 4$. Therefore, the permissible functions for the inclusion phase and the matrix phase denoted as f_1 and f_2, respectively, are expressed with unknown coefficients, $c_1(i)$ and $c_2(i)$ as

$$f_1(x,y) = c_1(11)x^4 + c_1(12)x^3y + c_1(7)x^3 + c_1(13)x^2y^2 + c_1(8)x^2y + c_1(4)x^2$$
$$+ c_1(14)xy^3 + c_1(9)xy^2 + c_1(5)xy + c_1(2)x + c_1(15)y^4 + c_1(10)y^3$$
$$+ c_1(6)y^2 + c_1(3)y + c_1(1), \qquad (4.49)$$

$$f_2(x,y) = (x^2 - a_1^2)(y^2 - a_2^2)(c_2(11)x^4 + c_2(12)x^3y + c_2(7)x^3 + c_2(13)x^2y^2$$
$$+ c_2(8)x^2y + c_2(4)x^2 + c_2(14)xy^3 + c_2(9)xy^2 + c_2(5)xy + c_2(2)x$$
$$+ c_2(15)y^4 + c_2(10)y^3 + c_2(6)y^2 + c_2(3)y + c_2(1)). \qquad (4.50)$$

The unknown coefficients, $c_1(i)$ and $c_2(i)$, in Equations (4.49) and (4.50) can be determined by imposing the continuity conditions of Equations (4.47) and (4.48) along $x^2 + y^2 = a^2$. When x and y are on the circle, $x^2 + y^2 = a^2$,

$$x = a\cos\theta, \quad y = a\sin\theta,$$

can be substituted into Equations (4.49) and (4.50) and rearranging the result in independent terms of $\cos m\theta$ and $\sin n\theta$ yields the following formula:

$$f_1 - f_2 = a^8(-c_2(11))\sin(2\theta)\cos(6\theta) - a^8 c_2(12)\sin(3\theta)\cos(5\theta)$$
$$- a^8 c_2(13)\sin(4\theta)\cos(4\theta) - a^8 c_2(14)\sin(5\theta)\cos(3\theta) - a^8 c_2(15)\sin(6\theta)\cos(2\theta)$$
$$- a^7 c_2(7)\sin(2\theta)\cos(5\theta)$$
$$- a^7 c_2(8)\sin(3\theta)\cos(4\theta) - a^7 c_2(9)\sin(4\theta)\cos(3\theta) - a^7 c_2(10)\sin(5\theta)\cos(2\theta)$$
$$+ \sin(2\theta)\cos(4\theta)(a^6 a_1^2 c_2(11) + a^6 a_2^2 c_2(13) + a^6(-c_2(4))) + \sin(3\theta)\cos(3\theta)$$
$$(a^6 a_1^2 c_2(12) + a^6 a_2^2 c_2(14) + a^6(-c_2(5))) + \sin(4\theta)\cos(2\theta)$$
$$(a^6 a_1^2 c_2(13) + a^6 a_2^2 c_2(15) + a^6(-c_2(6))) + a^6 a_1^2 c_2(15)\sin(6\theta)$$
$$+ a^6 a_1^2 c_2(14)\sin(5\theta)\cos(\theta) + a^6 a_2^2 c_2(12)\sin(\theta)\cos(5\theta) + \sin(2\theta)\cos(3\theta)$$
$$(a^5 a_1^2 c_2(7) + a^5 a_2^2 c_2(9) + a^5(-c_2(2))) + \sin(3\theta)\cos(2\theta)$$
$$(a^5 a_1^2 c_2(8) + a^5 a_2^2 c_2(10) + a^5(-c_2(3))) + a^5 a_1^2 c_2(10)\sin(5\theta)$$
$$+ a^5 a_1^2 c_2(9)\sin(4\theta)\cos(\theta) + a^5 a_2^2 c_2(8)\sin(\theta)\cos(4\theta) + \sin(4\theta)$$
$$(-a^4 a_1^2 a_2^2 c_2(15) + a^4 a_1^2 c_2(6) + a^4 c_1(15)) + \sin(\theta)\cos(3\theta)$$
$$(-a^4 a_1^2 a_2^2 c_2(12) + a^4 a_2^2 c_2(5) + a^4 c_1(12)) + \sin(2\theta)\cos(2\theta)$$
$$(-a^4 a_1^2 a_2^2 c_2(13) + a^4 a_1^2 c_2(4) + a^4 a_2^2 c_2(6) + a^4 c_1(13) - a^4 c_2(1)) + \sin(3\theta)\cos(\theta)$$
$$(-a^4 a_1^2 a_2^2 c_2(14) + a^4 a_1^2 c_2(5) + a^4 c_1(14)) + \sin(3\theta)$$

$$(-a^3 a_1^2 a_2^2 c_2(10) + a^3 a_1^2 c_2(3) + a^3 c_1(10))$$
$$+ \sin(\theta)\cos(2\theta)(-a^3 a_1^2 a_2^2 c_2(8) + a^3 a_1^2 c_2(3) + a^3 c_1(8)) + \sin(2\theta)\cos(\theta)$$
$$(-a^3 a_1^2 a_2^2 c_2(9) + a^3 a_1^2 c_2(2) + a^3 c_1(9)) + \sin(2\theta)$$
$$(-a^2 a_1^2 a_2^2 c_2(6) + a^2 a_1^2 c_2(1) + a^2 c_1(6))$$
$$+ \sin(\theta)\cos(\theta)(a^2 c_1(5) - a^2 a_1^2 a_2^2 c_2(5)) + \sin(\theta)(ac_1(3) - aa_1^2 a_2^2 c_2(3)).$$

For $f_1(x, y) - f_2(x, y)$ to vanish along $x^2 + y^2 = a^2$, each coefficient of $\cos m\theta \sin n\theta$ must vanish. Therefore, a set of simultaneous equations for the unknowns, $c_1(i)$ and $c_2(i)$, can be obtained by equating each of the coefficients of the $\sin m\theta \cos n\theta$ terms to 0.

For the continuity condition for heat flux across the interface, $k_1 \frac{\partial f_1}{\partial n} - k_2 \frac{\partial f_2}{\partial n}$ must vanish along $x^2 + y^2 = a^2$. A similar procedure is possible to derive an additional set of simultaneous equations for $c_1(i)$ and $c_2(i)$.

For *Mathematica* coding, the following function, decomp, takes a polynomial on x and y, evaluates the polynomial along the circle, $x^2 + y^2 = r^2$, and isolates independent terms based on $\cos m\theta \sin n\theta$.

```
In[19]:= decomp[f_] := CoefficientList[TrigReduce[f /. {x -> r Cos[θ], y -> r Sin[θ]}] /.
         {Cos[i_. * θ] -> cosine^i, Sin[i_. * θ] -> sine^i}, {cosine, sine}] // Flatten
```

This is useful to derive a polynomial satisfying such conditions that its values or directional derivative vanish along a circle. By using this function, the following code automatically generates a set of permissible functions. It is assumed that the entire medium is defined as $\{(x, y), -a_1 < x < a_1, -a_2 < y < a_2\}$ and the radius of the circle is $r = a$.

```
In[96]:= terms = (order + 1) (order + 2) / 2;
        f1 = poly[order, c1];
        f2 = poly[order, c2] (x^2 - a1^2) (y^2 - a2^2);
        e1 = decomp[f1 - f2] /. r -> a;
        d[f_] := D[f, x] x / r + D[f, y] y / r
        e2 = decomp[(k1 d[f1] - k2 d[f2])] /. r -> a;
        eq1 = Map[# == 0 &, e1];
        eq2 = Map[# == 0 &, e2];
        sol = Solve[Flatten[{eq1, eq2}],
               Flatten[{Table[c1[i], {i, 1, terms}], Table[c2[i], {i, 1, terms}]}]][[1]]
        ff1 = f1 /. sol;
        ff2 = f2 /. sol;
        fff1 = Table[Coefficient[Expand[ff1], c1[i]], {i, terms}];
        fff2 = Table[Coefficient[Expand[ff2], c1[i]], {i, terms}];
        fff3 = Table[Coefficient[Expand[ff1], c2[i]], {i, terms}];
        fff4 = Table[Coefficient[Expand[ff2], c2[i]], {i, terms}];
        base1 = DeleteCases[{fff1, fff2} // Transpose, {0, 0}];
        base2 = DeleteCases[{fff3, fff4} // Transpose, {0, 0}];
        base = Join[base1, base2];
```

For instance, for the tenth-order polynomial, one of the permissible functions is as follows:

In[48]:= **base[[70]] // Simplify**

Out[48]= $\{\frac{1}{24\,k1\,(k1-k2)} a^2\,x\,y^2\,(-6\,a^8\,(7\,k1^2-20\,k1\,k2+13\,k2^2) +$
$a^6\,(k1-k2)\,(10\,a2^2\,(3\,k1-5\,k2) - 16\,a1^2\,k2 + 3\,(7\,k1-13\,k2)\,(3\,x^2+4\,y^2)) -$
$8\,(k1-k2)\,k2\,(x^2+y^2)\,(a2^2\,x^2\,(-x^2+y^2) + a1^2\,(x^4 - 4\,a2^2\,y^2 - x^2\,y^2)) -$
$2\,a^4\,(k1-k2)\,(a2^2\,(9\,k1\,x^2 - 21\,k2\,x^2 + 30\,k1\,y^2 - 50\,k2\,y^2) +$
$3\,a1^2\,(a2^2\,(5\,k1-9\,k2) - 4\,k2\,x^2 - 7\,k1\,y^2 + 3\,k2\,y^2) +$
$3\,y^2\,(-13\,k2\,(2\,x^2+y^2) + k1\,(16\,x^2+7\,y^2))) +$
$a^2\,(-3\,k1^2\,(x^2+y^2)\,(7\,x^4 - 10\,a2^2\,y^2 - 7\,x^2\,y^2) + 4\,k1\,k2\,(15\,x^6 - 20\,a2^2\,y^4 -$
$x^2\,(20\,a2^2\,y^2 + 11\,y^4)) + k2^2\,(-39\,x^6 + 50\,a2^2\,y^4 + x^2\,(34\,a2^2\,y^2 + 39\,y^4)) +$
$2\,a1^2\,(-y^2\,(15\,k1^2\,(x^2+y^2) - 24\,k1\,k2\,(x^2+y^2) + k2^2\,(9\,x^2+17\,y^2)) +$
$a2^2\,(9\,k1^2\,(x^2+y^2) - 4\,k1\,k2\,(9\,x^2+11\,y^2) + k2^2\,(27\,x^2+43\,y^2))))),$
$-\frac{1}{6\,(k1-k2)} x\,(a1^2-x^2)\,y^2\,(a2^2-y^2)\,(3\,k1\,(x^6+3\,a^4\,y^2 - 3\,x^2\,y^4 - 2\,y^6) +$
$k2\,(-3\,x^6 - 13\,a^4\,y^2 + 9\,x^2\,y^4 + 6\,y^6))\}$

It is possible that permissible functions for an elliptic inclusion can also be derived in a similar manner.

4.2.2 Finding Temperature Field Using Permissible Functions

By using the procedure described in the previous section to generate a set of permissible functions, the Galerkin method introduced in Section 4.1.1 can be applied to obtain the temperature field for inclusion problems.

The steady-state heat conduction equation with a heat source is expressed as

$$\nabla \cdot (k\nabla u) + q = 0, \quad (4.51)$$

where u is the unknown temperature, q is the given heat source, and k is the thermal conductivity, which is a function of the position. Following MWR in Section 4.1.1, the solution to Equation (4.51) can be assumed as

$$\tilde{u}_N(x) = \sum_{\alpha=1}^{N} u_\alpha f_\alpha(x), \quad (4.52)$$

where $\tilde{u}_N(x)$ is an N term approximation to $u(x)$, u_α is the unknown coefficient, and $f_\alpha(x)$ is the αth permissible function. By using the Galerkin method, the unknown coefficient vector, u_α, can be determined by solving a set of simultaneous equations as

$$\mathbf{Au} = \mathbf{d},$$

where **u** is a vector whose components are u_α ($\alpha = 1, 2, \ldots, N$), and A is a matrix whose components, a_{ij}, are expressed as

$$a_{\alpha\beta} = \int\int_{\text{whole}} \nabla \cdot (k\nabla f_\alpha) f_\beta dS$$

$$= \oint k\nabla f_\alpha f_\beta \mathbf{n} \, d\ell - \int\int_{\text{whole}} k\nabla f_\alpha \cdot \nabla f_\beta dS$$

$$= -\int\int_{\text{whole}} k\nabla f_\alpha \cdot \nabla f_\beta dS$$

$$= -\left\{\int\int_{\text{circle}} k_1 \nabla f_{\alpha_1} \cdot \nabla f_{\beta_1} dS + \int\int_{\text{matrix}} k_2 \nabla f_{\alpha_2} \cdot \nabla f_{\beta_2} dS\right\}$$

$$= -\left\{\left(k_1 \int\int_{\text{circle}} \nabla f_{\alpha_1} \cdot \nabla f_{\beta_1} dS - k_2 \int\int_{\text{circle}} \nabla f_{\alpha_2} \cdot \nabla f_{\beta_2} dS\right)\right.$$

$$\left. + k_2 \int\int_{\text{whole}} \nabla f_{\alpha_2} \cdot \nabla f_{\beta_2} dS\right\}. \tag{4.53}$$

where the term "whole," "circle," and "matrix" refer to the integration over the whole region, integration over the circle (inclusion) only, and integration over the matrix phase, respectively. The quantities, k_1 and k_2, refer to the thermal conductivity for the inclusion and the matrix, respectively. The functions, f_{α_1} and f_{α_2}, refer to the αth permissible function for the inclusion phase and for the matrix phase, respectively. Noting that the integration by parts is used, the line integral in the second line vanishes as f_β vanishes on the boundary.

The components of the vector **d** are expressed as

$$d_\alpha = \int\int_{\text{circle}} q f_\alpha dS. \tag{4.54}$$

To obtain $a_{\alpha\beta}$ and d_α, it is necessary to integrate polynomials over the rectangle or the circle. If the built-in `Integrate` function in *Mathematica* is used, it takes an extremely long time to perform such integrations, which makes the program execution impractical. This problem can be easily circumvented if an alternative to the `Integrate` function is used.

The following code integrates a polynomial on x and y over a circle of radius a.

```
intcircle[f_, a_]:= Module[{j1,j2},
j1=f r/.{x->r Cos[th], y->r Sin[th]}//TrigReduce//Expand;
j2=2 Pi j1/.{Cos[i_. th]->0,Sin[i_. th]->0}/.
{r^i_.->a^(i+1)/(i+1)}//Simplify;j2]
```

To integrate a polynomial over a rectangle $\{(x, y), -a \le x \le a, -b \le y \le b\}$, the following code can be used:

```
intrec[f_, a_, b_] := Module[{coeff, ii, jj, i, j},
  coeff = CoefficientList[f, {x, y}];
  {ii, jj} = Dimensions[coeff];
  Sum[coeff[[i, j]] (a^i - (-a)^i)/i (b^j - (-b)^j)/j ,
    {i, 1, ii}, {j, 1, jj}]]
```

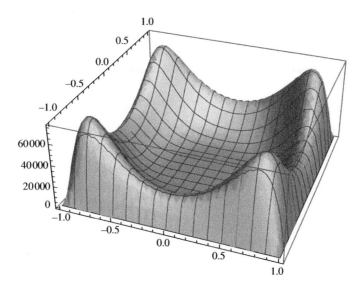

Figure 4.8 First base function

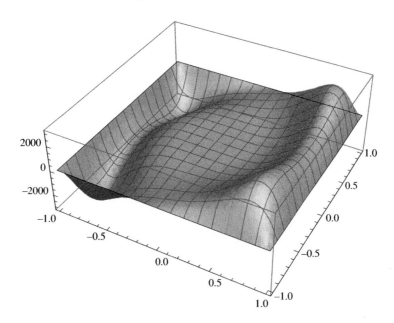

Figure 4.9 Second base function

It should be noted that this code is much faster than the Integrate command by a whopping 2000 times for the following example function:

```
f = Sum[ Random[Integer, {-15, 15}] x^i y^j, {i, 0, 25}, {j, 0, 32}];
j1 = Timing[intrec[f, 1, 1]]
j2 = Timing[Integrate[f, {x, -1, 1}, {y, -1, 1}]]
j2[[1]]/j1[[1]]
```

The execution time for $j1$ was 0.016 seconds and the time for $j2$ was 32.39 seconds (2000 times).

Figures 4.8–4.10 represent the first three permissible functions.

Figure 4.11 is the temperature distribution based on the following geometry:

$$k_1 = 10, \quad k_2 = 1, a = \frac{1}{3}.$$

Taking sixth-order polynomials yielded convergence.

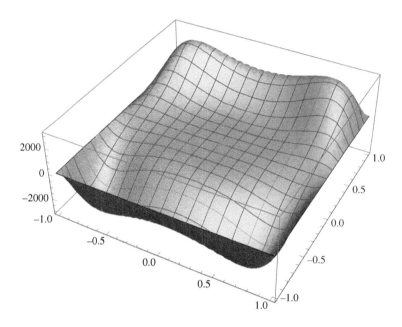

Figure 4.10 Third base function

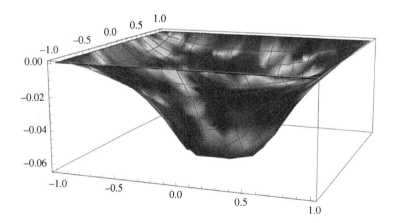

Figure 4.11 Temperature distribution

The *Mathematica* code to generate the base functions and the temperature distribution is as follows:

```
In[2]:= a = 1 / 3; k1 = 10; k2 = 1; order = 6;
    intcircle[f_, a_] := Module[{j1, j2},
      j1 = f r /. {x → r Cos[th], y → r Sin[th]} // TrigReduce // Expand;
      j2 = 2 Pi j1 /. {Cos[i_. th] → 0, Sin[i_. th] → 0} /.
        {r^i_. → a^(i+1) / (i+1)} // Simplify;
      j2]

    intrec[f_, a_, b_] :=
      If[f === 0, 0, Module[{coeff, ii, jj, i, j}, coeff = CoefficientList[f, {x, y}];
        {ii, jj} = Dimensions[coeff];
        Sum[
          coeff[[i, j]] (a^i - (-a)^i) / i (b^j - (-b)^j) / j, {i, 1, ii}, {j, 1, jj}]]]
In[8]:=
    laplace[f_] := D[f, {x, 2}] + D[f, {y, 2}];
    divdel[f_, g_] := D[f, x] D[g, x] + D[f, y] D[g, y];

    poly[n_, a_] := Sum[a[(i+j) (i+j+1) / 2 + i + 1] x^j y^i, {i, 0, n}, {j, 0, n-i}]
    decomp[f_] := CoefficientList[
      TrigReduce[f /. {x → r Cos[th], y → r Sin[th]}] /. {Cos[i_. * th] → cosine^i,
        Sin[i_. * th] → sine^i}, {cosine, sine}] // Flatten
In[9]:= terms = (order + 1) (order + 2) / 2;
    f1 = poly[order, c1];
    f2 = poly[order, c2] (x^2 - 1) (y^2 - 1);
    e1 = decomp[f1 - f2] /. r → a;
    d[f_] := D[f, x] x / r + D[f, y] y / r
    e2 = decomp[(k1 d[f1] - k2 d[f2])] /. r → a;
    eq1 = Map[# == 0 &, e1];
    eq2 = Map[# == 0 &, e2];
    sol = Solve[Flatten[{eq1, eq2}],
      Flatten[{Table[c1[i], {i, 1, terms}], Table[c2[i], {i, 1, terms}]}]][[1]]
    ff1 = f1 /. sol;
    ff2 = f2 /. sol;
    fff1 = Table[Coefficient[Expand[ff1], c1[i]], {i, terms}];
    fff2 = Table[Coefficient[Expand[ff2], c1[i]], {i, terms}];
    fff3 = Table[Coefficient[Expand[ff1], c2[i]], {i, terms}];
    fff4 = Table[Coefficient[Expand[ff2], c2[i]], {i, terms}];
    base1 = DeleteCases[{fff1, fff2} // Transpose, {0, 0}];
    base2 = DeleteCases[{fff3, fff4} // Transpose, {0, 0}];
    base = Join[base1, base2];
In[27]:= basefunc[i_, x_, y_] :=
      If[x^2 + y^2 < a^2, Evaluate[base[[i, 1]]], Evaluate[base[[i, 2]]]]

    max = Length[base];

    amat = Table[0, {i, max}, {j, max}];
    Do[amat[[i, j]] = -(intcircle[k1 divdel[base[[i, 1]], base[[j, 1]]] -
          k2 divdel[base[[i, 2]], base[[j, 2]]], a] + k2 intrec[
          divdel[base[[i, 2]], base[[j, 2]]], 1, 1]), {i, 1, max}, {j, i, max}]
    Do[amat[[j, i]] = amat[[i, j]], {i, 1, max}, {j, i+1, max}];
In[32]:= bvec = Table[0, {i, max}];
    Do[bvec[[i]] = intcircle[base[[i, 1]], a], {i, 1, max}]
    coeff = Inverse[amat].bvec;
    inclusion = Sum[coeff[[i]] base[[i, 1]], {i, 1, max}];
    matrix = Sum[coeff[[i]] base[[i, 2]], {i, 1, max}];
    temperature[x_, y_] := If[x^2 + y^2 < a^2, inclusion, matrix]
    Plot3D[temperature[x, y], {x, -1, 1}, {y, -1, 1}]
```

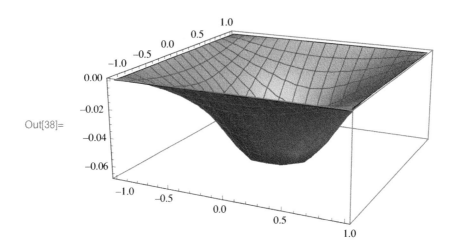
Out[38]=

4.3 Elastic Fields with Bounded Boundaries

In the previous section, the Galerkin method was used to derive a semianalytical expression for the solution of the elliptic partial differential equations (steady-state heat conduction equations) using a set of permissible functions satisfying both the homogeneous boundary conditions and associated continuity conditions.

It is, therefore, natural to presume that the same approach can be carried over to static elasticity as the characteristics for the steady-state heat conduction equation and the static displacement equilibrium equation in linear elasticity are similar. In steady-state heat conduction, the temperature is a scalar and the thermal conductivity is a second-rank tensor, while in static elasticity, the displacement is a first-rank tensor and the elastic constant is a fourth-rank tensor. The steady-state heat conduction equation is expressed as

$$(k_{ij}T_{,j})_{,i} = 0.$$

The static displacement equilibrium equation is expressed as

$$(C_{ijkl}u_{k,l})_{,j} = 0.$$

Therefore, the following correspondence can be established.

$$T \leftrightarrow u_i,$$

$$C_{ijkl} \leftrightarrow k_{ij}.$$

Most analytical methods that are developed for the steady-state heat conduction equation can be employed for the static equilibrium equation in elasticity by using the aforementioned correspondence. However, the amount of algebra involved in elasticity problems is much larger than in the steady-state heat conduction problems for the same geometry. Similar to the steady-state heat conduction problems, it is assumed that an inclusion is surrounded by a finite-sized matrix with different elastic moduli subject to the homogeneous boundary condition.

We restrict ourselves to 2-D analysis although it is possible to extend the procedure for 3-D media. The permissible functions for 2-D elasticity must satisfy the homogeneous boundary condition and the continuity conditions across the phase boundary for the displacement and traction force. The permissible functions are separately defined in the inclusion and matrix phases in both the x and y directions. Imposing the homogeneous boundary conditions and continuity conditions for the displacement and the traction force across the phases results in a set of underdetermined linear equations for the unknown coefficients, which can be solved by *Mathematica*.

First, a polynomial of the βth order, $h(\alpha, \beta)$, is defined as

$$h(\alpha, \beta) = x^\alpha y^{\beta-\alpha}.$$

The permissible function f_i^A in the phase A is defined as

$$f_i^A(x, y) = \sum_{\beta=0}^{n} \sum_{\alpha=0}^{\beta} d_i^{A\alpha\beta} h(\alpha, \beta),$$

where A refers to each distinct phase (inclusion and matrix phases) and i refers to either x or y.

The unknown coefficients, $d_i^{A\alpha\beta}$, can be determined to satisfy the following boundary conditions:

$$f_i^A - f_i^B = 0,$$
$$t_i^A - t_i^B = 0,$$

where t_i^A is the surface traction, which is related to the permissible function f_i by

$$t_i = \sigma_{ij} n_j$$
$$= C_{ijkl} f_{k,l} n_j$$
$$= (\mu(\delta_{ik}\delta_{jl} + \delta_{il}\delta_{jk}) + \lambda \delta_{ij}\delta_{kl}) f_{k,l} n_j$$
$$= \mu f_{i,j} n_j + \mu f_{j,i} n_j + \lambda f_{l,l} n_i.$$

The following *Mathematica* code generates permissible functions that satisfy the homogeneous boundary condition and the continuity conditions across the two phases. The geometry is the same as in the previous section, i.e., a circular inclusion is embedded in a square-shaped matrix.

In[1]:=

```
mat = {μm → 1, μf → 4, λm → 2, λf → 20}; mat2 = {μm → 1.0, μf → 4.0, λm → 2.0, λf → 20.0, a → 4

poly[n_] := Table[x^j y^i, {i, 0, n}, {j, 0, n-1}] // Flatten
order = 5; (*min=4*) terms = (order + 1) (order + 2) / 2;
g = poly[order];
(*
ff is the permissible function in the x direction in
 the inclusion. fm is the permissbiel function in the matrix.
  gf is the permissible function in the y direction in the inclusion.gm
  is the permissible function in the y direction in the matrix.
*)
ff = Sum[cf[α] g[[α]], {α, 1, terms}];
fm = Sum[cm[α] g[[α]], {α, 1, terms}] (x^2 - a^2) (y^2 - a^2);
gf = Sum[df[α] g[[α]], {α, 1, terms}];
gm = Sum[dm[α] g[[α]], {α, 1, terms}] (x^2 - a^2) (y^2 - a^2);
(*
tf[i]is traction inside inclusion, tm[i] is traction inside matrix
```

$\sigma_{ij} = 2\mu\varepsilon_{ij} + \lambda\delta_{ij}\varepsilon_{kk} = \mu u_{i,j} + \mu u_{j,i} + \lambda\delta_{ij}u_{k,k}$

$t_i = \sigma_{ij}n_j = \mu u_{i,j}n_j + \mu u_{j,i}n_i + \lambda u_{k,k}n_i$

```
*)
a = 0;
n = {(x - a), y}; xx = {x, y};
(* traction in x direction *)
txf = (2 μf + λf) D[ff, x] n[[1]] + μf (D[ff, y] n[[2]] + D[gf, x] n[[2]]) + λf D[gf, y] n[[1]]
txm = (2 μm + λm) D[fm, x] n[[1]] + μm (D[fm, y] n[[2]] + D[gm, x] n[[2]]) + λm D[gm, y] n[[1]]
(* traction in y direction *)
tyf = (2 μf + λf) D[gf, y] n[[2]] + μf (D[gf, x] n[[1]] + D[ff, y] n[[1]]) + λf D[ff, x] n[[2]]
tym = (2 μm + λm) D[gm, y] n[[2]] + μm (D[gm, x] n[[1]] + D[fm, y] n[[1]]) + λm D[fm, x] n[[2]]
(**)
eq1 = DeleteCases[
   CoefficientList[TrigReduce[(ff - fm) /. {x → a + r Cos[th], y → r Sin[th]} /. r → 1] /.
    {Cos[i_. th] → cosine^i, Sin[j_. th] → sine^j}, {cosine, sine}] // Flatten, 0];
eq2 = DeleteCases[CoefficientList[TrigReduce[
     (gf - gm) /. {x → a + r Cos[th], y → r Sin[th]} /. r → 1] /.
    {Cos[i_. th] → cosine^i, Sin[j_. th] → sine^j}, {cosine, sine}] // Flatten, 0];
eq3 = DeleteCases[CoefficientList[TrigReduce[
     (txf - txm) /. {x → a + r Cos[th], y → r Sin[th]} /. r → 1] /.
    {Cos[i_. th] → cosine^i, Sin[j_. th] → sine^j}, {cosine, sine}] // Flatten, 0];
eq4 = DeleteCases[CoefficientList[TrigReduce[
     (tyf - tym) /. {x → a + r Cos[th], y → r Sin[th]} /. r → 1] /.
    {Cos[i_. th] → cosine^i, Sin[j_. th] → sine^j}, {cosine, sine}] // Flatten, 0];
eqall = Flatten[{eq1, eq2, eq3, eq4}](*/.mat*);
eqall = Map[#1 == 0 &, eqall];
varall = {Table[cf[j], {j, 1, terms}], Table[cm[j], {j, 1, terms}],
    Table[df[j], {j, 1, terms}], Table[dm[j], {j, 1, terms}]} // Flatten;
sol = Solve[eqall, varall][[1]];
Fm = Table[Coefficient[fm /. sol, varall[[i]]], {i, 1, 4*terms}];
Ff = Table[Coefficient[ff /. sol, varall[[i]]], {i, 1, 4*terms}];
Gm = Table[Coefficient[gm /. sol, varall[[i]]], {i, 1, 4*terms}];
Gf = Table[Coefficient[gf /. sol, varall[[i]]], {i, 1, 4*terms}];
base1 = Transpose[{Ff, Fm}]; (* base functions in the x1 direction *);
base2 = Transpose[{Gf, Gm}]; (* base functions in the x2 direction *);
j1 = Transpose[{base1, base2}];
(*base[[i]] is the i-th trial function. base[[i,1]] is the i-
  th permissible function in the x direction,
base[[i,2]] is the i-th permissible function in the y-direction.*)
(*base[[i,1,1]] is the i-th permissible function in the x-
  direction in the inclusion. base[[i,1,2]] is the i-
  th permissible function in the x-direction in the matrix*)
base = DeleteCases[j1, {{0, 0}, {0, 0}}];
```

Inclusions in Finite Matrix

As the full expressions of the derived permissible functions are rather lengthy, the Short command can be used to display a partial expression as

In[37]:= **Short[base // Simplify, 20]**

Out[37]//Short= $\{\{\{y(-1+x^2+y^2)^2, 0\}, \{0, 0\}\}, \{\{\frac{1}{2}(-1+x^2+y^2)^2, 0\}, \{0, 0\}\},$

$\{\{\frac{1}{2}x(-1+x^2+y^2)^2, 0\}, \{0, 0\}\}, \{\{(x(\lambda m+2\mu m)(2y^2\mu f-(-1+x^2)^2\mu m+y^4(-2\mu f+\mu m)))/$
$(2(\mu f(-3\lambda m+\mu f-\mu m)+\lambda f(\mu f+2\mu m))),$
$x^3 y^2 (x^2+y^2+(-\lambda f(\mu f+2\mu m)+\mu f(4\lambda m-\mu f+3\mu m))/(\mu f(-3\lambda m+\mu f-\mu m)+\lambda f(\mu f+2\mu m)))\},$
$\{(y(\lambda m+2\mu m)(((-1+y^2)^2+x^2(-1+2y^2))\mu f-(-1+y^2)(-1+x^2+y^2)\mu m))/$
$(\mu f(-3\lambda m+\mu f-\mu m)+\lambda f(\mu f+2\mu m)), x^2 y^3$
$(x^2+y^2+(-\lambda f(\mu f+2\mu m)+\mu f(4\lambda m-\mu f+3\mu m))/(\mu f(-3\lambda m+\mu f-\mu m)+\lambda f(\mu f+2\mu m)))\}\},$

$\{\{0, x^2 y^2(-1+x^2+y^2)^2\}, \{0, 0\}\}, \ll 6\gg, \{\{0, 0\}, \{\frac{1}{2}(-1+x^2+y^2)^2, 0\}\},$

$\{\{0, 0\}, \{\frac{1}{2}x(-1+x^2+y^2)^2, 0\}\},$

$\{\{0, 0\}, \{0, x^2 y^2(-1+x^2+y^2)^2\}\},$

$\{\{0, 0\}, \{0, \frac{1}{2}x^3 y^2(-1+x^2+y^2)^2\}\},$

$\{\{0, 0\}, \{0, \frac{1}{2}x^2 y^3(-1+x^2+y^2)^2\}\}\}$

For specific material and geometrical constants, the full expression of the permissible functions is manageable as

In[35]:= **base = base /. mat;**
Short[base, 10]

Out[36]//Short= $\{\{\{y-2x^2 y+x^4 y-2y^3+2x^2 y^3+y^5, 0\}, \{0, 0\}\},$

$\{\{\frac{1}{2}-x^2+\frac{x^4}{2}-y^2+x^2 y^2+\frac{y^4}{2}, 0\}, \{0, 0\}\},$

$\{\{\frac{x}{2}-x^3+\frac{x^5}{2}-x y^2+x^3 y^2+\frac{x y^4}{2}, 0\}, \{0, 0\}\},$

$\{\{0, (-a^2+x^2)(-a^2+y^2)(1-2x^2+x^4-2y^2+2x^2 y^2+y^4)\}, \{0, 0\}\},$

$\{\{-\frac{(72-228 a^2+200 a^4) y}{1248}-\frac{(-72+340 a^2-312 a^4) x^2 y}{1248}-$
$\frac{(-144+340 a^2-312 a^4) y^3}{1248}+\frac{5x^2 y^3}{156}-\frac{3 y^5}{52},$
$(-a^2+x^2)(-a^2+y^2)(-\frac{71 y}{78}+x^2 y+y^3)\},$

$\{-\frac{(-184+456 a^2-400 a^4) x}{2496}-\frac{(184-340 a^2+312 a^4) x^3}{1248}+\frac{23 x^5}{312}-$
$\frac{(112-340 a^2+312 a^4) x y^2}{1248}+\frac{5 x y^4}{312}, (-a^2+x^2)(-a^2+y^2)(\frac{71 x}{78}-x^3-x y^2)\}\},$

$\ll 7\gg, \{\{0, 0\}, \{\frac{x}{2}-x^3+\frac{x^5}{2}-x y^2+x^3 y^2+\frac{x y^4}{2}, 0\}\},$

$\{\{0, 0\}, \{0, (-a^2+x^2)(-a^2+y^2)(\frac{x}{2}-x^3+\frac{x^5}{2}-x y^2+x^3 y^2+\frac{x y^4}{2})\}\},$

$\{\{0, 0\}, \{0, (-a^2+x^2)(-a^2+y^2)(\frac{y}{2}-x^2 y+\frac{x^4 y}{2}-y^3+x^2 y^3+\frac{y^5}{2})\}\},$

$\{\{0, 0\}, \{0, (-a^2+x^2)(-a^2+y^2)(1-2x^2+x^4-2y^2+2x^2 y^2+y^4)\}\}\}$

By using the permissible functions derived, the displacement fields in 2-D with body force must satisfy the displacement equilibrium equations as

$$\mu \Delta u + (\mu + \lambda)(u_{,xx} + v_{,xy}) + b_x = 0, \quad (4.55)$$

$$\mu \Delta v + (\mu + \lambda)(u_{,xy} + v_{,yy}) + b_y = 0, \quad (4.56)$$

where u and v are the displacements in the x and y directions, respectively. Approximate solutions to Equations (4.55) and (4.56) are sought as a linear combination of the permissible functions, $f_\alpha(x, y)$ and $g_\alpha(x, y)$, that satisfy the displacement and traction continuity conditions at the phase interface as

$$\tilde{u} = \sum_{\alpha=1}^{m} c_\alpha f_\alpha(x, y), \quad (4.57)$$

$$\tilde{v} = \sum_{\alpha=1}^{n} d_\alpha g_\alpha(x, y), \quad (4.58)$$

where $\tilde{u}$ and $\tilde{v}$ are approximation to u and v, respectively. Substitution of Equations (4.57) and (4.58) into Equations (4.55) and (4.56) yields the following equation for c_α and d_α.

$$\mu \sum_{\alpha=1}^{m} c_\alpha \Delta f_\alpha + (\mu + \lambda) \sum_{\alpha=1}^{m} c_\alpha f_{\alpha,xx} + (\mu + \lambda) \sum_{\alpha=1}^{n} d_\alpha g_{\alpha,xy} + b_x = 0, \quad (4.59)$$

$$\mu \sum_{\alpha=1}^{n} d_\alpha \Delta g_\alpha + (\mu + \lambda) \sum_{\alpha=1}^{m} c_\alpha f_{\alpha,xy} + (\mu + \lambda) \sum_{\alpha=1}^{n} d_\alpha g_{\alpha,yy} + b_y = 0. \quad (4.60)$$

As seen from Equations (4.59) and (4.60), the displacement fields are coupled, and hence, in order to determine the unknown coefficients, c_α and d_α, using the Galerkin method, Equation (4.59) must be multiplied by f_β and g_β separately and then integrated over the whole material points. The same applies to Equation (4.60) as well. This generates $2(m+n)$ equations for $m+n$ unknowns; hence, suggesting an overspecified system of equations where the solution may not exist. Thus, the Galerkin method fails to produce a solution for the elasticity equilibrium equation.

As an alternative to the Galerkin method, the Rayleigh–Ritz method can be used, which is also an approximation technique that reduces a continuous system to a discretized model. The Rayleigh–Ritz method is a widely used approximation technique that is based on the principle of minimum potential energy, which states that for conservative systems, of all the kinematically admissible displacement fields, those corresponding to equilibrium extremize the total potential energy, I, and if the extremum condition is minimum, the equilibrium state is stable. In other words, the displacement field that extremizes I is the one that satisfies the equilibrium equation. The Galerkin method and the Rayleigh–Ritz method usually yield the same set of equations. However, this is an example where the Rayleigh–Ritz method works properly while the Galerkin method does not.

The potential energy, I, for static elasticity with body force is defined as

$$I = \int L\,dx\,dy$$
$$= \int \left(\frac{1}{2}\sigma_{ij}\epsilon_{ij} - b_i u_i\right) dx\,dy$$
$$= \int \left(\frac{1}{2}C_{ijkl}u_{i,j}u_{k,l} - b_i u_i\right) dx\,dy, \qquad (4.61)$$

where L represents the Lagrangian.

By extremizing Equation (4.61), the Euler–Lagrange equation is derived as

$$\frac{\partial L}{\partial u_k} = \left(\frac{\partial L}{\partial u_{k,l}}\right)_{,l},$$

which is equivalent to

$$(C_{ijkl}u_{i,j})_{,l} + b_k = 0.$$

The strain energy is expressed using Equations (4.57) and (4.58) as

$$C_{ijkl}u_{k,l}u_{i,j} = \mu(u_{i,j}u_{i,j} + u_{i,j}u_{j,i}) + \lambda u_{k,k}u_{l,l}$$
$$= (2\mu + \lambda)(u_{,x}^2 + v_{,y}^2) + \mu(u_{,y}^2 + v_{,x}^2 + 2u_{,y}v_{,x}) + 2\lambda u_{,x}v_{,y}$$
$$= (2\mu + \lambda)\left(\sum_{\alpha,\beta} c_\alpha c_\beta f_{\alpha,x} f_{\beta,x} + \sum_{\alpha,\beta} d_\alpha d_\beta g_{\alpha,y} g_{\beta,y}\right)$$
$$+ \mu\left(\sum_{\alpha,\beta} c_\alpha c_\beta f_{\alpha,y} f_{\beta,y} + \sum_{\alpha,\beta} d_\alpha d_\beta g_{\alpha,x} g_{\beta,x} + 2\sum_{\alpha,\beta} c_\alpha d_\beta f_{\alpha,y} g_{\beta,x}\right)$$
$$+ 2\lambda \sum_{\alpha,\beta} c_\alpha d_\beta f_{\alpha,x} g_{\beta,y}. \qquad (4.62)$$

Differentiating Equation (4.62) with respect to c_α and d_α yields

$$(2\mu + \lambda)f_{\alpha,x}f_{\beta,x}c_\beta + \mu(f_{\alpha,y}f_{\beta,y}c_\beta + f_{\alpha,y}g_{\beta,x}d_\beta) + \lambda f_{\alpha,x}g_{\beta,y}d_\beta = b_x f_\alpha,$$
$$(2\mu + \lambda)g_{\alpha,y}g_{\beta,y}d_\beta + \mu(g_{\alpha,x}g_{\beta,x}d_\beta + f_{\alpha,y}g_{\beta,x}c_\beta) + \lambda f_{\alpha,x}g_{\beta,y}c_\beta = b_y g_\alpha.$$

or

$$\mathbf{Ac} + \mathbf{Bd} = \int\int b_x f_\alpha(x,y)\,dx\,dy, \qquad (4.63)$$
$$\mathbf{Bc} + \mathbf{Dd} = \int\int b_y g_\alpha(x,y)\,dx\,dy, \qquad (4.64)$$

where

$$A_{\alpha\beta} = \iint (2\mu + \lambda) f_{\alpha,x} f_{\beta,x} dx dy + \iint \mu f_{\alpha,y} f_{\beta,y} dx dy,$$

$$B_{\alpha\beta} = \iint \mu f_{\alpha,y} g_{\beta,x} dx dy + \iint \lambda f_{\alpha,x} g_{\beta,y} dx dy,$$

$$D_{\alpha\beta} = \iint (2\mu + \lambda) g_{\alpha,y} g_{\beta,y} dx dy + \iint \mu g_{\alpha,x} g_{\beta,x} dx dy.$$

Once the matrices, $A_{\alpha\beta}$, $B_{\alpha\beta}$, and $D_{\alpha\beta}$ are computed, the unknown coefficients, **c** and **d**, can be solved from Equations (4.63) and (4.64).

4.4 Numerical Examples

In this section, numerical examples based on the formulations developed in the previous section are shown for (1) a homogeneous medium and (2) a medium with one inclusion. For simplicity, only the Dirichlet-type homogeneous boundary condition where the displacement is prescribed on the boundary is covered although other boundary conditions such as the Neumann condition can be handled in a similar manner.

All the *Mathematica* code used for the calculation in this section is available from the companion web page as they are too lengthy to be printed.

4.4.1 Homogeneous Medium

A 2-D elastic medium shown in Figure 4.12 is considered. It is square shaped defined by $\{(x, y), -d < x < d, -d < y < d\}$ subject to a uniform body force, $\mathbf{b} = (b_x, b_y)$ throughout the body, and the Dirichlet-type boundary condition is imposed in which the displacement vanishes on the boundary. The body is isotropic and there is no inclusion. The following polynomial function of the nth order can be used as a permissible function as

$$f_n(x, y) = \sum_{i=0}^{n} c_i^n x^{n-i} y^i (x^2 - d^2)(y^2 - d^2).$$

This function, $f_n(x, y)$, consists of $n + 1$ terms that cover all the polynomials from the zeroth order to the nth order. The coefficient c_i^n needs to be determined by the Galerkin method. In Figure 4.12, the values for the geometry (d) and the material properties (μ and λ) were chosen at will as

$$d = 4, \quad \mu = \lambda = 1, \quad b_x = b_y = 1.$$

By using those values and $f_n(x, y)$, Equations (4.63) and (4.64) were solved. A tenth-order polynomial turned out to be sufficient to provide a more than reasonable convergence.

Figures 4.13 and 4.14 show the cross-sectional displacement profiles for $u(x, y)$ and $v(x, y)$ along the x and y axes, respectively, and their comparison with the respective finite element (FEM) solutions.

Figures 4.15 and 4.16 depict a 3-D view of the displacement, u_x and u_y.

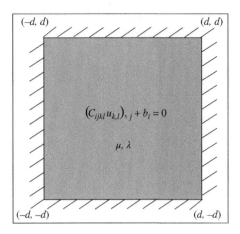

Figure 4.12 Homogeneous medium

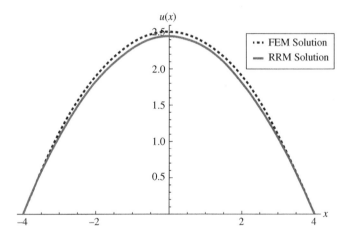

Figure 4.13 Comparison between the Rayleigh–Ritz solution and an FEM solution for the x-component of displacement in a homogeneous medium

The displacement contour plots for $u(x, y)$ and $v(x, y)$ are also presented as in Figures 4.17 and 4.18.

The results from Figures 4.13–4.18 conform to a tenth-order approximation. The FEM model was constructed in ANSYS using the PLANE183 elements characterized by their two-dimensional geometry with eight nodes. The material constants, namely, the Young modulus, E, and the Poisson ratio, ν, were calculated from the Lamé constants to be $E = 2.5$ and $\nu = 0.25$ corresponding to $\mu = 1$ and $\lambda = 1$. In order to accommodate the application of body forces, the system was modeled as a plane stress representation with the thickness such that the volume of the system ($8 \times 8 \times 0.015625$) equals one unit. The density, ρ, and the acceleration due to gravity, g, were also chosen to be one unit each, so that the body force ($\rho \times V \times g$) equals one unit, as chosen for the present model. Hence, the FEM model, so

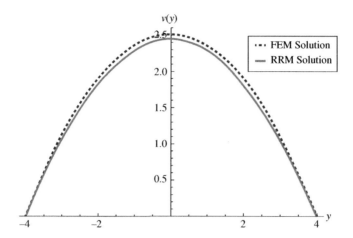

Figure 4.14 Comparison between the Rayleigh–Ritz solution and an FEM solution for the y-component of displacement in a homogeneous medium

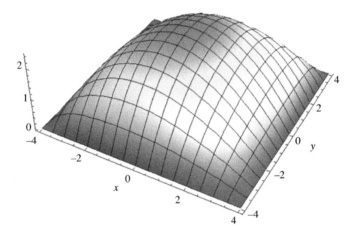

Figure 4.15 3-D profile for $u(x, y)$ in the homogeneous medium

constructed, only serves as a reasonable approximation to the original 2-D stress equilibrium equation. This explains the slim disparity in the solutions obtained from the FEM model to that procured from the present solution; and the same trend will be observed for the heterogeneous cases as well. However, it should be noted that the FEM solutions only serve to corroborate the present solutions in terms of the behavior of the displacement profiles, which can be noticed from the cross-sectional plots depicted next, to be consistent with those obtained from the semianalytical model and do not serve as the ultimate basis of comparison.

4.4.2 Single Inclusion

An elastic medium that contains a single inclusion is shown in Figure 4.19. The geometry constitutes a centrally located elliptical inclusion (with the Lamé constants of μ_1 and λ_1) embedded

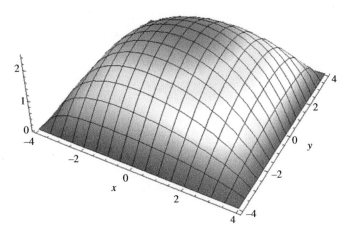

Figure 4.16 3-D profile for $v(x, y)$ in the homogeneous medium

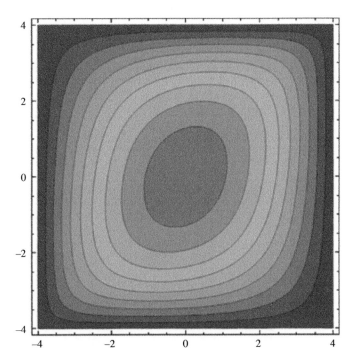

Figure 4.17 Contour plot of $u(x, y)$ in the homogeneous medium

in a square-shaped matrix medium (with the Lamé constants of μ_2 and λ_2) subjected to the zero displacement boundary condition (the homogeneous Dirichlet boundary condition).

The displacement solutions were obtained for the single elliptical inclusion problem with the material constants arbitrarily chosen as $\mu_1 = 30$, $\lambda_1 = 40$, $\mu_2 = 1$, and $\lambda_2 = 1$. The size of the elliptical inclusion was chosen as $a = 1$ and $b = 1.5$ inside a square-shaped matrix

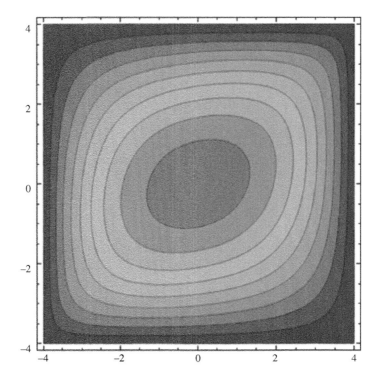

Figure 4.18 Contour plot of $v(x, y)$ in the homogeneous medium

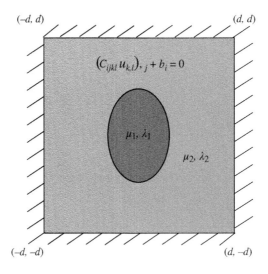

Figure 4.19 A medium with a single inclusion

characterized by $d = 4$. Also, the body forces in the x and y directions were chosen to be $b_x = b_y = 1$. Furthermore, the results for a single circular inclusion problem are also presented retaining the same material and geometrical parameters as indicated with the exception that $b = 1$.

Figures 4.20 and 4.21 depict the cross-sectional displacement plots for $u(x, y)$ and $v(x, y)$ along the x-axis and y-axis, respectively, for the single elliptical inclusion problem. The corresponding FEM solutions are also provided on the same plots.

Figures 4.22–4.25 present the 3-D displacement profiles and the displacement contour plots for the same problem.

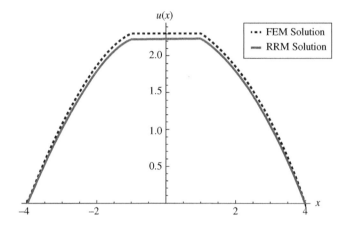

Figure 4.20 Comparison between the Rayleigh–Ritz method and FEM solution for the x-component of displacement in the single inclusion problem

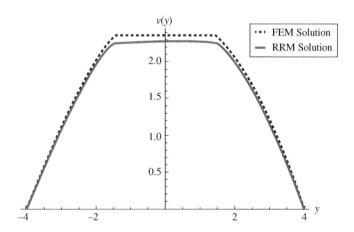

Figure 4.21 Comparison between the Rayleigh–Ritz method and FEM solution for the y-component of displacement in the single inclusion problem

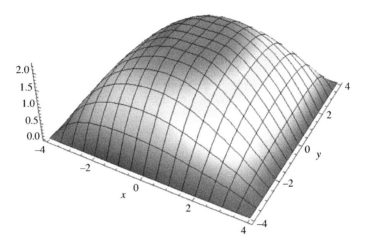

Figure 4.22 3-D profile of the x-component of displacement for the single inclusion problem

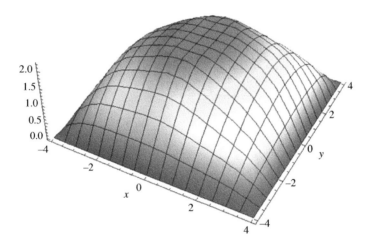

Figure 4.23 3-D profile of the y-component of displacement for the single inclusion problem

Figures 4.26–4.31 demonstrate a similar study of the displacement fields for the single circular inclusion problems.

It can be observed that the present semianalytical solution is well corroborated by the purely numerical solution, for both types of problems. Figures 4.32 and 4.33 show the effect of varying inclusion surface areas on the x and y displacement components for the single inclusion problem.

A similar study was carried out to investigate the effect of varying aspect ratios of the inclusion and material constants of the individual phases on the displacement fields. In the first instance (Figures 4.34 and 4.35), the aspect ratios of the inclusion are varied while retaining a constant inclusion surface area.

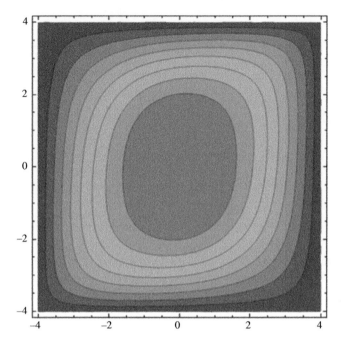

Figure 4.24 Contour plot of the *x*-component of displacement for the single inclusion problem

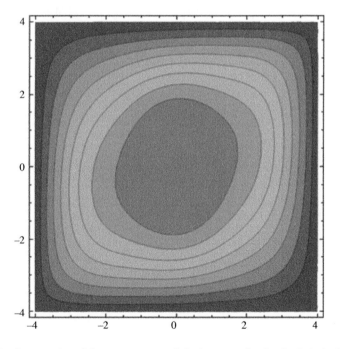

Figure 4.25 Contour plot of the *y*-component of displacement for the single inclusion problem

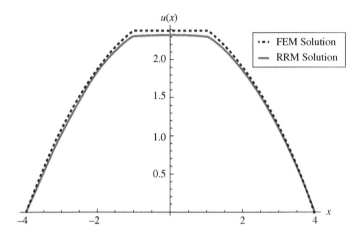

Figure 4.26 Comparison between the Rayleigh–Ritz method and an FEM solution for the x-component of displacement in the single inclusion problem

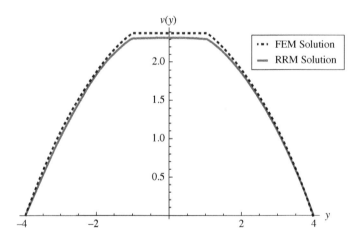

Figure 4.27 Comparison between the Rayleigh–Ritz method and an FEM solution for the y-component of displacement in the single inclusion problem

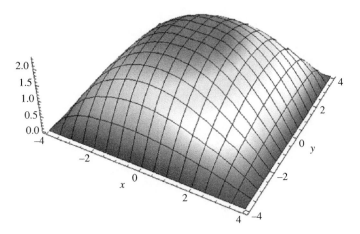

Figure 4.28 3-D profile of the x-component of displacement for the single inclusion problem

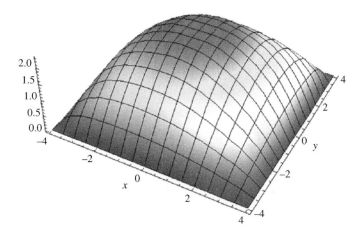

Figure 4.29 3-D profile of the y-component of displacement for the single inclusion problem

It is interesting to observe that as the aspect ratios become larger, the displacements become smaller irrespective of the inclusion area being a constant. Also, it can be seen that the decrease in displacement is more pronounced in $u(x)$ than in $v(y)$. This can be attributed to the geometry and orientation of the elliptical inclusion, i.e., $a > b$ and also that a lies along the x axis. In the second case (Figures 4.36 and 4.37), while retaining the same geometrical parameters throughout, only the material properties of the inclusion and the matrix are varied. Here, a noticeable feature is the profound effect on the displacement field due to the change in the material constants of the matrix phase, compared to the change in the material properties of the inclusion.

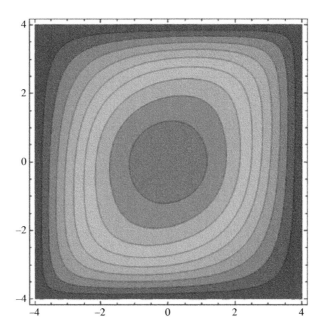

Figure 4.30 Contour plot of the x-component of displacement for the single inclusion problem

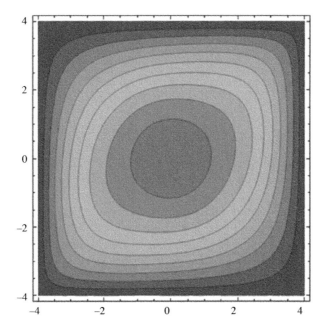

Figure 4.31 Contour plot of the y-component of displacement for the single inclusion problem

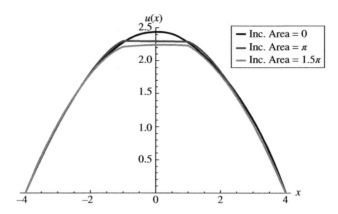

Figure 4.32 Effect of varying inclusion surface areas on the x-component of displacement for the single inclusion problem

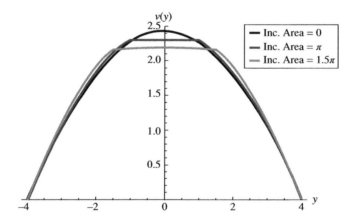

Figure 4.33 Effect of varying inclusion surface areas on the y-component of displacement for the single inclusion problem

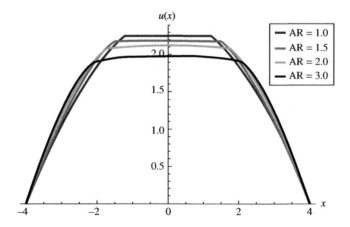

Figure 4.34 Effect of varying aspect ratios of the inclusion on the x-component of displacement for the single inclusion problem

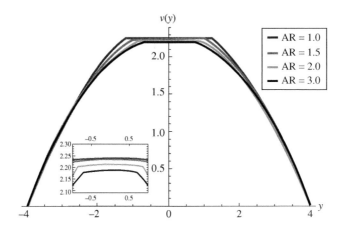

Figure 4.35 Effect of varying aspect ratios of the constituent phases on the y-component of displacement for the single inclusion problem

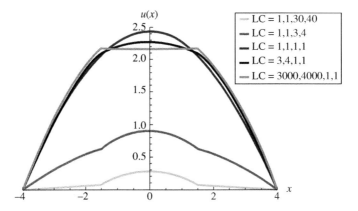

Figure 4.36 Effect of varying material constants of the constituent phases on the x-component of displacement for the single inclusion problem

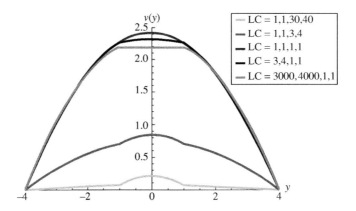

Figure 4.37 Effect of varying material constants of the constituent phases on the y-component of displacement for the single inclusion problem

4.5 Exercises

1. Derive a two-term approximate solution to the following differential equation:
$$-u'' = \cos x, \quad u(0) = 0, \quad u(1) = 0,$$
using
$$e_1(x) = x(x-1), \quad e_2(x) = x^2(x-1),$$
as the permissible functions.
   (a) The collocation method.
   (b) The least squares method.
   (c) The Galerkin method.
   (d) Draw three curves obtained from (a) to (c) for $0 < x < 1$ along with the exact solution in the same graph.

2. (a) Solve the following Laplace equation using the eigenfunction expansion method.

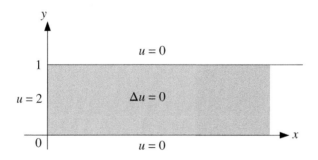

   (b) Plot the profile of $u(x, y)$ at $y = 0.5$ for $0 < x < 2$. You can take the first five terms.

3. Find one-term approximate solution to the differential equation,
$$Lu = 1, \quad Lu \equiv \frac{d^2u}{dx^2} + u, \quad u(0) = u(1) = 0,$$
using the Galerkin method. Use $e(x) = x(x-1)$ as a base function.

4. Obtain permissible functions, $f(x, y)$, that satisfy the following mixed boundary conditions along the square-shaped boundary:
$$f = 0, \quad x = -a, \quad -b < y < b \quad \text{and} \quad y = -b, \quad -a < x < a,$$
and
$$\frac{\partial f}{\partial n} = 0 \quad x = a, \quad -b < y < b \quad \text{and} \quad y = b, \quad -a < x < a.$$

5. Under what condition do the Galerkin method and the Rayleigh–Ritz method yield the same set of equations?

6. Derive permissible functions, $f(x, y)$, for a square plate that contains two circular inclusions.
Note: Refer www.wiley.com/go/nomura0615 for solutions for the Exercise section.

References

Finlayson BA 1972 *The Method of Weighted Residuals and Variational Principles: With Application in Fluid Mechanics, Heat and Mass Transfer*, vol. 87. Academic Press.

Greenberg M 1971 *Application of Green's Functions in Science and Engineering*, vol. 30. Prentice-Hall.

Hassani S 2013 Sturm-Liouville systems, *Mathematical Physics*. Springer-Verlag, pp. 563–602.

Pathapalli T 2013 *A Semi-Analytical Approach to Obtain Physical Fields in Heterogeneous Materials*. Dissertation. University of Texas, Arlington, TX.

Appendix A

Introduction to *Mathematica*

In this appendix, a brief introduction to *Mathematica* is presented so that the first-time reader can follow and understand the *Mathematica* code used in this book and even modify the code to suit the needs if necessary. Implementation of the analytical methods presented in this book would not have been possible had it not been for software that is capable of symbolic manipulation along with numerical capabilities.

For in-depth understanding of the detailed programming techniques and the syntax of *Mathematica*, many books or online tutorials are available although the most important reference is the one written by the creator of the software (Wolfram 1999), which is now part of the software distribution whose entire contents can be accessed through *Mathematica*'s online help. This reference book is not only a comprehensive reference to all the available *Mathematica* commands and statements but also an excellent introductory tutorial on *Mathematica*.

As is with any software originally written by a single person, *Mathematica* has its own peculiarities in which some of the conventions and syntax seem odd first compared with other scientific software applications. Here is a list of some of the peculiarities in *Mathematica*:

- All the statements and functions are case-sensitive. The first character of all the built-in functions and constants is always capitalized.

Examples
- `Sin[x]` for $\sin(x)$.
- `Exp[x]` for $\exp(x)$.
- `Integrate[Exp[-x] Cos[x], {x, 0, Infinity}]` for

$$\int_0^\infty e^{-x} \cos x \, dx.$$

- `Pi` for π.
- `I` for $i = \sqrt{-1}$.

- The square brackets ([···]), not the parentheses, are used for function arguments in functions.

Examples

- Exp[2 + 3 I] for e^{2+3i}.
- Log[x] for log x.

• Multiplications are entered by either an * or a space.

Examples

- x y for $x \times y$.
- x*y for $x \times y$ as well.

• Most of the built-in function and constant names are spelled out. The first letter is capitalized as well.

Examples

- Integrate[Sin[x], x] for integration.
- Infinity for ∞.
- Denominator[6/7] for the denominator of 6/7.

• A range can be specified by the braces, { and }.

Examples

- Plot[Cos[x], {x, -4, 4}] for plotting cos x over the interval between $[-4, 4]$.
- Sum[1/n, {n, 1, 100}] for

$$\sum_{n=1}^{100} \frac{1}{n}.$$

- Plot3D[Sin[x y], {x,-2,1},{y,-3,3}] for plotting a 3-D graph of sin xy over $-2 < x < 1$ and $-3 < y < 3$.

• Arrays and matrices are entered by (nested) braces, { and }.

Examples

- vec={x, y, z} to define a vector, **vec** $= (x, y, z)$.
- mat={{1,2,3},{4,5,6},{7,8,9}} to define a matrix,

$$mat = \begin{pmatrix} 1 & 2 & 3 \\ 4 & 5 & 6 \\ 7 & 8 & 9 \end{pmatrix}.$$

• A single square bracket ([···]) is used for the placeholder for function arguments while double square brackets ([[···]]) are used to refer to components in a list.

Examples

- Sin[x]+Exp[-x] for $\sin(x) + \exp(-x)$.
- m[[2, 3]] = 12 for $m_{23} = 12$.

• Free format. Ending a statement with a ";" (semicolon) suppresses output echo.

Appendix A: Introduction to *Mathematica*

Examples
- a=Expand[(x+y)^2]; assigns expansion of $(x+y)^2$ to a but the result is not displayed.
- m[[2, 3]] = 12 sets $m_{23} = 12$ and echoes back the result.

Mathematica has been in existence since 1988, and the current version as of this book is Version 10. Although many enhancements and new functionalities have been added, the backward compatibility has been always maintained and most of the *Mathematica* commands that worked in Version 2 still work in Version 10 without modifications. All of the *Mathematica* codes in this book should work in any version of *Mathematica*.

Mathematica is often compared with MATLAB as both are developed for the sole mission of solving scientific/engineering problems, are exclusively used by engineers and scientists, and many *Mathematica* functions are also available in MATLAB. However, there is a fundamental difference between *Mathematica* and MATLAB - the symbolic capabilities. In *Mathematica*, variables and functions can be both symbolic and numeric and can be mixed seamlessly. MATLAB uses the Maple engine with Symbolic Math Toolbox to perform symbolic computations, and it is not possible to mix symbolic variables with numerical variables. It is not possible to do computations presented in this book by programming in MATLAB alone.

Mathematica runs on many platforms including Windows, Mac, and Linux, with almost identical interface throughout different environments. Although the statements and commands used in the appendix were based on the Windows version, there should be little problem in trying those commands on different platforms.

A.1 Essential Commands/Statements

When *Mathematica* is launched first, it is in input mode waiting for commands to be entered from the keyboard.

In[3]:= **Expand[(x + y) ^5]**

Out[3]= $x^5 + 5 x^4 y + 10 x^3 y^2 + 10 x^2 y^3 + 5 x y^4 + y^5$

The input line from the keyboard begins with

In[3] =

and the output line from *Mathematica* begins with

Out[3] =

To recall from the immediately output result, a percentage symbol (%) can be used instead of retyping as

In[4]:= **%3**

Out[4]= $x^5 + 5 x^4 y + 10 x^3 y^2 + 10 x^2 y^3 + 5 x y^4 + y^5$

You can enter any *Mathematica* statement directly. The following are some examples of *Mathematica* statements that can be tried:

In[5]:= **Integrate[x Sin[x], x]**

Out[5]= $-x \cos[x] + \sin[x]$

In[6]:= **Expand[(x+y)^12]**

Out[6]= $x^{12} + 12 x^{11} y + 66 x^{10} y^2 + 220 x^9 y^3 + 495 x^8 y^4 + 792 x^7 y^5 +$
$924 x^6 y^6 + 792 x^5 y^7 + 495 x^4 y^8 + 220 x^3 y^9 + 66 x^2 y^{10} + 12 x y^{11} + y^{12}$

In[7]:= **Series[Sin[x], {x, 0, 10}]**

Out[7]= $x - \dfrac{x^3}{6} + \dfrac{x^5}{120} - \dfrac{x^7}{5040} + \dfrac{x^9}{362\,880} + O[x]^{11}$

In[8]:= **Plot[Sin[x] / x, {x, -2 Pi, 2 Pi}]**

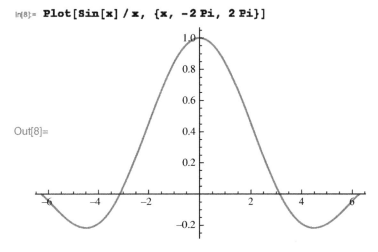

Out[8]=

As is seen from the aforementioned examples, most of the statements are self-explanatory. After entering the statement from the keyboard, it is necessary to press the Shift and Enter keys simultaneously or the Enter key in the 10-key pad for execution in Windows. Pressing the Enter key does not execute the statement but advances the cursor to the next line for additional entry of statements. All of the built-in functions in *Mathematica* such as **Integral** must be spelled out and the first letter must be capitalized. Function parameters must be enclosed by the square brackets ([···]) as in **f[x]** instead of the parentheses ((···)). The curly brackets ({···}) are used to specify the range of a variable. They are also used for components of a list.

A.2 Equations

Mathematica can handle many types of equations both analytically and numerically. The Solve function and its variations can do most of the jobs as

Appendix A: Introduction to *Mathematica*

In[1]:= `sol = Solve[a x^2 - b x - c == 0, x]`

Out[1]= $\left\{\left\{x \to \dfrac{b - \sqrt{b^2 + 4 a c}}{2 a}\right\}, \left\{x \to \dfrac{b + \sqrt{b^2 + 4 a c}}{2 a}\right\}\right\}$

The aforementioned example solves a quadratic equation, $ax^2 - bx - c = 0$, for x and outputs the two roots. Two equal signs (==) are needed to define the equation. The output from the `Solve` command is stored in the variable `sol` and also output to the screen in the list format with two elements each of which is enclosed by the curly brackets ({ ··· }). The right arrow symbol ($\to$) is the substitution rule (replace) that replaces what is to the left of the arrow by what is to the right of the arrow. By using this rule, it is possible to evaluate an expression that contains x.

In[3]:= `ex1 = x^2 + x - 1 /. sol`

Out[3]= $\left\{-1 + \dfrac{b - \sqrt{b^2 + 4 a c}}{2 a} + \dfrac{\left(b - \sqrt{b^2 + 4 a c}\right)^2}{4 a^2}, -1 + \dfrac{b + \sqrt{b^2 + 4 a c}}{2 a} + \dfrac{\left(b + \sqrt{b^2 + 4 a c}\right)^2}{4 a^2}\right\}$

In the aforementioned example, $x^2 + x - 1$ is evaluated for the two roots of the equation, $ax^2 - bx - c = 0$, whose two roots are stored by the `Solve` command to the variable, `sol`. As there are two roots, the evaluation of $x^2 + x - 1$ also results in two possible values for each of the roots; hence, the output is in the list format.

In *Mathematica*, to apply a rule to an expression, "/." is used immediately after the expression.

In[2]:= `x^2 + x - 1 /. x -> b`

Out[2]= $-1 + b + b^2$

In the aforementioned example, every occurrence of x is replaced by b. The replacement rule can be applied to more than one variable as

In[3]:= `x^2 - 4 x y + y^2 /. {x -> a, y -> b}`

Out[3]= $a^2 - 4 a b + b^2$

Each element in a list can be referenced by double square brackets `[[···]]` as in `a[[2]]`.

In[4]:= `ex1[[2]]`

Out[4]= $-1 + \dfrac{b + \sqrt{b^2 + 4 a c}}{2 a} + \dfrac{\left(b + \sqrt{b^2 + 4 a c}\right)^2}{4 a^2}$

`Solve` can recognize many types of equations. For an algebraic equation, it can solve up to the fifth-order equation exactly as

In[1]:= `Solve[x^3 - x - 1 == 0, x]`

Out[1]= $\left\{\left\{x \to \frac{1}{3}\left(\frac{27}{2}-\frac{3\sqrt{69}}{2}\right)^{1/3} + \frac{\left(\frac{1}{2}(9+\sqrt{69})\right)^{1/3}}{3^{2/3}}\right\},\right.$

$\left\{x \to -\frac{1}{6}(1+i\sqrt{3})\left(\frac{27}{2}-\frac{3\sqrt{69}}{2}\right)^{1/3} - \frac{(1-i\sqrt{3})\left(\frac{1}{2}(9+\sqrt{69})\right)^{1/3}}{2\times 3^{2/3}}\right\},$

$\left.\left\{x \to -\frac{1}{6}(1-i\sqrt{3})\left(\frac{27}{2}-\frac{3\sqrt{69}}{2}\right)^{1/3} - \frac{(1+i\sqrt{3})\left(\frac{1}{2}(9+\sqrt{69})\right)^{1/3}}{2\times 3^{2/3}}\right\}\right\}$

However, in many instances, numerical values are more desirable than lengthy exact expressions. The function N can convert a symbolic expression into its numerical value as

In[2]:= `N[Solve[x^3 - x - 1 == 0, x]]`

Out[2]= `{{x → 1.32472}, {x → -0.662359 + 0.56228 i}, {x → -0.662359 - 0.56228 i}}`

Alternatively, it is also possible to use the function NSolve to directly solve the equation numerically as

In[3]:= `NSolve[x^3 - x - 1 == 0, x]`

Out[3]= `{{x → -0.662359 - 0.56228 i}, {x → -0.662359 + 0.56228 i}, {x → 1.32472}}`

A set of simultaneous equations can be solved using a list as

In[4]:= `Solve[{ 2 x - y + z == 4, 4 x + y + z == -1, x - y - z == -4}, {x, y, z}]`

Out[4]= $\left\{\left\{x \to -1, y \to -\frac{3}{2}, z \to \frac{9}{2}\right\}\right\}$

The Solve command is not effective for nonlinear equations. The FindRoot function can solve nonlinear equations using the Newton–Raphson method as

In[5]:= `FindRoot[Exp[-x^2] - Sin[x] == 0, {x, 2}]`

Out[5]= `{x → 3.14154}`

The aforementioned statement solves a nonlinear equation,

$$e^{-x^2} - \sin x = 0,$$

with an initial guessing value of x as 2. As the Newton–Raphson method requires an initial guessing value to start, the FindRoot function uses a list, with {x, 2} as the initial value. The FindRoot function can also solve a set of nonlinear simultaneous equations as

In[19]:= `FindRoot[{ Exp[x - 2] + x^2 == y - 1, Sin[y] == 1/ (x^2 + 1)}, {{x, -1}, {y, 2}}]`

Out[19]= `{x → -1.3154, y → 2.7666}`

for

$$e^{x-2} + x^2 = y - 1, \quad \sin y = \frac{1}{x^2 + 1},$$

with initial guessing values of $x = -1, y = 2$.

Appendix A: Introduction to *Mathematica*

Differential equations (initial value problems) can be solved using `DSolve` command as

In[20]:= `DSolve[ y''[x] + 3 y'[x] + 2 y[x] == 3, y[x], x]`

Out[20]= $\left\{\left\{y[x] \to \frac{3}{2} + e^{-2x} C[1] + e^{-x} C[2]\right\}\right\}$

for

$$y'' + 3y' + 2y = 3.$$

The output is a rule in the list format as there may be multiple solutions available. The aforementioned solution is expressed with two integral constants, `C[1]` and `C[2]`.

The `DSolve` function has three components. The first component is the differential equation to be solved, the second component, $y[x]$, shows that the unknown function, y, is a function of x, and the third component, x, is the variable for differentiation.

The `DSolve` function can also solve a set of simultaneous differential equations.

In[12]:= `DSolve[{x'[t] - 2 y'[t] == Exp[t],`
`2 x'[t] + 3 y'[t] == 1, x[0] == 1, y[0] == 2}, {x[t], y[t]}, t]`

Out[12]= $\left\{\left\{x[t] \to \frac{1}{7}\left(4 + 3 e^t + 2 t\right), y[t] \to \frac{1}{7}\left(16 - 2 e^t + t\right)\right\}\right\}$

This statement solves the following simultaneous differential equations with the initial conditions supplied.

$$x'(t) - 2y'(t) = e^t, \quad 2x'(t) + 3y'(t) = 1, \quad x(0) = 1, \quad y(0) = 2.$$

The initial conditions can be specified as part of the equation entry.

If it is not possible to solve a differential equation analytically, `NDSolve` can be used to solve the differential equation numerically as

In[56]:= `sol = NDSolve[{y''[x] + Sin[y[x]] == 0, y[0] == 2, y'[0] == 1}, y[x], {x, 0, 10}]`
`Plot[y[x] /. sol, {x, 0, 10}]`

Out[56]= $\left\{\left\{y[x] \to \text{InterpolatingFunction}\left[\boxed{\begin{array}{l}\text{Domain: }\{\{0., 10.\}\}\\ \text{Output: scalar}\end{array}}\right][x]\right\}\right\}$

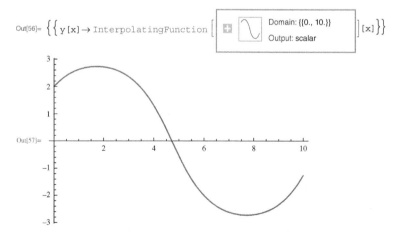

Out[57]=

The `NDSolve` function numerically solves the following differential equation over the interval of $0 < x < 10$.

$$y'' + \sin y = 0, \quad y(0) = 2, \quad y'(0) = 1.$$

The solution stored in the variable sol is in the format of numerically defined interpolating function. To plot the result, it is necessary to apply the rule sol to y[x].

In[60]:= `y[x] /. sol /. x → 2`

Out[60]= {2.71349}

If the solution at a specific x value is required, the substitution command can be used.

A.3 Differentiation/Integration

For differentiation of a function, the D function is used. This is one of a few built-in *Mathematica* functions that are not spelled out. Other functions that are not spelled out include Abs (the absolute value) Im (the imaginary part of a complex function) and Re (the real part of a complex function).

To differentiate $x^{100}\sin x$ with respect to x, use

In[1]:= `D[x^100 Sin[x], x]`

Out[1]= x^{100} Cos[x] + 100 x^{99} Sin[x]

Higher order derivatives can be specified by a list as (the second-order derivative)

In[2]:= `D[x Exp[x], {x, 2}]`

Out[2]= $2\,e^x + e^x\,x$

Indefinite integration of a function is evaluated by the Integrate command as

In[3]:= `Integrate[ x Sin[x], x]`

Out[3]= $-x\,$Cos[x] + Sin[x]

To carry out definite integration of a function, it is necessary to use a list for the lower and upper bounds as

In[5]:= `Integrate[ Exp[-x] Sin[x], {x, 0, Infinity}]`

Out[5]= $\dfrac{1}{2}$

If a function cannot be integrated analytically, it can be integrated numerically using NIntegrate as

In[6]:= `NIntegrate[ 1/ (Cos[x]^2 + 2), {x, 0, 20}]`

Out[6]= 8.13209

A.4 Matrices/Vectors/Tensors

In *Mathematica*, there is no distinction between vectors and matrices. In fact, there is no distinction between matrices and tensors. They are all represented by a list or nested lists. Hence, there is no distinction between column vectors and row vectors either.

Appendix A: Introduction to *Mathematica*

A vector can be entered as a flat list as

In[7]:= **vec = {v1, v2, v3}**

Out[7]= {v1, v2, v3}

To make a reference to a specific component, double square brackets ([[···]]) are used as

In[2]:= **vec[[2]]**

Out[2]= v2

The inner product between vectors can be computed by a dot (·) instead of a space or an asterisk (*) as

In[5]:= **vec.{a1, a2, a3}**

Out[5]= a1 v1 + a2 v2 + a3 v3

A 3 × 3 matrix can be entered as a nested list as

In[2]:= **mat = {{a11, a12, a13}, {a21, a22, a23}, {a31, a32, a33}}**

Out[2]= {{a11, a12, a13}, {a21, a22, a23}, {a31, a32, a33}}

To make a reference to a specific component, double square brackets ([[···]]) are used as

In[7]:= **mat[[2, 3]]**

Out[7]= a23

The product between a matrix and a vector or between matrices must use a dot (·) instead of a space or an asterisk (*) as

In[11]:= **mat.vec**

Out[11]= {a11 v1 + a12 v2 + a13 v3, a21 v1 + a22 v2 + a23 v3, a31 v1 + a32 v2 + a33 v3}

Almost all operations in linear algebra are available in *Mathematica*. The inverse of a matrix can be computed by Inverse as

In[8]:= **Inverse[mat]**

Out[8]= {{(-a23 a32 + a22 a33) /
      (-a13 a22 a31 + a12 a23 a31 + a13 a21 a32 - a11 a23 a32 - a12 a21 a33 + a11 a22 a33),
      (a13 a32 - a12 a33) / (-a13 a22 a31 + a12 a23 a31 + a13 a21 a32 -
      a11 a23 a32 - a12 a21 a33 + a11 a22 a33), (-a13 a22 + a12 a23) /
      (-a13 a22 a31 + a12 a23 a31 + a13 a21 a32 - a11 a23 a32 - a12 a21 a33 + a11 a22 a33)},
   {(a23 a31 - a21 a33) / (-a13 a22 a31 + a12 a23 a31 + a13 a21 a32 -
      a11 a23 a32 - a12 a21 a33 + a11 a22 a33), (-a13 a31 + a11 a33) /
      (-a13 a22 a31 + a12 a23 a31 + a13 a21 a32 - a11 a23 a32 - a12 a21 a33 + a11 a22 a33),
      (a13 a21 - a11 a23) / (-a13 a22 a31 + a12 a23 a31 + a13 a21 a32 -
      a11 a23 a32 - a12 a21 a33 + a11 a22 a33)}, {(-a22 a31 + a21 a32) /
      (-a13 a22 a31 + a12 a23 a31 + a13 a21 a32 - a11 a23 a32 - a12 a21 a33 + a11 a22 a33),
      (a12 a31 - a11 a32) / (-a13 a22 a31 + a12 a23 a31 + a13 a21 a32 -
      a11 a23 a32 - a12 a21 a33 + a11 a22 a33), (-a12 a21 + a11 a22) /
      (-a13 a22 a31 + a12 a23 a31 + a13 a21 a32 - a11 a23 a32 - a12 a21 a33 + a11 a22 a33)}}

In[10]:= **Simplify[Inverse[mat].mat]**

Out[10]= {{1, 0, 0}, {0, 1, 0}, {0, 0, 1}}

It is necessary to apply the Simplify function to simplify the result for AA^{-1}.

The eigenvalues and eigenvectors of a matrix can be computed using the `Eigensystem` function as

In[5]:= `{eval, evec} = Eigensystem[{{1, 2, 3}, {2, 3, 4}, {3, 4, 5}}]`

Out[5]= $\{\{\frac{1}{2}(9+\sqrt{105}), \frac{1}{2}(9-\sqrt{105}), 0\}, \{\{-\frac{-11-\sqrt{105}}{2(10+\sqrt{105})}, -\frac{-31-3\sqrt{105}}{4(10+\sqrt{105})}, 1\},$
$\{-\frac{11-\sqrt{105}}{2(-10+\sqrt{105})}, -\frac{31-3\sqrt{105}}{4(-10+\sqrt{105})}, 1\}, \{1, -2, 1\}\}\}$

The three eigenvalues for the matrix,

$$\begin{pmatrix} 1 & 2 & 3 \\ 2 & 3 & 4 \\ 3 & 4 & 5 \end{pmatrix}$$

are stored in `eval` and the corresponding eigenvectors are stored in `evec`. For example, the eigenvector corresponding to the second eigenvalue is `evec[[2]]` as

In[6]:= `evec[[2]]`

Out[6]= $\{-\frac{11-\sqrt{105}}{2(-10+\sqrt{105})}, -\frac{31-3\sqrt{105}}{4(-10+\sqrt{105})}, 1\}$

A.5 Functions

The first letter of all the built-in *Mathematica* functions is always capitalized as exemplified by `Integrate`, `Log`, `Series`, and so on. This helps to distinguish user-defined functions from the built-in functions.

To create a user-defined function, the following syntax is used:

`func[x_, y_, z_] := definition of the function of x, y and z.`

where x, y, z are the variables of the function, `func`. Note that an underscore follows after each variable and a colon and an equal symbol (:=) are used instead of an equal symbol. The definition of the function must be given to the right of the ":=" containing the variables with the underscores within the function argument. To define a function, $f(x) = x \sin x$, use the syntax:

In[7]:= `f[x_] := x Sin[x]`

In[8]:= `f[5]`
Out[8]= `5 Sin[5]`

In[9]:= `f[a+b]`
Out[9]= `(a+b) Sin[a+b]`

A function of several variables can be defined similarly as

Appendix A: Introduction to *Mathematica*

In[10]:= `f[x_, y_] := (x - y)^4`
`f[2, 4]`

Out[11]= `16`

In[12]:= `f[5.0]`

Out[12]= `-4.79462`

Mathematica remembers all the definitions of the function the user has input so the function `f` returns different results, depending on the number of the arguments. If there is only one argument (i.e., `f[5]`), it returns the result of $x \sin x$, but if there are two arguments, x and y, it returns $(x - y)^4$.

If the definition of a function is more involved than a single formula requiring several steps, `Module` can be used to define a function that requires multiple steps with local variables as

In[2]:= `f[x_, y_] := Module[{a, b}, a = x + y; b = x * y; a + b]`

In[5]:= `f[x, y]`

Out[5]= `x + y + x y`

In[6]:= `f[5, 2]`

Out[6]= `17`

The variables, `a` and `b`, inside the braces in the `Module` function are local variables inside `Module` that do not retain their values outside the function. `Module` returns the last statement `a + b`, when it is called.

A.6 Graphics

Mathematica has a variety of commands that can visualize functions or data points in 2-D and 3-D. To plot a graph of $\sin x/x$ over $[-3\pi, 3\pi]$, use the `Plot` function as

In[7]:= `Plot[Sin[x] / x, {x, -3 Pi, 3 Pi}]`

Out[7]=
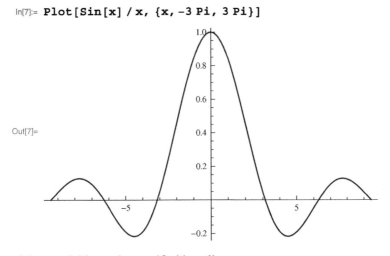

The range of the x variable can be specified by a list.

To plot multiple graphs together, use a list of functions as

In[11]:= `Plot[{Sin[x], Cos[x]}, {x, -2 Pi, 2 Pi}]`

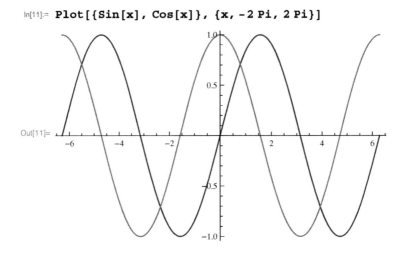

Out[11]=

The function `Plot3D` can be used to plot a function of two variables

In[11]:=
`Plot3D[ Exp[-x*y], {x, -4, 4}, {y, -5, 5}]`

Out[11]=

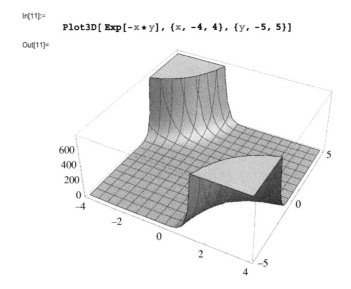

The plot range for x and y can be specified by the lists.
To plot a sequence of numbers, `ListPlot` can be used.

In[1]:= `a = Table[i, {i, 1, 20}]`
Out[1]= {1, 2, 3, 4, 5, 6, 7, 8, 9, 10, 11, 12, 13, 14, 15, 16, 17, 18, 19, 20}

Appendix A: Introduction to *Mathematica*

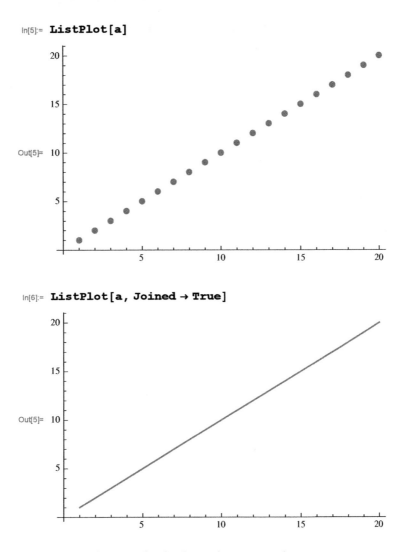

The option `Joined` must be `True` for the dots to be connected.

A.7 Other Useful Functions

Some of the useful functions in *Mathematica* are as follows. The usage of most of the functions is self-explanatory.

`Series[f[x], {x, a, n}]` returns Taylor's series expansion of `f[x]` about $x = a$ up to and including the nth order.

In[1]:= **Series[Sin[x], {x, 0, 10}]**

Out[1]= $x - \dfrac{x^3}{6} + \dfrac{x^5}{120} - \dfrac{x^7}{5040} + \dfrac{x^9}{362880} + O[x]^{11}$

Sum[] and NSum[] can sum up a series exactly and numerically, respectively, as

In[1]:= **Sum[1/i!, {i, 1, 100}]**

Out[1]= 2 717 978 646 037 002 453 111 381 514 508 703 728 041 213 733 930 788 487 528 512 238 851 383 ⸳ 964 663 670 015 423 712 227 949 076 805 203 622 200 194 928 470 090 819 123 541 234 409 774 741 771 022 426 508 339 / 1 581 800 261 761 765 299 689 817 607 733 333 906 622 304 546 853 925 787 603 270 574 495 213 ⸳ 559 207 286 705 236 295 999 595 873 191 292 435 557 980 122 436 580 528 562 896 896 000 000 ⸳ 000 000 000 000 000 000

In[2]:= **NSum[1/i!, {i, 1, 100}]**

Out[2]= 1.71828

Both compute

$$\sum_{i=1}^{100} \dfrac{1}{i!}.$$

The Apart function can be used to perform partial fraction of a rational function as

In[1]:= **Apart[1/(x^2 - 4)]**

Out[1]= $\dfrac{1}{4(-2+x)} - \dfrac{1}{4(2+x)}$

One of the most useful functions in *Mathematica* is the Manipulate function that demonstrates the power of *Mathematica*. Here is an example of its basic usage:

In[7]:= **Manipulate[Plot[x^a, {x, 0, 1}], {a, 1, 5}]**

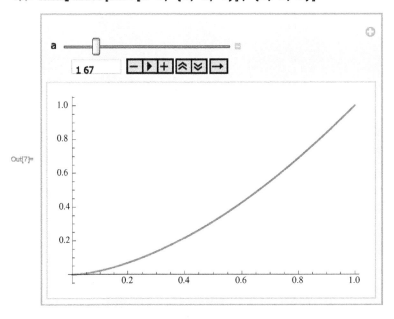

Appendix A: Introduction to *Mathematica*

The Manipulate function works just like the Table function except that the variable a in the aforementioned example can be varied by the slider in the output. By dragging the slider in the output changing the value of a, the graph of x^a is updated in real time.

A.8 Programming in *Mathematica*

By default, *Mathematica* is launched in interactive mode waiting for commands from the keyboard. The statements can be entered directly into the Notebook cells, and each Notebook session can be saved as a separate file (the extension of the file is nb, which stands for notebook) that can be loaded later.

However, for a long program, it is possible to prepare a file that contains *Mathematica* statements and run it thorough a batch process. For example, a text file with the following content can be prepared and saved in the c:\temp directory.

```
f[k1_, k2_] := Module[{y, c, eq, sol, u, range},
  y[1] = c[1] x + c[2]; y[2] = -x^2/2/k1 + c[3] x + c[4];
  y[3] = c[5] x + c[6];
  eq[1] = (y[1] /. x -> -1) == 0;
  eq[2] = ((y[1] - y[2]) /. x -> -1/3) == 0;
  eq[3] = ((y[2] - y[3]) /. x -> 1/3) == 0;
  eq[4] = (y[3] /. x -> 1) == 0;
  eq[5] = ((k2 D[y[1], x] - k1 D[y[2], x]) /. x -> -1/3) == 0;
  eq[6] = ((k1 D[y[2], x] - k2 D[y[3], x]) /. x -> 1/3) == 0;

  sol = Solve[Table[eq[i], {i, 1, 6}], Table[c[i], {i, 1, 6}]][[1]];
  u = Table[y[i], {i, 1, 3}] /. sol;
  range = {Boole[ x > -1 && x < -1/3], Boole[ x > -1/3 && x < 1/3],
    Boole[ x > 1/3 && x < 1]};
  Plot[u.range, {x, -1, 1}, PlotRange -> {0, 0.4}]]

Manipulate[f[k1, k2], {k1, 1, 100}, {k2, 1, 100}]
```

Once the file is saved, it can be loaded later into a *Mathematica* session. First, it is necessary to specify the working directory in which the file is saved by using the SetDirectory function as

In[8]:= **SetDirectory["c:/temp"]**

Out[8]= c:\temp

This will set the directory, c:\temp, as the default directory in which files are saved/loaded. Note that the forward slash character (/) must be used as the Windows directory delimiter instead of the backslash (\) character. To load the file into a *Mathematica* Notebook session, use the following syntax:

<<filename

In[2]:= **SetDirectory["c:/tmp"]**

Out[2]= c:\tmp

In[4]:= **<< batch_example.txt**

Out[4]=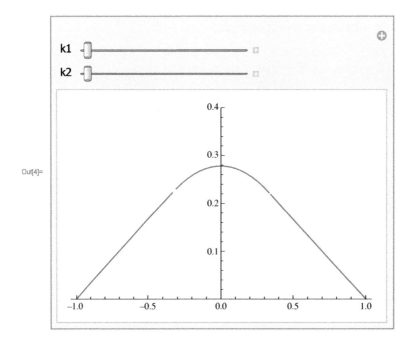

A.8.1 Control Statements

Programming in *Mathematica* requires conditional statements for repetitive computations or branching out to different routines. The following are important conditional statements that are useful in *Mathematica* programming.

A.8.1.1 Do

The syntax is Do[body, range] where the contents of body are repeated for the range specified by range. The following code prints all elements of the matrix a defined as

$$a = \begin{pmatrix} 1 & 2 & 3 \\ 4 & 5 & 6 \\ 7 & 8 & 9 \end{pmatrix}$$

```
In[1]:= a = {{1, 2, 3}, {4, 5, 6}, {7, 8, 9}}
Out[1]= {{1, 2, 3}, {4, 5, 6}, {7, 8, 9}}

In[2]:= Do[Print[a[[i, j]]], {i, 3}, {j, 3}]
1
2
3
4
5
6
7
8
9
```

The iteration range for the indices can be entered by the braces, and if omitted, the lower bound is assumed to be 1.

A.8.1.2 Table

The syntax of the `Table` function is `Table[ expr, range]` where the contents of `expr` are evaluated to generate a list for the range specified by `range`.

The `Table` function is used to generate a list. The following command generates a list of $\cos i\pi$ when i is increased from 1 to 10.

```
In[3]:= Table[ Cos[Pi i], {i, 1, 10}]
Out[3]= {-1, 1, -1, 1, -1, 1, -1, 1, -1, 1}
```

A.8.1.3 For

The syntax of the `For` function is `For[init, final, incr, expr]` where `expr` is kept evaluated while `final` is true with the initial counter value specified in `int` and the counter value is incremented by `incr` after each evaluation of `expr`, similar to the *for* loop in C. The following code prints i from $i = 1$ to $i = 9$.

```
In[1]:= For[i = 1, i < 10, i++, Print[i]]
```
1
2
3
4
5
6
7
8
9

A.8.1.4 If

The syntax of the If function is If[test, body1, body2] in which if test is true, body1 is executed, and if test is false, body2 is executed. The following code defines a function that returns the absolute value of x (same as Abs[x]).

```
In[2]:= abs[x_] := If[x > 0, x, -x]

In[4]:= abs[-5]
Out[4]= 5

In[5]:= abs[1]
Out[5]= 1
```

A.8.2 Tensor Manipulations

Mathematica itself does not support tensors in native mode, implying that such operations as summation convention are not automatically applied for the product of tensors. However, *Mathematica* is flexible in manipulating data types and using nested lists, any rank of tensors can be defined.

A first-rank tensor can be defined by a simple list as

```
In[6]:= v1 = {x, y, z}
Out[6]= {x, y, z}
```

A second-rank tensor can be defined by a double-nested list as

```
In[7]:= v2 = Table[v[i, j], {i, 1, 3}, {j, 1, 3}]
Out[7]= {{v[1, 1], v[1, 2], v[1, 3]}, {v[2, 1], v[2, 2], v[2, 3]}, {v[3, 1], v[3, 2], v[3, 3]}}
```

A third-rank tensor can be defined by a triple-nested list as

In[8]:= **v3 = Table[i + j + k, {i, 3}, {j, 3}, {k, 3}]**

Out[8]= {{{3, 4, 5}, {4, 5, 6}, {5, 6, 7}},
{{4, 5, 6}, {5, 6, 7}, {6, 7, 8}}, {{5, 6, 7}, {6, 7, 8}, {7, 8, 9}}}

In[9]:= **v3[[1, 1, 2]]**

Out[9]= 4

The Kronecker delta, δ_{ij}, can be defined as a function instead of a list as

In[1]:= **delta[i_, j_] := If[i == j, 1, 0]**

In[2]:= **delta[1, 1]**

Out[2]= 1

In[3]:= **delta[2, 3]**

Out[3]= 0

As the summation convention is not implemented in *Mathematica*, it is necessary to explicitly apply the **Sum** function as

In[4]:= **Sum[delta[i, i], {i, 1, 3}]**

Out[4]= 3

to compute $\delta_{ii} = 3$.

It is possible to define the Kronecker delta as a list instead of a function as

In[5]:= **Clear[delta]**
delta = IdentityMatrix[3]

Out[6]= {{1, 0, 0}, {0, 1, 0}, {0, 0, 1}}

In[7]:= **delta[[1, 1]]**

Out[7]= 1

In[8]:= **delta[[1, 2]]**

Out[8]= 0

For instance, an elastic modulus is a fourth-rank tensor, and when the properties are isotropic, it can be expressed as

$$C_{ijkl} = \mu(\delta_{ik}\delta_{jl} + \delta_{il}\delta_{jk}) + \lambda\delta_{ij}\delta_{kl},$$

which can be implemented as

```
In[1]:= delta[i_, j_] := If[i == j, 1, 0]

In[2]:= c = Table[0, {i, 3}, {j, 3}, {k, 3}, {l, 3}];
    Do[c[[i, j, k, l]] = mu (delta[i, k] delta[j, l] + delta[i, l] delta[j, k]) +
        lambda delta[i, j] delta[k, l], {i, 3}, {j, 3}, {k, 3}, {l, 3}]

In[4]:= c[[1, 1, 1, 1]]

Out[4]= lambda + 2 mu

In[5]:= c[[1, 2, 1, 2]]

Out[5]= mu

In[6]:= c[[1, 1, 1, 2]]

Out[6]= 0
```

The summation convention needs to be implemented by the Sum command as

```
In[7]:= strain = Table[epsilon[i, j], {i, 3}, {j, 3}]

Out[7]= {{epsilon[1, 1], epsilon[1, 2], epsilon[1, 3]},
    {epsilon[2, 1], epsilon[2, 2], epsilon[2, 3]},
    {epsilon[3, 1], epsilon[3, 2], epsilon[3, 3]}}

In[8]:= stress = Table[Sum[c[[i, j, k, l]] strain[[k, l]], {k, 3}, {l, 3}], {i, 3}, {j, 3}]

Out[8]= {{(lambda + 2 mu) epsilon[1, 1] + lambda epsilon[2, 2] + lambda epsilon[3, 3],
    mu epsilon[1, 2] + mu epsilon[2, 1], mu epsilon[1, 3] + mu epsilon[3, 1]},
    {mu epsilon[1, 2] + mu epsilon[2, 1],
    lambda epsilon[1, 1] + (lambda + 2 mu) epsilon[2, 2] + lambda epsilon[3, 3],
    mu epsilon[2, 3] + mu epsilon[3, 2]},
    {mu epsilon[1, 3] + mu epsilon[3, 1], mu epsilon[2, 3] + mu epsilon[3, 2],
    lambda epsilon[1, 1] + lambda epsilon[2, 2] + (lambda + 2 mu) epsilon[3, 3]}}

In[9]:= stress[[1, 1]]

Out[9]= (lambda + 2 mu) epsilon[1, 1] + lambda epsilon[2, 2] + lambda epsilon[3, 3]

In[10]:= stress[[1, 2]]

Out[10]= mu epsilon[1, 2] + mu epsilon[2, 1]
```

Other ranks of tensors can be defined and manipulated.

References

Wolfram S 1999 *The Mathematica Book*. Cambridge university press.

Index

Airy's stress function, 155
Analytic function, 157
Anisotropy, 50
Antisymmetrization operator, 49
Aspect ratio, 88

Balance of moment, 29
Base vectors, 69
Bianchi formulas, 50
Biharmonic differential equation, 155
Biharmonic equation, 157
Boundary condition, 30
Bulk modulus, 55

Cartesian tensors, 11, 13
Cauchy strain, 43
Christoffel symbol, 70
Collocation Method, 192
Compatibility condition, 49, 156
Concentric inclusions, 81
Constitutive Relation, 50
Continuity equation, 62
Contravariant quantity, 66
Covariant derivatives, 70
Covariant quantity, 66
Curvilinear coordinate system, 2

Deviator, 35, 54
Deviatoric part, 108
Differentiation, 19

Differentiation of Cartesian tensors, 21
Dirac delta function, 83
Dirichlet boundary condition, 214
Divergence theorem, 59
Dummy index, 2

Effective properties, 172
Eigenstrain, 82, 85
Eigenvalue, 205
Eigenvalues, 32
Eigenvectors, 32
Einstein's notation, 2
Elastic compliance, 55
Elastic modulus, 54
Ellipsoid, 88
Equation of continuity, 62
Equation of energy, 63
Equation of motion, 62
Equilibrium, 29
Equilibrium equation, 141
Eshelby, 82
Eshelby tensor, 86
Eshelby tensors, 87
Euler equation, 203
Euler strain, 43

Finite element method, 238
Finite matrix, 191
Fluids, 58
Four-phase materials, 132

Micromechanics with Mathematica, First Edition. Seiichi Nomura.
© 2016 John Wiley & Sons, Ltd. Published 2016 by John Wiley & Sons, Ltd.
Companion Website: www.wiley.com/go/nomura0615

Fourier transforms, 84
Free index, 2

Galerkin's Method, 194
Gauss theorem, 59
Green strain, 42
Green's function, 83

Heat flow, 146
Heat source, 138
Hermitian operators, 83
Hill's condition, 173
Hydrostatic component, 35
Hydrostatic part, 54, 108

Implicit differentiation, 139
Inclusion, 95
Index notation, 1, 2
Inelastic source, 85
Inhomogeneity, 95
Invariance of tensor equations, 22
Isotropic, 19
Isotropic fluids, 65
Isotropic solids, 65
Isotropy, 50, 52

Kronecker delta, 6, 52

Lagrangian derivative, 60
Lagrangian strain, 42
Lamé constants, 54, 108
Laplace equation, 147
Laplacian, 76
Laurent series, 82, 158, 163
Layered inclusions, 81
Least Squares Method, 193
Leibniz integral rule, 61

Many-body problem, 172
Material derivative, 60
Matrix product, 10
Metric tensor, 69
micromech.m, 98, 177
Mohr's circle, 38
Moment, 29

Momentum, 62
Multilayered inclusions, 104
Multiphase inclusion problems, 163
Multiphase materials, 111
MWR, 192

Navier equation, 65
Navier's equation, 83, 105, 108
Navier–Stokes equation, 66
Neumann boundary condition, 215

Oblate spheroid, 88
Orthotropy, 56

Permutation symbol, 9
Physical components, 73
Poisson equation, 138
Polar coordinate system, 70, 158
Principal stresses, 31
Prolate spheroid, 88
Pure shear, 47, 48

Quotient rule, 22, 25, 29, 111

Rayleigh–Ritz method, 203
Reuss bound, 173
Ritz method, 204

Scalar, 13
Self-consistent method, 175
Shear deformation, 36, 47
Signature, 9
Simple shear, 47, 48
Steady-state heat conduction, 83, 213
Stokes fluid, 58
Stokes' hypothesis, 58
Strain, 40
Strain energy, 36
Strain proportionality factor, 98
Strain–displacement relation, 75
Stress, 25
Stress deviator, 35
Stress equilibrium equation, 73
Stress invariant, 33
Sturm–Liouville System, 205

Summation convention, 2, 5, 105
Surface traction, 25
Symmetrization operator, 49

Taylor series, 82, 163
Temperature effect, 137
Tensor components, 73
Thermal conductivities, 82
Thermal effect, 58
Thermal effects, 66
Thermal expansion, 146
Thermal expansion coefficient, 58
Thermal expansion coefficients, 82
Thermal stress, 82, 137
Three-phase composites, 175

Three-phase materials, 123
Trace, 47
Transformation matrix, 12
Transverse isotropy, 57
Transversely-isotropic medium, 91
Two-phase materials, 116

Uniform heat flow, 138

Variational principle, 203
Vector, 14
Voight bounds, 173
von Mises failure criterion, 37

Weighted residuals, 192

Printed and bound by CPI Group (UK) Ltd, Croydon, CR0 4YY
09/06/2025

14685996-0001